Classics in Game Theory

FRONTIERS OF ECONOMIC RESEARCH

Series Editors

David M. Kreps   Thomas J. Sargent

Classics in Game Theory

Edited by Harold W. Kuhn

PRINCETON UNIVERSITY PRESS

PRINCETON, NEW JERSEY

Library of Congress Cataloging-in-Publication Data

Classics in game theory / edited by Harold W. Kuhn
    p. cm. — Frontiers of Economic Research
    Includes bibliographical references and index.
    ISBN 0-691-01193-1 (cloth: alk. paper)—ISBN 0-691-01192-3
(pbk.: alk. paper)
    1. Game theory. 2. Mathematical economics. I. Kuhn, Harold W.
(Harold William), 1925– . II. Series.
HB144.C53   1997
519.3—dc20   96-20693

This book has been composed in 10/13 Times Roman

Princeton University Press books are printed on acid-free paper
and meet the guidelines for permanence and durability of the
Committee on Production Guidelines for Book Longevity of the
Council on Library Resources

Printed in the United States of America
10 9 8 7 6 5 4 3 2 1

10 9 8 7 6 5 4 3 2 1
(Pbk.)

CONTENTS

PERMISSIONS

1. JOHN F. NASH, JR.
   Equilibrium Points in n-Person Games. Reprinted from *PNAS* 36 (1950) 48–49, by permission of the author.

2. JOHN F. NASH, JR.
   The Bargaining Problem. Reprinted from *Econometrica* 18 (1950) 155–162, by permission of The Econometric Society.

3. JOHN NASH
   Non-Cooperative Games. Reprinted from *Annals of Mathematics Journal* 54 (1951) 286–295, by permission of the *Annals of Mathematics Journal*.

4. JULIA ROBINSON
   An Iterative Method of Solving a Game. Reprinted from *Annals of Mathematics Journal* 54 (1951) 296–301, by permission of the *Annals of Mathematics Journal*.

5. F. B. THOMPSON
   Equivalence of Games in Extensive Form. January 1952, RM-759, 12pp. Used by Permission of RAND.

6. H. W. KUHN
   Extensive Games and the Problem of Information. Reprinted from *Contributions to the Theory of Games* II (1953) 193–216, by permission of Princeton University Press.

7. L. S. SHAPLEY
   A Value for n-Person Games. Reprinted from *Contributions to the Theory of Games* II (1953) 307–317, by permission of Princeton University Press.

8. L. S. SHAPLEY
   Stochastic Games. *PNAS* 39 (1953) 1095–1100, by permission of *PNAS* and L. S. Shapley.

9. H. EVERETT
   Recursive Games. Reprinted from *Contributions to the Theory of Games* III (1957) 47–78, by permission of Princeton University Press.

10. R. J. AUMANN AND B. PELEG
    Von Neumann–Morgenstern Solutions to Cooperative Games with-

out Side Payments. Reprinted from *Bulletin of the American Mathematical Society* 66 (1960) 173–179, by permission of the American Mathematical Society.

11. GERARD DEBREU AND HERBERT E. SCARF

    A Limit Theorem on the Core of an Economy. Reprinted from *The International Economic Review* 4 (1963) 235–246, by permission of the International Economic Review.

12. ROBERT J. AUMANN AND MICHAEL MASCHLER

    The Bargaining Set for Cooperative Games. Reprinted from *Advances in Game Theory* (1964) 443–477, by permission of Princeton University Press.

13. ROBERT J. AUMANN

    Existence of Competitive Equilibria in Markets with a Continuum of Traders. Reprinted from *Econometrica* 34 (1966) 1–17, by permission of The Econometric Society.

14. HERBERT E. SCARF

    The Core of an *n*-Person Game. Reprinted from *Econometrica* 35 (1967) 50–69, by permission of The Econometric Society.

15. Reprinted by permission, JOHN C. HARSANYI, Games with Incomplete Information Played by "Bayesian" Players. Part I: The Basic Model, *Management Science* 14 (1967) 159–182; Part II: Bayesian Equilibrium Points, *Management Science* 14 (1968) 320–334; Part III: The Basic Probability Distribution of the Game, *Management Science* 14 (1968) 486–502, The Institute of Management Sciences (currently INFORMS), 2 Charles Street, Suite 300, Providence, RI.

16. DAVID BLACKWELL AND T. S. FERGUSON

    The Big Match. Reprinted from *Annals Mathematical Statistics* 39 (1968) 159–163, by permission of the Institute for Mathematical Statistics.

17. LLOYD S. SHAPLEY AND MARTIN SHUBIK

    On Market Games. Reprinted from the *Journal of Economic Theory* 1 (1969) 9–25, by permission of Academic Press, Inc., P.O. Box 860630, Orlando, FL 32886-0630

18. R. SELTEN

    Reexamination of the Perfectness Concept for Equilibrium Points in Extensive Games. *International Journal of Game Theory* 4 (1975) 25–55. Reprinted by permission of Physica-Verlag, GmbH & Co. KG, TiergartenstraBe 17, D-69121 Heidelberg, Germany.

FOREWORD

In 1988 several colleagues in the Economics Department proposed that I select a group of papers in game theory that would be published as a set of readings to supplement the small group of textbooks in the subject for the burgeoning courses in both mathematics and economics departments. The original proposal was that the papers to be included would be drawn from the *Annals of Mathematics Studies* devoted to game theory (Contributions to the Theory of Games, Vols. I–IV, and Advances in Game Theory). However, the more thought that I gave to the project, the more I believed that this would be unduly restrictive, and that papers should be sought from other sources. Accordingly, I asked the advice of a number of friends and colleagues whose judgment I trust, enclosing a tentative list of papers to be included. There were some differences of opinion, but nevertheless a great deal of agreement about what should be considered the "classics in game theory." Although no one of them bears responsibility for the final list, I am happy to thank the following for their advice and comments: Ken Arrow, Paolo Caravani, V. P. Crawford, Gerard Debreu, Avinash Dixit, Sergiu Hart, Ehud Kalai, Roger Meyerson, Hervé Moulin, Guillermo Owen, John Roberts, Herbert Scarf, David Schmeidler, Martin Shubik, William Thompson, Robert Willig, Robert Wilson, and Peyton Young.

The title "Classics in Game Theory" will suggest different things to different people, but the heart of this volume is the basic building blocks on which the current edifice of game theory is built. With a confirmed procrastinator (myself) in charge of the project, the final list of papers was chosen by 1990, but the most urgent efforts of Jack Repcheck, the economics editor at Princeton University Press, could not shake it loose from me. At one time I intended to prepare an introductory essay that would incorporate some "prehistoric" excerpts (say, by Montmort, Zermelo, and von Neumann). The essay would have also given me the opportunity to give the volume some historical perspective and, incidentally, to explain some of the criteria used in the selection of the papers. However, that was not to be.

The project assumed a new urgency upon the announcement of the 1994 Nobel Memorial Prize in Economic Science, which recognized the central importance of the theory of games for economic theory by honoring John Nash, John Harsanyi, and Reinhard Selten. I am pleased to say that the key works for which they were honored were five of the final list of eighteen papers that was ready but not acted upon in 1990. At this point Peter Dougherty, publisher in social science and public affairs at Princeton University Press, entered the fray. After discussions with David Kreps and Ariel Rubinstein, who were active supporters of the collection, they kindly agreed to relieve me of the responsibility of writing the introductory essay. They have kindly written the "appreciation" that follows this foreword.

To all of the people who have kept my procrastination from preventing this collection's reaching the light of day, I hereby express my sincerest thanks. I can only hope that their patience has been worth the effort.

Harold W. Kuhn
Princeton, New Jersey

_____

The Bank of Sweden Prize in Economic Sciences in Memory of Alfred Nobel, 1994, was awarded jointly to John C. Harsanyi, John F. Nash, and Reinhard Selten, "for their pioneering analysis of equilibria in the theory of noncooperative games." In so doing, the Royal Swedish Academy of Sciences took note of a revolutionary change in the language and style of analysis of economics and economists, in which game-theoretic ideas have become commonplace. Game theory has provided economists with a flexible language for discussing many issues central to economic inquiry, from two-person bargaining, to highly personal, repeated, and long-run exchange, to the theoretical foundations of economic models of monopoly and perfect competition. Most fields in economics and certainly economic theory have been dramatically affected by these ideas.

But game theory is not only a subfield of economics. It analyzes abstractly conflicts of interest. Thus game theory has expanded far beyond economics. We find a growing tendency to use game-theoretic concepts and models in a variety of fields: Political scientists use game theory to examine political institutions. Philosophers find game theory a tool for reexamination of norms and social institutions. Biologists find game theory a framework to analyze the conflicting interests between creatures in nature.

The history of game theory in economics may be told as follows. Basic game-theoretic concepts of equilibrium, in the context of competition with a small number of participants, were developed all but formally by Cournot, von Stackelberg, and Bertrand. Mathematical models of games of strategy had been broached by the French mathematician Borel, the Polish mathematician Steinhaus, and the German mathematician Zermelo (for the special case of chess). The first attempt at a general theory with a solution concept for zero-sum two-person games and the associated minimax theorem was published by von Neumann in 1928. However, formal game theory reached a larger audience with the publication of von Neumann and Morgenstern's treatise *Theory of*

Games and Economic Behavior, first published in 1944. Following the Second World War (and in some instances evolving out of work done during the war), the subject underwent explosive development in the 1950s and 1960s. Concepts which today form the heart of game theory as it is applied to economics and other disciplines—Nash equilibrium, the theory of extensive form games, axiomatic bargaining theory, the Shapley value, the core and its connections to competitive equilibrium, most other cooperative game-theory solution concepts, the folk theorem, games of incomplete information, and basic notions of perfection —all were created during this heroic period. A few fundamental developments came a bit later, most notably Selten's further development of the idea of perfection and foundational issues of common knowledge. But by the end of the 1960s, the tools were essentially in place. The revolution in economics occurred quite shortly thereafter. Beginning in the 1970s, and especially in the context of Industrial Organization, the language and techniques of game theory moved from being an esoteric and limited tool of microeconomic theorists to become part of the mainstream language of the discipline.

In this volume, Prof. Harold Kuhn has collected eighteen papers from the heroic era of game theory. Included here you will find the following:

Nash equilibrium. If any concept has achieved primacy in game theory, it is Nash's equilibrium concept regarding the model of a game in strategic form. Chapters 1 and 3 contain the first formal statements of this concept in the literature.

Evolution of equilibrium. In the late 1940s, the RAND Corporation mathematician George Brown proposed an adaptive behavior algorithm (called "fictitious play") for solving zero-sum two-person games. It was not known whether the method converged (a money prize was offered by the RAND Corporation to settle this question) until the publication of Julia Robinson's paper (Chapter 3). With convergence established by her elegant arguments, the model of fictitious play has become a cornerstone of recent work on evolutionary and adaptive learning models.

Extensive form games and perfect recall. Whereas the strategic form of a game lacks a dynamic structure, the extensive form allows the analysis of dynamic considerations. Thompson's paper (Chapter 5) proposes a set of transformations that connects strategically equivalent extensive form games together and to their equivalent strategic form counterpart.

His paper uses Kuhn's model for extensive form games (Chapter 6) which has become the standard model. Kuhn analyzes games with and without perfect information, and introduces the concepts of subgame and perfect recall. This paper contains fundamental insights on the role of perfect recall in answering the question of when games can be solved with "behavioral" modes of play as effectively as by mixed strategies.

Games with incomplete information. How can game-theoretic models be used to analyze competitive situations in which some parties have information that others lack, about their own capabilities or tastes, or about the underlying state of nature? In Chapter 15, John Harsanyi provides the standard answer to this question. After Nash equilibrium, Harsanyi's definition of games with incomplete information is perhaps the single most important innovation from the point of view of modern economic applications.

Perfect equilibrium. In 1965, Selten first formulated the notion of sub-game perfection, a criterion for studying whether a Nash equilibrium of an extensive form game is based on credible threats/strategies. Sub-game perfection had a somewhat limited reach, and in his 1975 paper, reprinted here as Chapter 18, Selten reformulated subgame perfection, giving a notion of perfection that applies to all extensive form games with perfect recall. The concept itself is a crucial tool for studying dynamic competitive interactions but, as importantly or more so, it is a cornerstone of Selten's seminal program for studying competitive dynamics as something more than simple static and simultaneous choice of strategies.

Repeated games. Several works in this volume are the pioneering work regarding the analysis of the model of dynamic zero-sum games: Shapley (Chapter 8), Everett (Chapter 9), and Blackwell–Ferguson (Chapter 11) give seminal analyses of different varieties of these dynamic games, suggesting some of the most innovative mathematical techniques used in game theory.

Games and markets. Game theory has been used to study the foundations of economic (price-mediated) equilibrium in markets, giving insights into the sources of market power and the implications of "competitive" agents for the existence and character of market equilibrium. Two of the seminal works in this line of thought are provided here: Aumann's paper on the existence of competitive equilibrium with (truly) competitive agents (Chapter 13), and Shapley and Shubik's model and analysis of market games (Chapter 17).

Cooperative game theory. The Nobel Prize for 1994 was granted for achievements in noncooperative game theory. Cooperative game theory is the other "half" of the subject. A cooperative game specifies the details about what a coalition (and not just a single player) can achieve without the agreement of the other players. This part of game theory may have made less of an impact on the broad community of economists so far, but we believe that its impact will be felt in the future. In Chapter 10, Aumann and Peleg analyze the von Neumann–Morgenstern Solution for games without side payments; in Chapter 12, Aumann and Maschler define the Bargaining Set.

The core. The central solution concept in cooperative game theory, and especially in cooperative game theory as it has been applied to economics, is the core. The core is especially important for its connection to competitive equilibria in "large" economies. In Chapter 14, Scarf discusses fundamental issues connected with the core; in Chapter 11, Debreu and Scarf formalize the intuition of Edgeworth, that in the competitive equilibrium of a large economy, every agent gets just what they "contribute" to the society.

Nash bargaining solution. In "small" economies, where bargaining can take place, game theory has developed axioms for saying what is a fair and/or reasonable outcome. In the context of two-person bargaining, Nash (Chapter 2) is the seminal reference. This paper sets up the general problem and then solves it axiomatically, paving the way for the huge literature of axiomatic bargaining theory that followed. It is remarkable in that it was written originally as a paper for an undergraduate course before Nash was familiar with the work of von Neumann and Morgenstern.

Shapley value. Along with the Nash Bargaining Solution, the Shapley Value, defined and axiomatized in Chapter 7, is the premier axiomatic standard of "equity" in game theory, the subject of intensive reaxiomatization and (more importantly) of enormous use in applications from cost allocation to recent work in corporate finance.

These eighteen papers include virtually all the foundation stones of game theory. They have also established the formal style of game theory. The emphasis on clear definition and well-knit proofs sets the tone for the further development of game theory.

Over the past five years or so, a number of excellent advanced textbooks in game theory have appeared. With the benefit of hindsight, they cover many of these foundation stones without some of the awkwardness that can accompany original efforts. But to our mind, this collection of papers stands out for their clarity and vision. To see these ideas as they were first expressed is a pleasure for us, the beneficiaries of the heroic era of game theory, and we are pleased to sit back with you and enjoy again these classics.

David Kreps and Ariel Rubinstein

Classics in Game Theory

EQUILIBRIUM POINTS IN n-PERSON GAMES

JOHN F. NASH, JR.*

PRINCETON UNIVERSITY

Communicated by S. Lefschetz, November 16, 1949

ONE MAY define a concept of an n-person game in which each player has a finite set of pure strategies and in which a definite set of payments to the n players corresponds to each n-tuple of pure strategies, one strategy being taken for each player. For mixed strategies, which are probability distributions over the pure strategies, the pay-off functions are the expectations of the players, thus becoming polylinear forms in the probabilities with which the various players play their various pure strategies.

Any n-tuple of strategies, one for each player, may be regarded as a point in the product space obtained by multiplying the n strategy spaces of the players. One such n-tuple counters another if the strategy of each player in the countering n-tuple yields the highest obtainable expectation for its player against the $n - 1$ strategies of the other players in the countered n-tuple. A self-countering n-tuple is called an equilibrium point.

The correspondence of each n-tuple with its set of countering n-tuples gives a one-to-many mapping of the product space into itself. From the definition of countering we see that the set of countering points of a point is *convex*. By using the continuity of the pay-off functions we see that the graph of the mapping is closed. The closedness is equivalent to saying: if $P_1, P_2, \ldots$ and $Q_1, Q_2, \ldots, Q_n, \ldots$ are sequences of points in the product space where $Q_n \to Q$, $P_n \to P$ and Q_n counters P_n then Q counters P.

Since the graph is closed and since the image of each point under the mapping is convex, we infer from Kakutani's theorem[1] that the mapping has a fixed point (i.e., point contained in its image). Hence there is an equilibrium point.

---

*The author is indebted to Dr. David Gale for suggesting the use of Kakutani's theorem to simplify the proof and to the A. E. C. for financial support.

[1] Kakutani, S., *Duke Math. J.*, **8**, 457–459 (1941).

In the two-person zero-sum case the "main theorem"[2] and the existence of an equilibrium point are equivalent. In this case any two equilibrium points lead to the same expectations for the players, but this need not occur in general.

---

[2] Von Neumann, J., and Morgenstern, O., *The Theory of Games and Economic Behaviour*, Chap. 3, Princeton University Press, Princeton, 1947.

THE BARGAINING PROBLEM[1]

John F. Nash, Jr.

A new treatment is presented of a classical economic problem, one which occurs in many forms, as bargaining, bilateral monopoly, etc. It may also be regarded as a nonzero-sum two-person game. In this treatment a few general assumptions are made concerning the behavior of a single individual and a group of two individuals in certain economic environments. From these, the solution (in the sense of this paper) of the classical problem may be obtained. In the terms of game theory, values are found for the game.

INTRODUCTION

A two-person bargaining situation involves two individuals who have the opportunity to collaborate for mutual benefit in more than one way. In the simpler case, which is the one considered in this paper, no action taken by one of the individuals without the consent of the other can affect the well-being of the other one.

The economic situations of monopoly versus monopsony, of state trading between two nations, and of negotiation between employer and labor union may be regarded as bargaining problems. It is the purpose of this paper to give a theoretical discussion of this problem and to obtain a definite "solution"—making, of course, certain idealizations in order to do so. A "solution" here means a determination of the amount of satisfaction each individual should expect to get from the situation, or, rather, a determination of how much it should be worth to each of these individuals to have this opportunity to bargain.

This is the classical problem of exchange and, more specifically, of bilateral monopoly as treated by Cournot, Bowley, Tintner, Fellner, and others. A different approach is suggested by von Neumann and Morgenstern in *Theory of Games and Economic Behavior*[2] which permits the

---

[1]The author wishes to acknowledge the assistance of Professors von Neumann and Morgenstern who read the original form of the paper and gave helpful advice as to the presentation.

[2]John von Neumann and Oskar Morgenstern, *Theory of Games and Economic Behavior*, Princeton: Princeton University Press, 1944 (Second Edition, 1947), pp. 15–31.

identification of this typical exchange situation with a nonzero-sum two-person game.

In general terms, we idealize the bargaining problem by assuming that the two individuals are highly rational, that each can accurately compare his desires for various things, that they are equal in bargaining skill, and that each has full knowledge of the tastes and preferences of the other.

In order to give a theoretical treatment of bargaining situations we abstract from the situation to form a mathematical model in terms of which to develop the theory.

In making our treatment of bargaining we employ a numerical utility, of the type developed in *Theory of Games*, to express the preferences, or tastes, of each individual engaged in bargaining. By this means we bring into the mathematical model the desire of each individual to maximize his gain in bargaining. We shall briefly review this theory in the terminology used in this paper.

UTILITY THEORY OF THE INDIVIDUAL

The concept of an "anticipation" is important in this theory. This concept will be explained partly by illustration. Suppose Mr. Smith knows he will be given a new Buick tomorrow. We may say that he has a Buick anticipation. Similarly, he might have a Cadillac anticipation. If he knew that tomorrow a coin would be tossed to decide whether he would get a Buick or a Cadillac, we should say that he had a $\frac{1}{2}$ Buick, $\frac{1}{2}$ Cadillac anticipation. Thus an anticipation of an individual is a state of expectation which may involve the certainty of some contingencies and various probabilities of other contingencies. As another example, Mr. Smith might know that he will get a Buick tomorrow and think that he has half a chance of getting a Cadillac too. The $\frac{1}{2}$ Buick, $\frac{1}{2}$ Cadillac anticipation mentioned above illustrates the following important property of anticipations: if $0 \leq p \leq 1$ and A and B represent two anticipations, there is an anticipation, which we represent by $pA + (1 - p)B$, which is a probability combination of the two anticipations where there is a probability p of A and $1 - p$ of B.

By making the following assumptions we are enabled to develop the utility theory of a single individual:

1. An individual offered two possible anticipations can decide which is preferable or that they are equally desirable.

2. The ordering thus produced is transitive; if A is better than B and B is better than C then A is better than C.

3. Any probability combination of equally desirable states is just as desirable as either.

4. If A, B, and C are as in assumption (2), then there is a probability combination of A and C which is just as desirable as C. This amounts to an assumption of continuity.

5. If $0 \le p \le 1$ and A and B are equally desirable, then $pA + (1 - p)C$ and $pB + (1 - p)C$ are equally desirable. Also, if A and B are equally desirable, A may be substituted for B in any desirability ordering relationship satisfied by B.

These assumptions suffice to show the existence of a satisfactory utility function, assigning a real number to each anticipation of an individual. This utility function is not unique, that is, if u is such a function then so also is $au + b$, provided $a > 0$. Letting capital letters represent anticipations and small ones real numbers, such a utility function will satisfy the following properties:

(a) $u(A) > u(B)$ is equivalent to A is more desirable than B, etc.

(b) If $0 \le p \le 1$ then $u[pA + (1 - p)B] = pu(A) + (1 - p)u(B)$.

This is the important linearity property of a utility function.

TWO-PERSON THEORY

In *Theory of Games and Economic Behavior* a theory of n-person games is developed which includes as a special case the two-person bargaining problem. But the theory there developed makes no attempt to find a value for a given n-person game, that is, to determine what it is worth to each player to have the opportunity to engage in the game. This determination is accomplished only in the case of the two-person zero-sum game.

It is our viewpoint that these n-person games should have values; that is, there should be a set of numbers which depend continuously upon the set of quantities comprising the mathematical description of the game and which express the utility to each player of the opportunity to engage in the game.

We may define a two-person anticipation as a combination of two one-person anticipations. Thus we have two individuals, each with a certain expectation of his future environment. We may regard the

one-person utility functions as applicable to the two-person anticipations, each giving the result it would give if applied to the corresponding
one-person anticipation which is a component of the two-person anticipation. A probability combination of two two-person anticipations is
defined by making the corresponding combinations for their components. Thus if $[A, B]$ is a two-person anticipation and $0 \leq p \leq 1$, then

$$p[A, B] + (1 - p)[C, D]$$

will be defined as

$$[pA + (1 - p)C, pB + (1 - p)D].$$

Clearly the one-person utility functions will have the same linearity
property here as in the one-person case. From this point onwards when
the term anticipation is used it shall mean two-person anticipation.

In a bargaining situation one anticipation is especially distinguished;
this is the anticipation of no cooperation between the bargainers. It is
natural, therefore, to use utility functions for the two individuals which
assign the number zero to this anticipation. This still leaves each
individual's utility function determined only up to multiplication by a
positive real number. Henceforth any utility functions used shall be
understood to be so chosen.

We may produce a graphical representation of the situation facing
the two by choosing utility functions for them and plotting the utilities
of all available anticipations in a plane graph.

It is necessary to introduce assumptions about the nature of the set of
points thus obtained. We wish to assume that this set of points is
compact and convex, in the mathematical senses. It should be convex
since an anticipation which will graph into any point on a straight line
segment between two points of the set can always be obtained by the
appropriate probability combination of two anticipations which graph
into the two points. The condition of compactness implies, for one
thing, that the set of points must be bounded, that is, that they can all
be inclosed in a sufficiently large square in the plane. It also implies
that any continuous function of the utilities assumes a maximum value
for the set at some point of the set.

We shall regard two anticipations which have the same utility for any
utility function corresponding to either individual as equivalent so that
the graph becomes a complete representation of the essential features

of the situation. Of course, the graph is only determined up to changes of scale since the utility functions are not completely determined.

Now since our solution should consist of *rational* expectations of gain by the two bargainers, these expectations should be realizable by an appropriate agreement between the two. Hence, there should be an available anticipation which gives each the amount of satisfaction he should expect to get. It is reasonable to assume that the two, being rational, would simply agree to that anticipation, or to an equivalent one. Hence, we may think of one point in the set of the graph as representing the solution, and also representing all anticipations that the two might agree upon as fair bargains. We shall develop the theory by giving conditions which should hold for the relationship between this solution point and the set, and from these deduce a simple condition determining the solution point. We shall consider only those cases in which there is a possibility that both individuals could gain from the situation. (This does not exclude cases where, in the end, only one individual could have benefited because the "fair bargain" might consist of an agreement to use a probability method to decide who is to gain in the end. Any probability combination of available anticipations is an available anticipation.)

Let u_1 and u_2 be utility functions for the two individuals. Let $c(S)$ represent the solution point in a set S which is compact and convex and includes the origin. We assume:

6. If α is a point in S such that there exists another point β in S with the property $u_1(\beta) > u_1(\alpha)$ and $u_2(\beta) > u_2(\alpha)$, then $\alpha \neq c(S)$.
7. If the set T contains the set S and $c(T)$ is in S, then $c(T) = c(S)$.

We say that a set S is symmetric if there exist utility operators u_1 and u_2 such that when (a, b) is contained in S, (b, a) is also contained in S; that is, such that the graph becomes symmetrical with respect to the line $u_1 = u_2$.

8. If S is symmetric and u_1 and u_2 display this, then $c(S)$ is a point of the form (a, a), that is, a point on the line $u_1 = u_2$.

The first assumption above expresses the idea that each individual wishes to maximize the utility to himself of the ultimate bargain. The third expresses equality of bargaining skill. The second is more complicated. The following interpretation may help to show the naturalness of this assumption: If two rational individuals would agree that $c(T)$ would

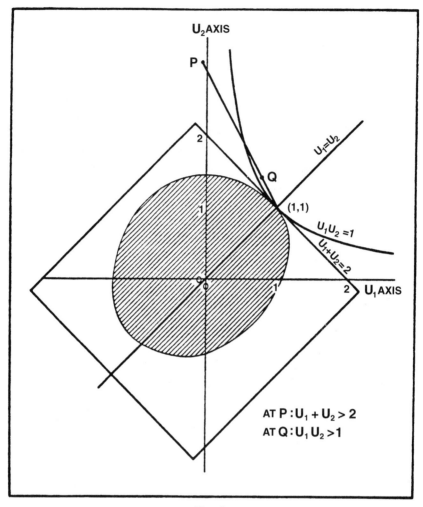

FIG. 1

be a fair bargain if T were the set of possible bargains, then they should be willing to make an agreement, of lesser restrictiveness, not to attempt to arrive at any bargains represented by points outside of the set S if S contained $c(T)$. If S were contained in T this would reduce their situation to one with S as the set of possibilities. Hence $c(S)$ should equal $c(T)$.

We now show that these conditions require that the solution be the point of the set in the first quadrant where u_1u_2 is maximized. We know some such point exists from the compactness. Convexity makes it unique.

Let us now choose the utility functions so that the above-mentioned point is transformed into the point $(1,1)$. Since this involves the multiplication of the utilities by constants, $(1,1)$ will now be the point of maximum $u_1 u_2$. For no points of the set will $u_1 + u_2 > 2$, now, since if there were a point of the set with $u_1 + u_2 > 2$ at some point on the line segment between $(1,1)$ and that point, there would be a value of $u_1 u_2$ greater than one (see Figure 1).

We may now construct a square in the region $u_1 + u_2 \leq 2$ which is symmetrical in the line $u_1 = u_2$, which has one side on the line $u_1 + u_2 = 2$, and which completely encloses the set of alternatives. Considering the square region formed as the set of alternatives, instead of the older set, it is clear that $(1,1)$ is the only point satisfying assumptions (6) and (8). Now using assumption (7) we may conclude that $(1,1)$ must also be the solution point when our original (transformed) set is the set of alternatives. This establishes the assertion.

We shall now give a few examples of the application of this theory.

EXAMPLES

Let us suppose that two intelligent individuals, Bill and Jack, are in a position where they may barter goods but have no money with which to facilitate exchange. Further, let us assume for simplicity that the utility to either individual of a portion of the total number of goods involved is the sum of the utilities to him of the individual goods in that portion. We give below a table of goods possessed by each individual with the utility of each to each individual. The utility functions used for the two individuals are, of course, to be regarded as arbitrary.

Bill's goods	Utility to Bill	Utility to Jack
book	2	4
whip	2	2
ball	2	1
bat	2	2
box	4	1
Jack's goods		
pen	10	1
toy	4	1
knife	6	2
hat	2	2

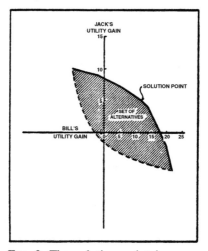

FIG. 2. The solution point is on a
rectangular hyperbola lying in the
first quadrant and touching the set
of alternatives at but one point.

The graph for this bargaining situation is included as an illustration
(Figure 2). It turns out to be a convex polygon in which the point where
the product of the utility gains is maximized is at a vertex and where
there is but one corresponding anticipation. This is:

Bill gives Jack: book, whip, ball, and bat,

Jack gives Bill: pen, toy, and knife.

When the bargainers have a common medium of exchange the
problem may take on an especially simple form. In many cases the
money equivalent of a good will serve as a satisfactory approximate
utility function. (By the money equivalent is meant the amount of
money which is just as desirable as the good to the individual with
whom we are concerned.) This occurs when the utility of an amount of
money is approximately a linear function of the amount in the range of
amounts concerned in the situation. When we may use a common
medium of exchange for the utility function for each individual the set
of points in the graph is such that that portion of it in the first quadrant

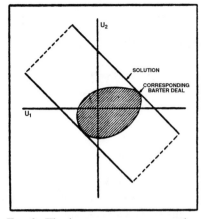

FIG. 3. The inner area represents the bargains possible without the use of money. The area between parallel lines represents the possibilities allowing the use of money. Utility and gain measured by money are here equated for small amounts of money. The solution must be formed using a barter-type bargain for which $u_1 + u_2$ is at a maximum and using also an exchange of money.

forms an isosceles right triangle. Hence the solution has each bargainer getting the same money profit (see Figure 3).

Princeton University

3

NON-COOPERATIVE GAMES

JOHN NASH
(RECEIVED OCTOBER 11, 1950)

INTRODUCTION

Von Neumann and Morgenstern have developed a very fruitful theory of two-person zero-sum games in their book *Theory of Games and Economic Behavior*. This book also contains a theory of *n*-person games of a type which we would call cooperative. This theory is based on an analysis of the interrelationships of the various coalitions which can be formed by the players of the game.

Our theory, in contradistinction, is based on the *absence* of coalitions in that it is assumed that each participant acts independently, without collaboration of communication with any of the others.

The notion of an *equilibrium point* is the basic ingredient in our theory. This notion yields a generalization of the concept of the solution of a two-person zero-sum game. It turns out that the set of equilibrium points of a two-person zero-sum game is simply the set of all pairs of opposing "good strategies."

In the immediately following sections we shall define equilibrium points and prove that a finite non-cooperative game always has at least one equilibrium point. We shall also introduce the notions of solvability and strong solvability of a non-cooperative game and prove a theorem on the geometrical structure of the set of equilibrium points of a solvable game.

As an example of the application of our theory we include a solution of a simplified three-person poker game.

FORMAL DEFINITIONS AND TERMINOLOGY

In this section we define the basic concepts of this paper and set up standard terminology and notation. Important definitions will be preceded by a subtitle indicating the concept defined. The non-cooperative idea will be implicit, rather than explicit, below:

Finite Game:

For us an *n-person game* will be a set of *n players*, or *positions*, each with an associated finite set of *pure strategies*; and corresponding to each player, i, a *payoff function*, p_i, which maps the set of all n-tuples of pure strategies into the real numbers. When we use the term *n-tuple* we shall always mean a set of n items, with each item associated with a different player.

Mixed Strategy, s_i:

A *mixed strategy* of player i will be a collection of non-negative numbers which have unit sum and are in one to one correspondence with his pure strategies.

We write $s_i = \sum_\alpha c_{i\alpha} \pi_{i\alpha}$ with $c_{i\alpha} \geq 0$ and $\sum_\alpha c_{i\alpha} = 1$ to represent such a mixed strategy, where the $\pi_{i\alpha}$'s are the pure strategies of player i. We regard the s_i's as points in a simplex whose vertices are the $\pi_{i\alpha}$'s. This simplex may be regarded as a convex subset of a real vector space, giving us a natural process of linear combination for the mixed strategies.

We shall use the suffixes i, j, k for players and α, β, γ to indicate various pure strategies of a player. The symbols s_i, t_i, and r_i, etc. will indicate mixed strategies; $\pi_{i\alpha}$ will indicate the ith player's α^{th} pure strategy, etc.

Payoff Function, p_i:

The payoff function, p_i, used in the definition of a finite game above, has a unique extension to the n-tuples of mixed strategies which is linear in the mixed strategy of each player [n-linear]. This extension we shall also denote by p_i, writing $p_i(s_1, s_2, \cdots, s_n)$.

We shall write $\mathfrak{s}$ or $\mathfrak{t}$ to denote an n-tuple of mixed strategies and if $\mathfrak{s} = (s_1, s_2, \cdots, s_n)$ then $p_i(\mathfrak{s})$ shall mean $p_i(s_1, s_2, \cdots, s_n)$. Such an n-tuple, $\mathfrak{s}$, will also be regarded as a point in a vector space, the product space of the vector spaces containing the mixed strategies. And the set of all such n-tuples forms, of course, a convex polytope, the product of the simplices representing the mixed strategies.

For convenience we introduce the substitution notation $(\mathfrak{s}; t_i)$ to stand for $(s_1, s_2, \cdots, s_{i-1}, t_i, s_{i+1}, \cdots, s_n)$ where $\mathfrak{s} = (s_1, s_2, \cdots, s_n)$. The effect of successive substitutions $((\mathfrak{s}; t_i); r_j)$ we indicate by $(\mathfrak{s}; t_i; r_j)$, etc.

Equilibrium Point:

An n-tuple $\mathfrak{s}$ is an *equilibrium point* if and only if for every i

$$p_i(\mathfrak{s}) = \max_{\text{all } r_i\text{'s}} [p_i(\mathfrak{s}; r_i)]. \tag{1}$$

Thus an equilibrium point is an n-tuple $\mathfrak{s}$ such that each player's mixed strategy maximizes his payoff if the strategies of the others are held fixed. Thus each player's strategy is optimal against those of the others. We shall occasionally abbreviate equilibrium point by eq. pt.

We say that a mixed strategy s_i *uses* a pure strategy $\pi_{i\alpha}$ if $s_i = \sum_\beta c_{i\beta}\pi_{i\beta}$ and $c_{i\alpha} > 0$. If $\mathfrak{s} = (s_1, s_2, \cdots, s_n)$ and s_i uses $\pi_{i\alpha}$ we also say that $\mathfrak{s}$ uses $\pi_{i\alpha}$.

From the linearity of $p_i(s_1, \cdots, s_n)$ in s_i,

$$\max_{\text{all } r_i\text{'s}} [p_i(\mathfrak{s}; r_i)] = \max_\alpha [p_i(\mathfrak{s}; \pi_{i\alpha})]. \tag{2}$$

We define $p_{i\alpha}(\mathfrak{s}) = p_i(\mathfrak{s}; \pi_{i\alpha})$. Then we obtain the following trivial necessary and sufficient condition for $\mathfrak{s}$ to be an equilibrium point:

$$p_i(\mathfrak{s}) = \max_\alpha p_{i\alpha}(\mathfrak{s}). \tag{3}$$

If $\mathfrak{s} = (s_1, s_2, \cdots, s_n)$ and $s_i = \sum_\alpha c_{i\alpha}\pi_{i\alpha}$ then $p_i(\mathfrak{s}) = \sum_\alpha c_{i\alpha}p_{i\alpha}(\mathfrak{s})$, consequently for (3) to hold we must have $c_{i\alpha} = 0$ whenever $p_{i\alpha}(\mathfrak{s}) < \max_\beta p_{i\beta}(\mathfrak{s})$, which is to say that $\mathfrak{s}$ does not use $\pi_{i\alpha}$ unless it is an optimal pure strategy for player i. So we write

$$\text{if } \pi_{i\alpha} \text{ is used in } \mathfrak{s} \text{ then } p_{i\alpha}(\mathfrak{s}) = \max_\beta p_{i\beta}(\mathfrak{s}) \tag{4}$$

as another necessary and sufficient condition for an equilibrium point.

Since a criterion (3) for an eq. pt. can be expressed by the equating of n pairs of continuous functions on the space of n-tuples $\mathfrak{s}$ the eq. pts. obviously form a closed subset of this space. Actually, this subset is formed from a number of pieces of algebraic varieties, cut out by other algebraic varieties.

EXISTENCE OF EQUILIBRIUM POINTS

A proof of this existence theorem based on Kakutani's generalized fixed point theorem was published in Proc. Nat. Acad. Sci. U.S.A., 36, pp. 48–49. The proof given here is a considerable improvement over that earlier version and is based directly on the Brouwer theorem. We proceed by constructing a continuous transformation T of the space of n-tuples such that the fixed points of T are the equilibrium points of the game.

THEOREM 1. *Every finite game has an equilibrium point.*

PROOF. Let $\mathfrak{s}$ be an n-tuple of mixed strategies, $p_i(\mathfrak{s})$ the corresponding payoff to player i, and $p_{i\alpha}(\mathfrak{s})$ the payoff to player i if he changes to his α^{th} pure strategy $\pi_{i\alpha}$ and the others continue to use their respective mixed strategies from $\mathfrak{s}$. We now define a set of continuous functions of $\mathfrak{s}$ by

$$\varphi_{i\alpha}(\mathfrak{s}) = \max(0, p_{i\alpha}(\mathfrak{s}) - p_i(\mathfrak{s}))$$

and for each component s_i of $\mathfrak{s}$ we define a modification s_i' by

$$s_i' = \frac{s_i + \sum_\alpha \varphi_{i\alpha}(\mathfrak{s})\pi_{i\alpha}}{1 + \sum_\alpha \varphi_{i\alpha}(\mathfrak{s})},$$

calling $\mathfrak{s}'$ the n-tuple $(s_1', s_2', s_3' \cdots s_n')$.

We must now show that the fixed points of the mapping $T: \mathfrak{s} \rightarrow \mathfrak{s}'$ are the equilibrium points.

First consider any n-tuple $\mathfrak{s}$. In $\mathfrak{s}$ the ith player's mixed strategy s_i will use certain of his pure strategies. Some one of these strategies, say $\pi_{i\alpha}$, must be "least profitable" so that $p_{i\alpha}(\mathfrak{s}) \leq p_i(\mathfrak{s})$. This will make $\varphi_{i\alpha}(\mathfrak{s}) = 0$.

Now if this n-tuple $\mathfrak{s}$ happens to be fixed under T the proportion of $\pi_{i\alpha}$ used in s_i must not be decreased by T. Hence, for all β's, $\varphi_{i\beta}(\mathfrak{s})$ must be zero to prevent the denominator of the expression defining s_i' from exceeding 1.

Thus, if $\mathfrak{s}$ is fixed under T, for any i and β $\varphi_{i\beta}(\mathfrak{s}) = 0$. This means no player can improve his payoff by moving to a pure strategy $\pi_{i\beta}$. But this is just a criterion for an eq. pt. [see (2)].

Conversely, if $\mathfrak{s}$ is an eq. pt. it is immediate that all φ's vanish, making $\mathfrak{s}$ a fixed point under T.

Since the space of n-tuples is a cell the Brouwer fixed point theorem requires that T must have at least one fixed point $\mathfrak{s}$, which must be an equilibrium point.

SYMMETRIES OF GAMES

An *automorphism*, or *symmetry*, of a game will be a permutation of its pure strategies which satisfies certain conditions, given below.

If two strategies belong to a single player they must go into two strategies belonging to a single player. Thus if ϕ is the permutation of the pure strategies it induces a permutation ψ of the players.

Each n-tuple of pure strategies is therefore permuted into another n-tuple of pure strategies. We may call χ the induced permutation of these n-tuples. Let ξ denote an n-tuple of pure strategies and $p_i(\xi)$ the payoff to player i when the n-tuple ξ is employed. We require that if

$$j = i^\psi \qquad \text{then } p_j(\xi^\chi) = p_i(\xi)$$

which completes the definition of a symmetry.

The permutation ϕ has a unique linear extension to the mixed strategies. If

$$s_i = \sum_\alpha c_{i\alpha} \pi_{i\alpha} \quad \text{we define} \quad (s_i)^\phi = \sum_\alpha c_{i\alpha} (\pi_{i\alpha})^\phi.$$

The extension of ϕ to the mixed strategies clearly generates an extension of χ to the n-tuples of mixed strategies. We shall also denote this by χ.

We define a *symmetric n-tuple* $\mathbf{s}$ of a game by $\mathbf{s}^\chi = \mathbf{s}$ for all χ's.

THEOREM 2. *Any finite game has a symmetric equilibrium point.*

PROOF. First we note that $s_{i0} = \sum_\alpha \pi_{i\alpha} / \sum_\alpha 1$ has the property $(s_{i0})^\phi = s_{j0}$ where $j = i^\psi$, so that the n-tuple $\mathbf{s}_0 = (s_{10}, s_{20}, \cdots, s_{n0})$ is fixed under any χ; hence any game has at least one symmetric n-tuple.

If $\mathbf{s} = (s_1, \cdots, s_n)$ and $\mathbf{t} = (t_1, \cdots, t_n)$ are symmetric then

$$\frac{\mathbf{s} + \mathbf{t}}{2} = \left(\frac{s_1 + t_1}{2}, \frac{s_2 + t_2}{2}, \cdots, \frac{s_n + t_n}{2} \right)$$

is also symmetric because $\mathbf{s}^\chi = \mathbf{s} \leftrightarrow s_j = (s_i)^\phi$, where $j = i^\psi$, hence

$$\frac{s_j + t_j}{2} = \frac{(s_i)^\phi + (t_i)^\phi}{2} = \left(\frac{s_i + t_i}{2} \right)^\phi,$$

hence

$$\left(\frac{\mathbf{s} + \mathbf{t}}{2} \right)^\chi = \frac{\mathbf{s} + \mathbf{t}}{2}.$$

This shows that the set of symmetric n-tuples is a convex subset of the space of n-tuples since it is obviously closed.

Now observe that the mapping $T\colon \mathfrak{s} \to \mathfrak{s}'$ used in the proof of the existence theorem was intrinsically defined. Therefore, if $\mathfrak{s}_2 = T\mathfrak{s}_1$ and χ is derived from an automorphism of the game we will have $\mathfrak{s}_2^\chi = T\mathfrak{s}_1^\chi$. If $\mathfrak{s}_1$ is symmetric $\mathfrak{s}_1^\chi = \mathfrak{s}_1$ and therefore $\mathfrak{s}_2^\chi = T\mathfrak{s}_1 = \mathfrak{s}_2$. Consequently this mapping maps the set of symmetric n-tuples into itself.

Since this set is a cell there must be a symmetric fixed point $\mathfrak{s}$ which must be a symmetric equilibrium point.

SOLUTIONS

We define here solutions, strong solutions, and sub-solutions. A non-cooperative game does not always have a solution, but when it does the solution is unique. Strong solutions are solutions with special properties. Sub-solutions always exist and have many of the properties of solutions, but lack uniqueness.

S_1 will denote a set of mixed strategies of player i and $\mathfrak{S}$ a set of n-tuples of mixed strategies.

Solvability:

A game is *solvable* if its set, $\mathfrak{S}$, of equilibrium points satisfies the condition

$$(\mathfrak{t}; r_i) \in \mathfrak{S} \quad \text{and} \quad \mathfrak{s} \in \mathfrak{S} \to (\mathfrak{s}; r_i) \in \mathfrak{S} \qquad \text{for all } i\text{'s}. \qquad (5)$$

This is called the *interchangeability* condition. The *solution* of a solvable game is its set, $\mathfrak{S}$, of equilibrium points.

Strong Solvability:

A game is *strongly solvable* if it has a solution, $\mathfrak{S}$, such that for all i's

$$\mathfrak{s} \in \mathfrak{S} \quad \text{and} \quad p_i(\mathfrak{s}; r_i) = p_i(\mathfrak{s}) \to (\mathfrak{s}; r_i) \in \mathfrak{S}$$

and then $\mathfrak{S}$ is called a *strong solution*.

Equilibrium Strategies:

In a solvable game let S_i be the set of all mixed strategies s_i such that for some $\mathfrak{t}$ the n-tuple $(\mathfrak{t}; s_i)$ is an equilibrium point. [s_i is the ith component of some equilibrium point.] We call S_i the set of *equilibrium strategies* of player i.

Sub-solutions:

If $\mathfrak{S}$ is a subset of the set of equilibrium points of a game and satisfies condition (1); and if $\mathfrak{S}$ is maximal relative to this property then we call $\mathfrak{S}$ a *sub-solution*.

For any sub-solution $\mathfrak{S}$ we define the ith *factor set*, S_i, as the set of all s_i's such that $\mathfrak{S}$ contains $(\mathfrak{t}; s_i)$ for some $\mathfrak{t}$.

Note that a sub-solution, when unique, is a solution; and its factor sets are the sets of equilibrium strategies.

THEOREM 3. *A sub-solution,* $\mathfrak{S}$, *is the set of all n-tuples* $(s_1, s_2, \cdots, s_n)$ *such that each* $s_i \in S_i$ *where* S_i *is the* i^{th} *factor set of* $\mathfrak{S}$. *Geometrically,* $\mathfrak{S}$ *is the product of its factor sets.*

PROOF. Consider such an n-tuple $(s_1, s_2, \cdots, s_n)$. By definition $\exists \mathfrak{t}_1, \mathfrak{t}_2, \cdots, \mathfrak{t}_n$ such that for each i $(\mathfrak{t}_i; s_i) \in \mathfrak{S}$. Using the condition (5) $n - 1$ times we obtain successively $(\mathfrak{t}_1; s_1) \in \mathfrak{S}$, $(\mathfrak{t}_1; s_1; s_2) \in \mathfrak{S}, \cdots, (\mathfrak{t}_1; s_1; s_2; \cdots; s_n) \in \mathfrak{S}$ and the last is simply $(s_1, s_2, \cdots, s_n) \in \mathfrak{S}$, which we needed to show.

THEOREM 4. *The factor sets* $S_1, S_2, \cdots, S_n$ *of a sub-solution are closed and convex as subsets of the mixed strategy spaces.*

PROOF. It suffices to show two things: (a) if s_i and $s_i' \in S_i$ then $s_i^* = (s_i + s_i')/2 \in S_i$; (b) if $s_i^\#$ is a limit point of S_i then $s_i^\# \in S_i$.

Let $\mathfrak{t} \in \mathfrak{S}$. Then we have $p_j(\mathfrak{t}; s_i) \geq p_j(\mathfrak{t}; s_i; r_j)$ and $p_j(\mathfrak{t}; s_i') \geq p_j(\mathfrak{t}; s_i'; r_j)$ for any r_j, by using the criterion of (1) for an eq. pt. Adding these inequalities, using the linearity of $p_j(s_1, \cdots, s_n)$ in s_i, and dividing by 2, we get $p_j(\mathfrak{t}; s_i^*) \geq p_j(\mathfrak{t}; s_i^*; r_j)$ since $s_i^* = (s_i + s_i')/2$. From this we know that $(\mathfrak{t}; s_i)$ is an eq. pt. for any $\mathfrak{t} \in \mathfrak{S}$. If the set of all such eq. pts. $(\mathfrak{t}; s_i^*)$ is added to $\mathfrak{S}$ the augmented set clearly satisfies condition (5), and since $\mathfrak{S}$ was to be maximal it follows that $s_i^* \in S_i$.

To attack (b) note that the n-tuple $(\mathfrak{t}; s_i^\#)$, where $\mathfrak{t} \in \mathfrak{S}$, will be a limit point of the set of n-tuples of the form $(\mathfrak{t}; s_i)$ where $s_i \in S_i$, since $s_i^\#$ is a limit point of S_i. But this set is a set of eq. pts. and hence any point in its closure is an eq. pt., since the set of all eq. pts. is closed. Therefore $(\mathfrak{t}; s_i^\#)$ is an eq. pt. and hence $s_i^\# \in S_i$ from the same argument as for s_i^*.

Values:

Let $\mathfrak{S}$ be the set of equilibrium points of a game. We define

$$v_i^+ = \max_{\mathfrak{s} \in \mathfrak{S}} [p_i(\mathfrak{s})], \qquad v_i^- = \min_{\mathfrak{s} \in \mathfrak{S}} [p_i(\mathfrak{s})].$$

If $v_i^+ = v_i^-$ we write $v_i = v_i^+ = v_i^-$. v_i^+ is the *upper value* to player i of the game; v_i^- the *lower value*; and v_i the *value*, if it exists.

Values will obviously have to exist if there is but one equilibrium point.

One can define *associated values* for a sub-solution by restricting $\subseteq$ to the eq. pts. in the sub-solution and then using the same defining equations as above.

A two-person zero-sum game is always solvable in the sense defined above. The sets of equilibrium strategies S_1 and S_2 are simply the sets of "good" strategies. Such a game is not generally strongly solvable; strong solutions exist only when there is a "saddle point" in *pure* strategies.

SIMPLE EXAMPLES

These are intended to illustrate the concepts defined in the paper and display special phenomena which occur in these games.

The first player has the roman letter strategies and the payoff to the left, etc.

Ex. 1
5	$a\alpha$	-3
-4	$a\beta$	4
-5	$b\alpha$	5
3	$b\beta$	-4

Solution $\left(\dfrac{9}{16}a + \dfrac{7}{16}b, \dfrac{7}{17}a + \dfrac{10}{17}\beta\right)$

$v_1 = \dfrac{-5}{17}, v_2 = +\dfrac{1}{2}$

Ex. 2
1	$a\alpha$	1
-10	$a\beta$	10
10	$b\alpha$	-10
-1	$b\beta$	-1

Strong Solution (b, β)

$v_1 = v_2 = -1$

Ex. 3
1	$a\alpha$	1
-10	$a\beta$	-10
-10	$b\alpha$	-10
1	$b\beta$	1

Unsolvable; equilibrium points $(a, \alpha), (b, \beta)$,

and $\left(\dfrac{a}{2} + \dfrac{b}{2}, \dfrac{\alpha}{2} + \dfrac{\beta}{2}\right)$. The strategies in the last case have maxi-min and mini-max properties.

Ex. 4
1	$a\alpha$	1
0	$a\beta$	1
1	$b\alpha$	0
0	$b\beta$	0

Strong Solution: all pairs of mixed strategies.

$v_1^+ = v_2^+ = 1, v_1^- = v_2^- = 0.$

Ex. 5
1	$a\alpha$	2
-1	$a\beta$	-4
-4	$b\alpha$	-1
2	$b\beta$	1

Unsolvable; eq. pts. (a, α), (b, α) and

$\left(\dfrac{1}{4}a + \dfrac{3}{4}b, \dfrac{3}{8}\alpha + \dfrac{5}{8}\beta\right)$. However, empirical tests show a tendency toward (a, α).

Ex. 6
1	$a\alpha$	1
0	$a\beta$	0
0	$b\alpha$	0
0	$b\beta$	0

Eq. pts.: (a, α) and (b, β), with (b, β) an example of instability.

GEOMETRICAL FORM OF SOLUTIONS

In the two-person zero-sum case it has been shown that the set of "good" strategies of a player is a convex polyhedral subset of his strategy space. We shall obtain the same result for a player's set of equilibrium strategies in any solvable game.

THEOREM 5. *The sets $S_1, S_2, \cdots, S_n$ of equilibrium strategies in a solvable game are polyhedral convex subsets of the respective mixed strategy spaces.*

PROOF. An n-tuple $\mathfrak{s}$ will be an equilibrium point if and only if for every i

$$p_i(\mathfrak{s}) = \max_{\alpha} p_{i\alpha}(\mathfrak{s}) \tag{6}$$

which is condition (3). An equivalent condition is for every i and α

$$p_i(\mathfrak{s}) - p_{i\alpha}(\mathfrak{s}) \geqq 0. \tag{7}$$

Let us now consider the form of the set S_j of equilibrium strategies, s_j, of player j. Let $\mathfrak{t}$ be any equilibrium point, then $(\mathfrak{t}; s_j)$ will be an equilibrium point if and only if $s_j \in S_j$, from Theorem 2. We now apply conditions (2) to $(\mathfrak{t}; s_j)$, obtaining

$$s_j \in S_j \leftrightarrow \text{for all } i, \alpha \qquad p_i(\mathfrak{t}; s_j) - p_i\alpha(\mathfrak{t}; s_j) \geqq 0. \tag{8}$$

Since p_i is n-linear and $\mathfrak{t}$ is constant these are a set of linear inequalities of the form $F_{i\alpha}(s_j) \geqq 0$. Each such inequality is either satisfied for all s_j or for those lying on and to one side of some hyperplane passing through the strategy simplex. Therefore, the complete set [which is finite] of conditions will all be satisfied simultaneously on some convex polyhedral subset of player j's strategy simplex. [Intersection of half-spaces.]

As a corollary we may conclude that S_j is the convex closure of a finite set of mixed strategies [vertices].

DOMINANCE AND CONTRADICTION METHODS

We say that s_i' dominates s_i if $p_i(\mathfrak{t}; s_i') > p_i(\mathfrak{t}; s_i)$ for every $\mathfrak{t}$.

This amounts to saying that s_i' gives player i a higher payoff than s_i no matter what the strategies of the other players are. To see whether a strategy s_i' dominates s_i it suffices to consider only pure strategies for the other players because of the n-linearity of p_i.

It is obvious from the definitions that *no equilibrium point can involve a dominated strategy s_i*.

The domination of one mixed strategy by another will always entail other dominations. For suppose s_i' dominates s_i and t_i uses all of the pure strategies which have a higher coefficient in s_i than in s_i'. Then for a small enough ρ

$$t_i' = t_i + \rho(s_i' - s_i)$$

is a mixed strategy; and t_i dominates t_i' by linearity.

One can prove a few properties of the set of undominated strategies. It is simply connected and is formed by the union of some collection of faces of the strategy simplex.

The information obtained by discovering dominances for one player may be of relevance to the others, insofar as the elimination of classes of mixed strategies as possible components of an equilibrium point is concerned. For the t's whose components are all undominated are all that need be considered and thus eliminating some of the strategies of one player may make possible the elimination of a new class of strategies for another player.

Another procedure which may be used in locating equilibrium points is the contradiction-type analysis. Here one assumes that an equilibrium point exists having component strategies lying within certain regions of the strategy spaces and proceeds to deduce further conditions which must be satisfied if the hypothesis is true. This sort of reasoning may be carried through several stages to eventually obtain a contradiction indicating that there is no equilibrium point satisfying the initial hypothesis.

A THREE-MAN POKER GAME

As an example of the application of our theory to a more or less realistic case we include the simplified poker game given below. The rules are as follows:

(a) The deck is large, with equally many *high* and *low* cards, and a hand consists of one card.

(b) Two chips are used to ante, open, or call.

(c) The players play in rotation and the game ends after all have passed or after one player has opened and the others have had a chance to call.

(d) If no one bets the antes are retrieved.

(e) Otherwise the pot is divided equally among the highest hands which have bet.

We find it more satisfactory to treat the game in terms of quantities we call "behavior parameters" than in the normal form of *Theory of Games and Economic Behavior*. In the normal form representation two mixed strategies of a player may be equivalent in the sense that each makes the individual choose each available course of action in each particular situation requiring action on his part with the same frequency. That is, they represent the same behavior pattern on the part of the individual.

Behavior parameters give the probabilities of taking each of the various possible actions in each of the various possible situations which may arise. Thus they describe behavior patterns.

In terms of behavior parameters the strategies of the players may be represented as follows, assuming that since there is no point in passing with a *high* card at one's last opportunity to bet that this will not be done. The greek letters are the probabilities of the various acts.

	First Moves	Second Moves
I	α Open on *high* β Open on *low*	κ Call III on *low* λ Call II on *low* μ Call II and III on *low*
II	γ Call I on *low* δ Open on *high* ε Open on *low*	ν Call III on *low* ξ Call III and I on *low*
III	ζ Call I and II on *low* η Open on *low* θ Call I on *low* ι Call II on *low*	Player III never gets a second move

We locate all possible equilibrium points by first showing that most of the greek parameters must vanish. By dominance mainly with a little contradiction-type analysis β is eliminated and with it go γ, ζ, and θ by dominance. Then contradictions eliminate μ, ξ, ι, λ, κ, and ν in that order. This leaves us with α, δ, ε, and η. Contradiction analysis shows that none of these can be zero or one and thus we obtain a system of simultaneous algebraic equations. The equations happen to have but one solution with the variables in the range $(0, 1)$. We get

$$\alpha = \frac{21 - \sqrt{321}}{10}, \qquad \eta = \frac{5\alpha + 1}{4}, \qquad \delta = \frac{5 - 2\alpha}{5 + \alpha}, \qquad \varepsilon = \frac{4\alpha - 1}{\alpha + 5}.$$

These yields $\alpha = .308$, $\eta = .635$, $\delta = .826$, and $\varepsilon = .044$. Since there is only one equilibrium point the game has values; these are

$$v_1 = -.147 = -\frac{(1 + 17\alpha)}{8(5 + \alpha)}, \qquad v_2 = -.096 = -\frac{1 - 2\alpha}{4},$$

and

$$v_3 = .243 = \frac{79}{40}\left(\frac{1 - \alpha}{5 + \alpha}\right).$$

A more complete investigation of this poker game is published in Annals of Mathematics Study No. 24, *Contributions to the Theory of Games*. There the solution is studied as the ratio of ante to bet varies, and the potentialities of coalitions are investigated.

APPLICATIONS

The study of n-person games for which the accepted ethics of fair play imply non-cooperative playing is, of course, an obvious direction in which to apply this theory. And poker is the most obvious target. The analysis of a more realistic poker game than our very simple model should be quite an interesting affair.

The complexity of the mathematical work needed for a complete investigation increases rather rapidly, however, with increasing complexity of the game; so that analysis of a game much more complex than the example given here might only be feasible using approximate computational methods.

A less obvious type of application is to the study of cooperative games. By a cooperative game we mean a situation involving a set of players, pure strategies, and payoffs as usual; but with the assumption that the players can and will collaborate as they do in the von Neumann and Morgenstern theory. This means the players may communicate and form coalitions which will be enforced by an umpire. It is unnecessarily restrictive, however, to assume any transferability or even comparability of the payoffs [which should be in utility units] to different players. Any desired transferability can be put into the game itself instead of assuming it possible in the extra-game collaboration.

The writer has developed a "dynamical" approach to the study of cooperative games based upon reduction to non-cooperative form. One proceeds by constructing a model of the pre-play negotiation so that the steps of negotiation become moves in a larger non-cooperative game [which will have an infinity of pure strategies] describing the total situation.

This larger game is then treated in terms of the theory of this paper [extended to infinite games] and if values are obtained they are taken as the values of the cooperative game. Thus the problem of analyzing a cooperative game becomes the problem of obtaining a suitable, and convincing, non-cooperative model for the negotiation.

The writer has, by such a treatment, obtained values for all finite two-person cooperative games, and some special n-person games.

ACKNOWLEDGEMENTS

Drs. Tucker, Gale, and Kuhn gave valuable criticism and suggestions for improving the exposition of the material in this paper. David Gale suggested the investigation of symmetric games. The solution of the Poker model was a joint project undertaken by Lloyd S. Shapley and the author. Finally, the author was sustained financially by the Atomic Energy Commission in the period 1949–50 during which this work was done.

BIBLIOGRAPHY

(1) von Neumann, Morgenstern, Theory of Games and Economic Behavior, Princeton University Press, 1944.
(2) J. F. Nash, Jr., *Equilibrium Points in N-Person Games*, Proc. Nat. Acad. Sci. U.S.A. 36 (1950) 48–49.
(3) J. F. Nash, L. S. Shapley, A Simple Three-Person Poker Game, Annals of Mathematics Study No. 24, Princeton University Press, 1950.
(4) John Nash, *Two Person Cooperative Games*, to appear in Econometrica.
(5) H. W. Kuhn, *Extensive Games*, Proc. Nat. Acad. Sci. U.S.A. 36 (1950) 570–576.

AN ITERATIVE METHOD OF SOLVING A GAME

JULIA ROBINSON

(JULY 28, 1950)

A TWO-PERSON GAME[1] can be represented by its pay-off matrix $A = (a_{ij})$. The first player chooses one of the m rows and the second player simultaneously chooses one of the n columns. If the ith row and the jth column are chosen, then the second player pays the first player a_{ij}.

If the first player plays the ith row with probability x_i and the second player plays the jth column with probability y_j where $x_i \geq 0$, $\Sigma x_i = 1$, $y_j \geq 0$, and $\Sigma y_j = 1$, then the expectation of the first player is $\Sigma\Sigma a_{ij} x_i y_j$. Furthermore,

$$\min_j \sum_i a_{ij} x_i \leq \max_i \sum_j a_{ij} y_j, \tag{1}$$

since

$$\min_j \sum_i a_{ij} x_i \leq \sum_i \sum_j a_{ij} x_i y_j \leq \max_i \sum_j a_{ij} y_j.$$

The minimax theorem of game theory (see [1] page 153) asserts that for some set of probabilities $X = (x_1, \cdots, x_m)$ and $Y = (y_1, \cdots, y_n)$ the equality holds in (1). Such a pair (X, Y) is called a solution of the game. The value v of the game is defined by

$$v = \min_j \sum_i a_{ij} x_i = \max_i \sum_j a_{ij} y_j,$$

where (X, Y) is a solution of the game.

In this paper, we shall show the validity of an iterative procedure suggested by George W. Brown [2]. This method corresponds to each player choosing in turn the best pure strategy against the accumulated mixed strategy of his opponent up to them.

---

[1] More technically, a finite two-person zero-sum game. See [1] in the bibliography at the end of the paper.

Let $A = (a_{ij})$ be an $m \times n$ matrix. $A_{i.}$ will denote the ith row of A and $A_{.j}$, the jth column. Similarly, if $V(t)$ is a vector, then $v_j(t)$ is the jth component. Let $\max V(t) = \max_j v_j(t)$ and $\min V(t) = \min_j v_j(t)$. In this notation, (1) can be rewritten as follows:

$$\min_i \sum_i A_{i.} x_i \leq \max_j \sum_j A_{.j} y_j, \tag{2}$$

whenever $x_i \geq 0$, $\Sigma x_i = 1$, $y_j \geq 0$, and $\Sigma y_j = 1$.

DEFINITION 1. A system (U, V) consisting of a sequence of n-dimensional vectors $U(0), U(1), \cdots$ and a sequence of m-dimensional vectors $V(0), V(1), \cdots$ is called a *vector system* for A provided that

$$\min U(0) = \max V(0),$$

and

$$U(t + 1) = U(t) + A_{i.}, \qquad V(t + 1) = V(t) + A_{.j},$$

where i and j satisfy the conditions

$$v_i(t) = \max V(t), \qquad u_j(t) = \min U(t).$$

Thus a vector system for A can be formed recursively from a given $U(0)$ and $V(0)$. At each step, the row added to U is determined by a maximum component of V and the column added to V is determined by a minimum component of U.

An alternate notion of vector system is obtained if the condition on j in Definition 1 is replaced by

$$u_j(t + 1) = \min U(t + 1).$$

A vector system of this new type can also be built up recursively. The only difference is that here successive U and V are determined alternatively while in the other definition U and V could be obtained simultaneously. In all the following proofs and theorems, either definition may be used.

In the special case $U(0) = 0$ and $V(0) = 0$, we see that $U(t)/t$ is a weighted average of the rows of A and $V(t)/t$ is a weighted average of the columns. Hence for every t and t',

$$\frac{\min U(t)}{t} \leqq v \leqq \frac{\max V(t')}{t'}.$$

If for some t and t', these two bounds are equal, we have a solution of the game. Unfortunately, this is not always the case. However George Brown [2] conjectured that as t and t' tend to ∞, the two bounds approach v. The main result of this paper is to prove this for any vector system. In numerical examples, vector systems of the second kind appear to converge more rapidly than the first.

THEOREM.[2] *If (U, V) is a vector system for A, then*

$$\lim_{t \to \infty} \frac{\min U(t)}{t} = \lim_{t \to \infty} \frac{\max V(t)}{t} = v.$$

The proof will be divided into four lemmas.

LEMMA 1. *If (U, V) is a vector system for a matrix A, then*

$$\liminf_{t \to \infty} \frac{\max V(t) - \min U(t)}{t} \geqq 0.$$

PROOF. For each t,

$$V(t) = V(0) + t \sum_j y_j A_{\cdot j} \quad \text{where } y_j \geqq 0, \ \sum y_j = 1,$$

and

$$U(t) = U(0) + t \sum_i x_i A_{i \cdot} \quad \text{where } x_i \geqq 0, \ \sum x_i = 1.$$

Hence

$$\max V(t) \geqq \min V(0) + t \max \sum_j y_j A_{\cdot j} \geqq \min V(0) + tv,$$

[2] The solution to Problem 5 in the RAND Mathematical Problem Series II is contained as a special case of this theorem.

and

$$\min U(t) \leq \max U(0) + t \min_{i} \sum_{i} x_i A_{i.} \leq \max U(0) + tv.$$

Therefore,

$$\liminf_{t \to \infty} \frac{\max V(t) - \min U(t)}{t} \geq 0.$$

DEFINITION 2. If (U, V) is a vector system for A, then we say that the ith row is eligible in the interval (t, t') provided that there exists t_1 with

$$t \leq t_1 \leq t'$$

and

$$v_i(t_1) = \max V(t_1).$$

Similarly, the jth column is eligible in the interval (t, t') if there exists t_2 with

$$t \leq t_2 \leq t'$$

and

$$u_j(t_2) = \min U(t_2).$$

LEMMA 2. *Given a vector system* (U, V) *for* A, *then if all the rows and columns of* A *are eligible in the interval* $(s, s + t)$,

$$\max U(s + t) - \min U(s + t) \leq 2at$$

and

$$\max V(s + t) - \min V(s + t) \leq 2at,$$

where

$$a = \max_{i, j} |a_{ij}|.$$

PROOF. Let j be such that

$$u_j(s + t) = \max U(s + t).$$

Choose t' with $s \leq t' \leq s + t$ so that

$$u_j(t') = \min U(t').$$

Then

$$u_j(s + t) \leq u_j(t') + at = \min U(t') + at,$$

since the change in the ith component in t steps is not more than at. But

$$\min U(s + t) \geq \min U(t') - at,$$

so that

$$\max U(s + t) - \min U(s + t) \leq 2at.$$

Similarly,

$$\max V(s + t) - \min V(s + t) \leq 2at.$$

LEMMA 3. *If all the rows and columns of A are eligible in* $(s, s + t)$ *for a given vector system* (U, V), *then*

$$\max V(s + t) - \min U(s + t) \leq 4at.$$

PROOF. By Lemma 2,

$$\max V(s + t) - \min U(s + t) \leq 4at + \min V(s + t) - \max U(s + t).$$

Hence it is sufficient to show that $\min V(s + t) \leq \max U(s + t)$. Now applying (2) to the transpose of A, we have

$$\min_{j} \sum A_{\cdot j} y_j \leq \max_{i} \sum A_{i \cdot} x_i,$$

whenever $x_i \geq 0$, $\Sigma x_i = 1$, $y_j \geq 0$, and $\Sigma y_j = 1$. In particular, choose x_i and y_j satisfying

$$U(s + t) = U(0) + (s + t) \sum A_{i \cdot} x_i,$$

$$V(s + t) = V(0) + (s + t) \sum A_{\cdot j} y_j.$$

Then

$$\min V(s + t) \leq \max V(0) + (s + t) \min \sum A_{\cdot j} y_j$$

$$\leq \min U(0) + (s + t) \max \sum A_{i \cdot} x_i$$

$$\leq \max U(s + t).$$

LEMMA 4. *To every matrix A and $\varepsilon > 0$, there exists t_0 such that for any vector system (U, V),*

$$\max V(t) - \min U(t) < \varepsilon t \qquad for\ t \geq t_0.$$

PROOF. The theorem holds for matrices of order 1 since $U(t) = V(t)$ for all t. Assume the theorem holds for all submatrices of A, then we will show by induction that it holds for A. Choose t^* so that for any vector system (U', V') corresponding to a submatrix A' of A, we have

$$\max V'(t) - \min U'(t) < \tfrac{1}{2}\varepsilon t \qquad \text{whenever } t \geq t^*.$$

We shall prove that if in the given vector system (U, V) for A, some row or column is not eligible in the interval $(s, s + t^*)$, then

$$\max V(s + t^*) - \min U(s + t^*) < \max V(s) - \min U(s) + \tfrac{1}{2}\varepsilon t^*. \quad (3)$$

Suppose, for example, that the kth row is not eligible in the interval $(s, s + t^*)$. Then we can construct a vector system (U', V') for the matrix A' obtained by deleting the kth row of A, in the following way:

$$U'(t) = U(s + t) + C,$$

$$V'(t) = \text{Proj}_k V(s + t) \qquad \text{for } t = 0, 1, \cdots, t^*,$$

where C is the n-dimensional vector all of whose components are equal to $\max V(s) - \min U(s)$ and $\text{Proj}_k V$ is the vector obtained from V by omitting the kth component. The rows of A' will be numbered $1, 2, \cdots, k - 1, k + 1, \cdots, m$. Notice first that $\min U'(0) = \max V'(0)$. Furthermore, if

$$U(s + t + 1) = U(s + t) + A_{i \cdot}, \qquad V(s + t + 1) = V(s + t) + A_{\cdot j},$$

then

$$U'(t + 1) = U'(t) + A'_{i\cdot}, \qquad V'(t + 1) = V'(t) + A'_{\cdot j}.$$

Also $v_i(s + t) = \max V(s + t)$ if and only if $v_i'(t) = \max V'(t)$ and $u_j(s + t) = \min U(s + t)$ if and only if $u_j'(t) = \min U'(t)$ for $0 \le t \le t^*$. Hence we see that U' and V' must satisfy the recursive restrictions of the definition of a vector system for $0 \le t \le t^*$, since U and V do. Naturally, we may continue U' and V' indefinitely to form a vector system for A'.

Now by the choice of t^*, we know that

$$\max V'(t^*) - \min U'(t^*) < \tfrac{1}{2}\varepsilon t^*.$$

Hence

$$\max V(s + t^*) - \min U(s + t^*)$$
$$= \max V'(t^*) - \min U'(t^*) + \max V(s) - \min U(s)$$
$$< \max V(s) - \min U(s) + \tfrac{1}{2}\varepsilon t^*.$$

We can now show that given any vector system (U, V) for A,

$$\max V(t) - \min U(t) < \varepsilon t \qquad \text{for } t \ge 8at^*/\varepsilon.$$

Consider $t > t^*$. Let θ with $0 \le \theta < 1$ and q a positive integer be so chosen that $t = (\theta + q)t^*$.

CASE 1. Suppose there is a positive integer $s \le q$ so that all rows and columns of A are eligible in the interval $((\theta + s - 1)t^*, (\theta + s)t^*)$. Take the largest such s, then

$$\max V(t) - \min U(t)$$
$$\le \max V((\theta + s)t^*) - \min U((\theta + s)t^*) + \tfrac{1}{2}\varepsilon(q - s)t^*. \quad (4)$$

We obtain this inequality by repeated application of (3), since in each of the intervals

$$((\theta + r - 1)t^*, (\theta + r)t^*) \qquad \text{for } r = s + 1, \cdots, q,$$

some row or column of A is not eligible. From Lemma 3 and the choice of s, we have

$$\max V((\theta + s)t^*) - \min U((\theta + s)t^*) \leq 4at^*. \qquad (5)$$

From (4) and (5), we obtain

$$\max V(t) - \min U(t) \leq 4at^* + \tfrac{1}{2}\varepsilon(q - s)t^* < (4a + \tfrac{1}{2}\varepsilon q)t^*.$$

CASE 2. If there is no such s, then in each interval $((\theta + r - 1)t^*, (\theta + r)t^*)$ some row or column of A is not eligible. Hence

$$\max V(t) - \min U(t) < \max V(\theta t^*) - \min U(\theta t^*) + \tfrac{1}{2}\varepsilon q t^*$$

$$\leq 4a\theta t^* + \tfrac{1}{2}\varepsilon q t^*.$$

Therefore, in either case,

$$\max V(t) - \min U(t) < (4a + \tfrac{1}{2}\varepsilon q)t^*$$

$$\leq 4at^* + \tfrac{1}{2}\varepsilon t < \varepsilon t \qquad \text{for } t \geq 8at^*/\varepsilon.$$

From Lemmas 1 and 4, we see that

$$\lim_{t \to \infty} \frac{\max V(t) - \min U(t)}{t} = 0.$$

But from (1),

$$\limsup_{t \to \infty} \frac{\min U(t)}{t} \leq v,$$

$$\liminf_{t \to \infty} \frac{\max V(t)}{t} \geq v.$$

Hence

$$\lim_{t \to \infty} \frac{\min V(t)}{t} = \lim_{t \to \infty} \frac{\max V(t)}{t} = v,$$

which completes the proof of the theorem.

THE RAND CORPORATION,
SANTA MONICA, CALIFORNIA

BIBLIOGRAPHY

[1] J. von Neumann and O. Morgenstern, Theory of Games and Economic Behavior, Princeton University Press.

[2] G. W. Brown, *Some notes on computation of Games Solutions*, RAND Report P-78, April 1949, The RAND Corporation, Santa Monica, California.

5

EQUIVALENCE OF GAMES IN EXTENSIVE FORM

F. B. THOMPSON

Summary: Four simple transformations are characterized which are sufficient to carry any two equivalent games in extensive form one into the other. Application is made to the problem of simplification of a game in extensive form.

Informal Discussion: Given a game, each player has available to him a set of pure strategies. Considerations involving the playing of the game in its so-called *extensive form* can often be simplified by passing to its *normal form* in which each player is to choose a (pure) strategy in ignorance of the choices of his opponents, and the appropriate payoff determined for these choices. It may happen that among the strategies available to a given player there are several which would yield the same payoffs against all combinations of strategies of his opponents. In such cases it may be desirable to consider a *reduced normal form* in which each player chooses an equivalence class of strategies, two equivalent strategies giving the same results against all strategies of the opponents [1]. The reduced normal form matrix will consequently have no repeating rows or columns. Example can easily be found where the reduced normal form has far fewer choices open to the players and thus is much easier to deal with in carrying out computations involved in solving the game.

In considering the differences between games, it seems natural to distinguish between games which differ only in payoff function and games which differ in other ways. We shall consider here the notion of game structure defined so that, loosely speaking, two games will have the same game structure if they differ only in payoff functions. One can define the corresponding notions of normal form and reduced normal form for game structures. Here a selection of a strategy for each player determines a play of the game rather than a numerical payoff. McKinsey has considered a special class of games and their patterns of information, a notion closely related to our notion of game structure [2]. His considerations can be stated in terms of game structures as follows. Two game structures of his class are to be considered equivalent if for every payoff function for one of them there is defined a function for the other such that the two resulting games have the same value ("value" is used here in the sense of von Neumann [3]). It is then proved [4] that two structures are equivalent if and only if they have isomorphic reduced normal forms. This theorem is easily extended to general game structures, and in fact, the notion of equivalence of game structures used in both the Krentel-McKinsey-Quine paper [4] and by Dalkey in his paper [5] is essentially that of having the same reduced normal form.

The motivation behind the papers of Krentel-McKinsey-Quine and of Dalkey is this. Given a game in extensive form, its normal form may be very large relative to its reduced normal form, the number of repetitions of rows or columns in its normal form matrix being very large. "In such a case, it becomes desirable to transform the given game in extensive form, so as to reduce the number of repetitions in its matrix." [4] The best results in this direction have been achieved by Dalkey [5]. However, the transformations he considered involved changes in the information partition alone. He obtained an elegant characterization of those changes in information patterns which would transform a game structure into an equivalent game structure. If one carries out on a given game structure the process of deflation which he describes, one obtains an equivalent structure whose normal form will be a simplification of that of the original. Unfortunately, he was able to find examples of game structures which could not be transformed in this way into structures whose normal form matrix have no repeated rows and columns.

The last-mentioned fact raised a question which was put to me by McKinsey in conversation. Are there game structures where the step-wise deflation process of Dalkey can be carried out in two ways so that the resulting two completely deflated game structures do not have the same size matrix? The answer to this question is "yes," as seen from the following example (Figure 1). We shall simply exhibit in graphical form two games. That they are completely deflated in the sense of Dalkey and that they are equivalent can easily be verified. The matrix for the first is 2×16 while that for the second is 2×8.

The problem then still remains to find transformations on a game structure which will carry it into an equivalent structure whose normal form will in fact be its reduced normal form. In attacking this problem we have found it desirable to modify the notion of game structure in the following way. Two games will have the same structure if they differ

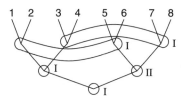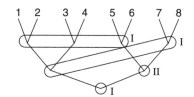

FIG. 1.

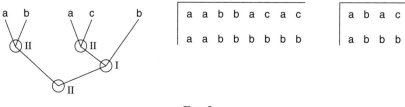

FIG. 2.

only in payoff function *and* if the payoff function for one of them assigns the same payoff to two plays then the payoff function for the other does likewise. Thus a game structure will involve an equivalence relation on the set of plays; if a game has a certain structure, then its payoff function assigns the same value to two plays if they are equivalent. The normal form for a game structure will be such that a selection of a strategy for each player determines an equivalence class of plays rather than a single play. A structure in this new sense can be presented in graphical form. For example, we exhibit in Figure 2 a structure and both its normal form and reduced normal form matrices, where "a," "b," "c" can be considered as names of equivalence classes of plays. We shall consider two games structures as equivalent if they have isomorphic reduced normal forms.

The two closely related problems we wish to attack are these: Find a simple set of transformations on game structures which:

(i) will carry one game structure into another if and only if they are equivalent;

(ii) will carry a game structure into another which is equivalent to it and whose normal form matrix is also its reduced normal form matrix.

Now it is easy to find transformations which will carry one game structure into an equivalent one. Consider the four transformations

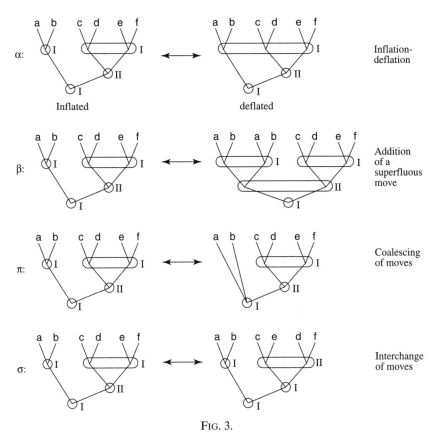

FIG. 3.

exemplified in Figure 3. One easily checks, either by inspection or by writing out their reduced normal form matrix, that the two structures in each of these pairs are equivalent. We shall characterize the four transformations of which the above examples are typical. It will then be shown that these four transformations are in fact sufficient to solve both of our problems. Indeed, given a game structure, we can transform it stepwise, using only transformations of the above types, into a structure whose normal form will be the reduced normal form of the original. We end this informal part of our paper by an example of such a reduction. See Figure 4. The original structure has a normal form matrix which is 16×4, while the normal form matrix of the simplified structure is 5×3.

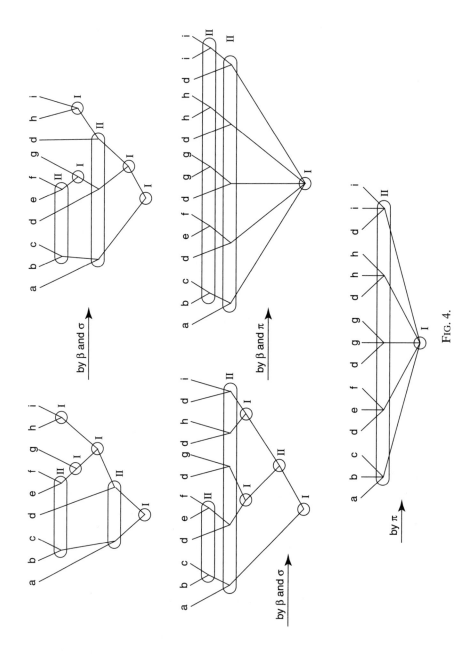

by β and σ

by β and π

by β and σ

by π

Fig. 4.

Formal Presentation: Let G be a finite, nonempty set partially ordered by a relation $\leq$ in such a manner that G has a least element and, for $a, b, c \in G$, $a \leq c$ and $b \leq c$ imply $a \leq b$ or $b \leq a$. For $a, b \in G$, a covers b if $b < a$ and $b < c \leq a$ implies $c = a$; let Aa be the set of all elements which cover a. If Aa is empty, then a will be called an "end-point"; otherwise, a will be called a "move." Let E be the set of end-points.

Let P be an equivalence relation on G such that the set of all end-points is an equivalence class. A "player" of the game is an equivalence class a/P for which a is a move. Let $\underline{P}^*$ denote the set of all players.

Let R be an equivalence relation on G such that, for all $a, b \in G$, $R \cap (Aa \times Ab)$ either is empty or is a biunique function on Aa onto Ab. R plays the role of both the alternative partition and the counter-clockwise ordering of alternatives which appear in the Kuhn formalization [1].

Let I be an equivalence relation on G such that (i) $I \subseteq P$, (ii) if aIb, then $R \cap (Aa \times Ab)$ is not empty, (iii) if aIb, then $a \nleq b$. If a is a move, then the equivalence class a/I which contains a will be called "an information set for the player a/P."

DEFINITION 1: $\underline{G} = \langle G, \leq, P, R, I \rangle$ is a game structure if $G, \leq, P, R, I$ are as described above.

DEFINITION 2: $\underline{G} = \langle\langle G, \leq, P, R, I, h \rangle\rangle$ is a game if $\langle G, \leq, P, R, I \rangle$ is a game structure and h is such a function on $E \times \underline{P}^*$ to the reals that for $a, b \in E$, aIb implies $h(a, p) = h(b, p)$ for all $p \in \underline{P}^*$. h is called the payoff function for the game. $\underline{G}$ will be called an n-person game if $\underline{P}^*$ has just n-elements.

DEFINITION 3: Let $\underline{G}$ be a game structure, $p \in \underline{P}^*$ a player. α is a strategy for p if α is such a function on p that:

(i) for $a \in p$, $\alpha(a) \in Aa$;
(ii) for $a, b \in p$, if aIb, then $\alpha(a)R\alpha(b)$.

Let S_p be the set of strategies for player p. Let $S = \Pi_{p \in P^*} S_p$ be the Cartesian product of the S_p's; S will be called the strategy space for $\underline{G}$.

Each element $\boldsymbol{\alpha}$ of S uniquely determines in a natural way an end-point $e = e(\boldsymbol{\alpha})$. In fact, let a_0 be the least element of G. Suppose $a_0, a_1, \ldots, a_k$ are defined. If a_k is an end-point, let $e(\boldsymbol{\alpha}) = a_k$. Otherwise, let $a_{k+1} = \boldsymbol{\alpha}_{a_k/p}(a_k)$. Clearly, $\{a_0, \ldots, a_k\}$ will be a chain, for each

k. Thus for some k, a_k will be an end-point. The set $\{a_0, \ldots, a_k\}$ is usually called a "play."

DEFINITION 4: Let e be the function defined above. e can be considered as the matrix of the game structure.

DEFINITION 5: Let G be a game with payoff h; let $p \in \underline{P}^*$. The payoff matrix of $\underline{G}$ for p is the function H_p on S such that for $\alpha \in S$, $H_p(\alpha) = h(e(\alpha), p)$.

DEFINITION 6: Let $\underline{G}$ be a game structure, $p_1 \in \underline{P}^*$, and $\sigma_1, \sigma_2 \in S_{p_1}$. Then $\sigma_1 \sim \sigma_2$ (σ_1 is equivalent to σ_2) if, for any $\alpha, \beta \in S$ such that $\alpha_{p_1} = \sigma_1$, $\beta_{p_1} = \sigma_2$, and $\alpha_p = \beta_p$ for $p \in \underline{P}^* - \{p_1\}$, then $e(\alpha) I e(\beta)$.

DEFINITION 7: Let $\underline{G}$ be a game structure. For $p \in \underline{P}^*$, let S_p^* be the family of equivalence classes of S_p under $\sim$. Thus an element of S_p^* is a set of equivalent strategies for p. Let S^* be the Cartesian product of the S_p^*'s for $p \in \underline{P}^*$. Then the reduced normal matrix of $\underline{G}$ is the function K on S^* such that for $\alpha^* \in S^*$ and $\alpha \in \alpha^*$, $K(\alpha^*) = e(\alpha)/I$.

A little reflection will convince one that Definition 7 formalizes the notion of reduced normal form matrix mentioned earlier.

DEFINITION 8: Two game structures $\underline{G}_1$ and $\underline{G}_2$ are equivalent, $\underline{G}_1 \sim \underline{G}_2$, if there are biunique functions u, v, w such that:

   i) u is on $\underline{P}_1^*$ onto $\underline{P}_2^*$;
   ii) v is such a function on $\underline{P}_1^*$ that for $p \in \underline{P}_1^*$, $v(p) = v_p$ maps in a one–one way the set S_p^* onto $S_{u(p)}^*$;
   iii) w is on the set E_1/I_1 onto E_2/I_2, where E_i/I_i is the family of all equivalence classes of elements of E_i under the equivalence relation I_i;
   iv) for $\alpha \in S_1^*$, $w(K_1(\alpha_1^*)) = K_2(\beta^*)$, where $v(\alpha_p^*) = \beta_{u(p)}^*$ for $p \in \underline{P}^*$.

DEFINITION 9: Two game structures $\underline{G}_1$ and $\underline{G}_2$ are isomorphic under the mapping Φ, $\underline{G}_1 \cong_\Phi \underline{G}_2$, if Φ maps $\underline{G}_1$ isomorphically onto $\underline{G}_2$ in the sense of general algebra, i.e., Φ maps G_1 one-to-one onto G_2 such that for a, b, G, $a \leq_1 b$ implies $\Phi(a) \leq_2 \Phi(b)$, etc.

DEFINITION 10: If $\underline{G} = \langle G, \leq, P, R, I \rangle$ is a game structure, $a \in G$, then $\underline{G}^a$ is the sequence $\langle G \cap \{b | a \leq b\}, \leq, P, R, I \rangle$.

LEMMA 11: If $\underline{G} = \langle G, \leq, P, R, I \rangle$ is a game structure, $a \in G$, then $\underline{G}^a$ is a game structure.

DEFINITION 12: Let $\underline{G}$ be a game structure, let $\underline{P}^*$ be ordered so that

$\underline{P}^* = \{p_1, p_2, \ldots, p_k\}$. Then $\underline{G}$ has normal form relative to this ordering if:

i) for all moves a and b, aPb implies aIb (i.e., each player has just one information set);

ii) for $a \in E$, $p \in \underline{P}^*$, there is a $b \in p$ such that $b < a$ (i.e., each player has a move in every play);

iii) if $a \in p_i$, $b \in p_j$, and $i < j$, then $a < b$;

iv) for $a \in G$, $b, c \in Aa$, there are $a' \in a/I$, $b', c' \in Aa'$ such that $b'Rb$, $c'Rc$, and $\underline{G}^{b'} \not\equiv \underline{G}^{c'}$.

THEOREM 13: *For every game structure $\underline{G}$, and every ordering of its players, there is a $\underline{G}'$ which has normal form relative to a fixed ordering of its players and $\underline{G} \sim \underline{G}'$ under mapping functions u, v, w, where u preserves the respective orderings of their players.*

THEOREM 14: *Let $\underline{G}, \underline{G}'$ be game structures in normal form relative to fixed ordering of their players. Then $\underline{G} \sim \underline{G}'$, under mapping functions u, v, w, where u preserves the respective orderings of their players, if and only if $\underline{G} \cong \underline{G}'$.*

The proofs of Theorems 13 and 14 are straightforward. If a game structure has normal form, then its matrix, the function e, has no "repeated rows or columns."

We have now defined equivalence classes of game structures essentially in terms of their reduced matrices (though this last notion has not been explicitly defined), and we have characterized certain canonical members of each class. We now turn to our major task: to characterize the equivalence classes of game structures in terms of four elementary transformations and independently of the notion of strategy.

DEFINITION 15: Let $\underline{G}$ be a game structure, $p \in \underline{P}^*$, $a, b \in G$. Then $a|_p b$ if there are $a', b' \in p$, $a'' \in Aa'$, $b'' \in Ab'$ such that $a'Ib'$, $a'' \leq a$, $b'' \leq b$ and not $a''Rb''$.

DEFINITION 16: Let $\underline{G}, \underline{G}'$ be game structures, $\underline{G} = \langle G, \leq, P, R, I \rangle$. Then:

i) (Inflation-deflation) $\underline{G} \sim_1 \underline{G}'$ if, for some $a, b \in G$,
   1) for $a'Ia$, $b'Ib$, $a'|_{a/p} b'$,
   2) $\underline{G}' \cong \langle G, \leq, P, R, I \cup \{\langle c, d \rangle | cIa \text{ and } dIb \text{ or } dIa \text{ and } cIb\} \rangle$;

ii) (Addition of superfluous moves) $\underline{G} \sim_2 \underline{G}'$ if, for some $a, b_1, b_2, \in G$ and some Φ:

1) $Aa = \{b_1, b_2\}$;

2) $\underline{G}^{b_1} \cong_\Phi \underline{G}^{b_2}$;

3) for $c \geq b_1$, $cI\Phi(c)$;

4) $\underline{G}' \cong \langle G - (\{a\} \cup \{c | b_2 \leq c\}), \leq, P, R, I \rangle$.

iii) (Coalescing of moves) $\underline{G} \sim_3 \underline{G}'$ if, for some $\{a_1, \ldots, a_k\} = a_1/I$, $\{b_1, \ldots, b_k\} = b_1/I$;

1) $b_i \in Aa_i$ for $1 \leq i \leq k$;

2) $b_i R b_j$ for $1 \leq i, j \leq k$;

3) $\underline{G}' \cong \langle G - \{b_1, \ldots, b_k\}, \leq, P, R, I \rangle$.

iv) (Interchange of moves) $\underline{G} \sim_4 \underline{G}'$ if, for some $a, b_1, b_2, c_1, c_2, d_1, d_2 \in G$, and some $\leq', P', I'$:

1) $Aa = \{b_1, b_2\}$, $Ab_1 = \{c_1, c_2\}$, $Ab_2 = \{d_1, d_2\}$;

2) $b_1 I b_2, c_1 R d_1, c_2 R d_2$;

3) $A'b_1 = \{c_1, d_1\}$, $A'b_2 = \{c_2, d_2\}$;

4) $(a/I - \{a\}) \cup \{b_1, b_2\} = b_1/I'$, $(b_1/I - \{b_1, b_2\}) \cup \{a\} = a/I'$;

5) $\underline{G}' = \langle G, \leq', P', R', I' \rangle$;

6) $\langle G' - G'^a, \leq', P', R', I' \rangle \cong_\Phi \langle G - G^a, \leq, P, R, I \rangle$, where Φ is the identity map;

7) $\underline{G}^{c_i} \cong G'^{c_i}$, $\underline{G}^{d_i} \cong G'^{d_i}$ for $i = 1, 2$.

In the remainder of this chapter we shall have need of two functions which we now define. Let $\underline{G}$ be a game structure, $\underline{U}$ the family of all of its information sets. (i) For $U \in \underline{U}$, let $A(U)$ be the number of alternatives at any move in U. We note that $A(U) \geq 2$. Let $\pi(\underline{G}) = \Sigma_{U \in \underline{U}}(A(U) - 2)$. (ii) Let V be an ordering of $\underline{U}$. Thus for some k, $\underline{U} = \{V_1, \ldots, V_k\}$. We shall say that $b \in G$ is out of order with respect to V if there are $i > j \geq k$, $a \in G$, such that $a \in V_j$, $b \in V_i$ and $a > b$. Let B be the set of elements which are out of order. For $a \in G$, let $\rho(a)$ be the number of elements which precede a, and $r = \max_{a \in G} \{\rho(a)\}$. Let $\sigma(\underline{G}, V) = \Sigma_{a \in B, a \in V_i} 4^{r+i-\rho(a)}$.

LEMMA 17: *Let $\underline{G}$ be a game structure. Then there are game structures $\underline{G}_1, \ldots, \underline{G}_t$ such that $\underline{G} \cong \underline{G}_1$, $\underline{G}_i \sim_3 \underline{G}_{i+1}$ for $1 \leq i < t$, and $\pi(\underline{G}_t) = 0$.*

PROOF: Let $\mathcal{J}(n)$ mean: If $\underline{G}$ is a game structure such that $\pi(\underline{G}) \leq n$, then there are game structures $\underline{G}_1, \ldots, \underline{G}_t$ such that $\underline{G} \cong \underline{G}_1$, $\underline{G}_i \sim_3 \underline{G}_{i+1}$ for $1 \leq i < t$, and $\pi(\underline{G}_t) = 0$. The lemma follows by induction on n.

LEMMA 18: *Let $\underline{G}$ be a game structure. Then there are game structures $\underline{G}_1, \ldots, \underline{G}_t$ such that $\underline{G} \cong \underline{G}_1$, $\underline{G}_i \sim_j \underline{G}_{i+1}$ for $1 \leq i < t$ and $j = 2, 3, 4$, $\pi(\underline{G}_t) = 0$ and, for some ordering V, $\sigma(\underline{G}_t, V) = 0$.*

PROOF: First we get a $\underline{G}'$ as given by Lemma 17. Let V be an ordering of its information sets. We complete our proof by supposing $\sigma(\underline{G}', V) \leq n$ and proceeding by induction on n.

THEOREM 19: *If $\underline{G}$ is a game structure, then there are game structures $\underline{G}_1, \ldots, \underline{G}_t$ such that $\underline{G} \cong \underline{G}_1$, $\underline{G}_i \sim_j \underline{G}_{i+1}$ or $\underline{G}_{i+1} \sim_j \underline{G}_i$ for $1 \leq i < t$ and $j = 1, 2, 3, 4$, and $\underline{G}_t$ is in normal form.*

PROOF: We first get a $\underline{G}'$ as given by Lemma 18. Using Definition 16 (ii), we get a $\underline{G}''$, preserving the properties of $\underline{G}'$ and such that each information set intersects every maximal chain. Now we inflate wherever possible. We easily check that several applications of Definition 16 (iii) give the desired result.

THEOREM 20: *If $\underline{G}_1, \ldots, \underline{G}_t$ are game structures such that $\underline{G}_1 \sim_j \underline{G}_{i+1}$ or $\underline{G}_{i+1} \sim_j \underline{G}_i$ for $1 \leq i < t$ and $j = 1, 2, 3, 4$, then $\underline{G}_1 \sim \underline{G}_t$.*

PROOF: It is clearly sufficient to prove that $\underline{G} \sim_j \underline{G}'$ implies $\underline{G} \sim \underline{G}'$ for $j = 1, 2, 3, 4$. The checking of the four cases is straightforward.

THEOREM 21: *$\underline{G} \sim \underline{G}'$ if and only if $\underline{G}'$ can be obtained from $\underline{G}$ by stepwise application of the four transformations of Definition 16.*

PROOF: This follows from Theorems 13, 14, 19, and 20.

REFERENCES

[1] Kuhn, H. W. "Extensive Games," *Proceedings of the National Academy of Sciences*, Vol. 36 (1950), pp. 570–576.
[2] McKinsey, J. C. C. "Notes on games in extensive form," RM-157.
[3] von Neumann, J. and Morgenstern, O. *Theory of Games and Economic Behavior*, Princeton, 1947.
[4] Krentel, W. D., McKinsey, J. C. C. and Quine, W. V. "A simplification of games in extensive form," P-140.
[5] Dalkey, N. "Equivalence of information patterns," D(L)-877.

6

EXTENSIVE GAMES AND THE PROBLEM
OF INFORMATION

H. W. Kuhn[1]

IN THE mathematical theory of games of strategy as described by von Neumann and Morgenstern,[2] the development is seen to proceed in two major steps: (1) the presentation of an all-inclusive formal characterization of a general n-person game, (2) the introduction of the concept of pure strategy which makes possible a radical simplification of this scheme, replacing an arbitrary game by a suitable prototype game. They called these two descriptions the *extensive* and the *normalized* forms of a game. As noted there, the normalized form is better suited to the derivation of general theorems (e.g., the main theorem of the zero-sum two-person game), while the extensive form exposes the characteristic differences between games and the decisive structural features which determine those differences. Since all games are found in extensive form, while it is practical to normalize but a few, it seems desirable to attack the completion of a general theory of games in extensive form.

First of all, this paper presents a new formulation of the extensive form which, while appearing quite natural intuitively, covers a larger class of games than that used by von Neumann. The use of a geometrical model reduces the amount of set theoretical equipment necessary and clarifies the delicate problem of information. After the definition of pure strategies, Theorem 1 removes the redundancy found in a direct definition. Theorems 2 and 3 characterize the properties of a natural decomposition of many games into a subgame and a number of difference games. They represent a generalization of the theorem that every zero-sum two-person game with perfect information has a solution in pure strategies. Theorem 4 presents a positive criterion that a game be solvable in behavior strategies, a method that presents extreme computational advantages over mixed strategies in many cases.

[1] The preparation of this paper was supported by the Office of Naval Research.

[2] von Neumann, J. and Morgenstern, O., *The Theory of Games and Economic Behavior*, 2nd ed., Princeton, 1947.

Throughout the paper, passages which serve as motivation, interpretation, or heuristic discussion are placed in brackets, [...]. This is done to emphasize the independence of the definitions and proofs of these sections.

§1. The Extensive Form of a Game

DEFINITION 1. A *game tree* K is a finite tree with a distinguished vertex 0 which is imbedded in an oriented plane.[3]

[The concept of a game tree is introduced as a natural geometric model of the essential character of a game as a successive presentation of alternatives. The distinguished vertex and the imbedding are devices which facilitate the arithmetization of the notion of strategy. Before proceeding to the definition of a game, it is necessary to present some general technical terms associated with a game tree; it is important to remark that, although these terms are taken from common parlance, their meaning is given by the definitions.]

TERMINOLOGY: The *alternatives* at a vertex $X \in K$ are the edges e incident at X and lying in components of K which do not contain 0 if we cut K at X. If there are j alternatives at X then they are indexed by the integers $1, \ldots, j$, circling X in the positive sense of the orientation. At the vertex 0, the first alternative may be assigned arbitrarily. If one circles a vertex $X \neq 0$ in the positive sense, the first alternative follows the unique edge at X which is not an alternative. The function thus defined which indexes the alternatives in K will be denoted by ν; thus, $\nu(e)$ is the index of the alternative e. Those vertices which possess alternatives will be called *moves*;[4] the remaining vertices will be called *plays*. The name play will also be used for the unique unicursal path from 0 to a play when no confusion results. The partition[5] of the moves into sets A_j, $j = 1, 2, \ldots$, where A_j contains all of the moves with j alternatives, will be called the *alternative partition*. A *temporal order* on K is defined by $X \leq Y$ if X lies on W_Y, the unicursal path joining 0 to

---

[3] A graphical representation of a game by a tree has been suggested by von Neumann, *loc. cit.*, p. 77; however, he only treats the case of games with perfect information.

[4] It is important to make a clear distinction between our "moves" and those of von Neumann. Precisely, a von Neumann move is the set of all moves of a given rank in our sense.

[5] In this paper a partition means an exhaustive decomposition into (possibly void) disjoint sets.

Y; it is a partial order. The *rank* of a move Y is the number of X such that $X \leq Y$ or, equivalently, the number of moves $X \in W_Y$.

DEFINITION 2. A *general n-person game* Γ is a game tree K with the following specifications:

(I) A partition of the moves into $n + 1$ indexed sets $P_0, P_1, \ldots, P_n$ which will be called the *player partition*. The moves in P_0 will be called *chance moves*; the moves in P_i will be called *personal moves of player i* for $i = 1, \ldots, n$.

(II) A partition of the moves into sets U which is a refinement of the player and alternative partitions (that is, each U is contained in $P_i \cap A_j$ for some i and j) and such that no U contains two moves on the same play. This partition is called the *information partition* and its sets will be called *information sets*.

(III) For each $U \subset P_0 \cap A_j$, a probability distribution on the integers $1, \ldots, j$, which assigns *positive* probability to each. Such information sets are assumed to be one-element sets.

(IV) An n-tuple of real numbers $h(W) = (h_1(W), \ldots, h_n(W))$ for each play W. The function h will be called the *payoff function*.

[How is this formal scheme to be interpreted? That is, how is a general n-person game played? To personalize the interpretation, one may imagine a number of people called *agents* isolated from each other and each in possession of the rules of the game. There is one agent for each information set and they are grouped into players in a natural manner, an agent belonging to the ith player if his information set lies in P_i. This seeming plethora of agents is occasioned by the possibly complicated state of information of our players who may be forced by the rules to forget facts which they knew earlier in a play.[6]

A play begins at the vertex 0. Suppose that is has progressed to the move X. If X is a personal move with j alternatives then the agent whose information set contains X chooses a positive integer not greater than j, knowing only that he is choosing an alternative at one of the moves in his information set. If X is a chance move, then an alternative is chosen in accordance with the probabilities specified by (III) for the information set containing X. In this manner, a path with initial point 0 is constructed. It is unicursal and, since K is finite, leads to a unique play W. At this point, player i is paid the amount $h_i(W)$ for $i = 1, \ldots, n$.

---

[6] It has been asserted by von Neumann that Bridge is a two-person game in exactly this manner.

The case in which K reduces to the vertex 0 is not excluded. Then Γ is a no-move game, no one does anything, and the payoff is $h(0)$.

The price of the intuition gained with the use of a geometric model is the introduction of a certain amount of redundancy. Suppose Γ_1 and Γ_2 are two games defined with game trees K_1 and K_2 such that:

(1) K_1 and K_2 are homeomorphic.

(2) The homeomorphic mapping σ of K_1 onto K_2 preserves the distinguished vertex and the specifications (I)–(IV).

(3) On each information set, σ effects a permutation of the indices of the alternatives. (To make this more precise, (3) requires the existence of a permutation τ_U of the integers $1, \ldots, j$ for each information set $U \subset A_j$ such that, if ν_1 and ν_2 denote the functions indexing the alternatives in K_1 and K_2, $\nu_2(\sigma(e)) = \tau_U(\nu_1(e))$ for all alternatives e at moves in U.) It is clear that in such a case the games Γ_1 and Γ_2 should be considered equivalent and that a proper definition would define a general n-person game as an equivalence class under this equivalence relation. However, this distinction need not be emphasized; if care is taken to frame definitions which apply to the equivalence class as well as to the representative, one can ignore it entirely in proofs and remark that all of the theorems hold for either class or representative.

Although the majority of the above formalization needs no justification, ample motivation appearing in the book of von Neumann and Morgenstern, several features deserve comment. The first concerns the finiteness of the game tree. Since the rules of most games include a Stop Rule which only insures that every play terminates after a finite number of choices it is not completely obvious that this entails that the game tree be finite. To demonstrate this, following König,[7] we will assume that there are an infinite number of possible plays and then contradict the Stop Rule by constructing a unicursal path starting at 0 and containing an infinite number of edges. Since the choice at 0 is made from a finite set of alternatives, there must be an infinite number of plays with the same first edge e_1. We proceed by induction; assume that edges $e_1, \ldots, e_l$ have been chosen such that $e_1 \ldots e_l$ is the beginning segment of an infinite number of plays. Then, since the next choice is made from a finite set, an infinite subset of these plays must continue with the same edge, say e_{l+1}. This completes the proof.

[7] König, D. "Über eine Schlussweise aus dem Endlichen ins Unendliche," *Acta Szeged* 3 (1927), pp. 121–130.

A more crucial question is that of formalizing the state of information of a player at the occasion of a decision, i.e., at a move. If one examines the information given to a player by von Neumann's "patterns of information" it is found to consist of several parts. First of all, he is told that it is his move and is told the number of alternatives. In our terminology, this says that the move lies in $P_i \cap A_j$ for some fixed i and j. Secondly, he is told that the move lies on one of a certain set of plays and has been preceded by a fixed number of choices. From this he can deduce that his move lies in a set of moves which all have the same rank. It is this set of moves which forms a U in the information partition; however, we have weakened the requirement that all moves in U have the same rank to the condition that no U contains two moves on the same play.]

§2. COMPARISON WITH THE FORMULATION OF VON NEUMANN

[The object of this section is to clarify the relation between our "general n-person game" and a "von Neumann n-person game." Incidentally, it will be shown that the formulation given above is more general than von Neumann's but this is of minor importance beside the major purpose of throwing light on the points of agreement. The best way to do this seems to be to describe the method of transition from one extensive description to the other. Accordingly, we commence by deriving a "general n-person game" from a "von Neumann n-person game."[8]

Take as vertices of K the non-void subsets A_κ of the partitions $\mathscr{A}_\kappa$, $\kappa = 1, \ldots, \nu + 1$. A vertex A_κ is joined to a vertex $A_{\kappa+1}$ by an edge if $A_{\kappa+1} = A_\kappa \cap C_\kappa$ for some C_κ. Remark that A_κ has j alternatives if it is contained in a D_κ which contains j sets C_κ and that the moves on K are the vertices A_κ, $\kappa = 1, \ldots, \nu$, while the plays are the vertices $A_{\nu+1}$. The player partition of the moves is defined by $P_k = \{A_\kappa | A_\kappa \subset \text{some} \ B_\kappa(k)\}$ for $k = 0, 1, \ldots, n$. The information partition of the moves is defined by $U = \{A_\kappa | A_\kappa \subset D_\kappa\}$ with one U defined for each $D_\kappa \in \mathscr{D}_\kappa(k)$, $\kappa = 1, \ldots, \nu$ and $k = 1, \ldots, n$; the chance moves $A_\kappa \subset B_\kappa(0)$ form one-element sets in the information partition. For each $U \subset P_0 \cap A_j$, that is, for each $A_\kappa \subset B_\kappa(0)$, where A_κ contains j sets C_κ of $\mathscr{C}_\kappa(0)$, the probability assigned to the alternative corresponding to $A_\kappa \cap C_\kappa$ is $p_\kappa(C_\kappa)$. Finally, the functions h_k are defined on the plays by $h_k(A_{\nu+1}) = \mathscr{F}_k(A_{\nu+1})$ for all plays $A_{\nu+1}$ and $k = 1, \ldots, n$.

[8] For the comparison, the notation of von Neumann, *loc. cit.*, pp. 73–75, is followed.

The imbedding has been left to the last, as it may be, since it is independent of the other specifications. Imbed 0 arbitrarily. Suppose the imbedding has progressed as far as the moves A_κ. We imbed the alternatives at all of the A_κ of a fixed U in the same order in the orientation. This is possible since each of these has the same number of alternatives (the number of C_κ contained in the fixed set D_κ which defines U).

There are two restrictive conditions which are fulfilled on the general n-person game thus obtained.

(A) All plays contain the same number of moves ν.

(B) All moves in a fixed information set U (defined by D_κ) have the same rank (κ).

Condition (A) is trivial and can be fulfilled in all of our games by filling out short plays with "dummy" chance moves with one alternative. Condition (B) is not trivial and will be discussed after the derivation of a "von Neumann n-person game" from a "general n-person game" *which satisfies conditions (A) and (B).*

(10:A:a) The number ν is the number given by condition (A).

(10:A:b) The finite set Ω is the set of plays in K.

(10:A:c) For every $i = 1, \ldots, n$: The function $\mathscr{F}_i(W) = h_i(W)$ for $W \in \Omega$.

(10:A:d) For every $r = 1, \ldots, \nu$: The partition $\mathscr{A}_r$ in Ω contains one set A_r for each move X of rank r which is defined by $A_r = \{W | W > X\}$. The partition $\mathscr{A}_{\nu+1}$ consists of the one-element sets $\{W\}$.

(10:A:e) For every $r = 1, \ldots, \nu$: The partition $\mathscr{B}_r$ in Ω contains one set $B_r(i)$ for each $i = 0, 1, \ldots, n$ which is defined by $B_r(i) = \{W | W > X$ where X is of rank r and $X \in P_i\}$.

(10:A:f) For every $r = 1, \ldots, \nu$ and every $i = 0, 1, \ldots, n$: The partition $\mathscr{C}_r(i)$ in $B_r(i)$ contains one set C_r for each alternative e at an information set $U \subset P_i$ which is of rank r. It is defined by $C_r = \{W | W$ follows some $X \in U$ by the alternative $e\}$.

(10:A:g) For every $r = 1, \ldots, \nu$ and every $i = 1, \ldots, n$: The partition $\mathscr{D}_r(i)$ in $B_r(i)$ contains one set D_r for each $U \subset P_i$ which contains only moves of rank r. It is defined by $D_r = \{W | W > X \in U\}$.

(10:A:h) For every $r = 1, \ldots, \nu$ and every $C_r \in \mathscr{C}_r(0)$: The number $p_r(C_r)$ is the probability assigned to the alternative e by specification (III).

The proofs that the games thus derived actually satisfy the requirements placed on them (that is, by Definition 1 and 2 in the first case and by (10:1:a)–(10:1:j) in the second) are omitted. If they are not always

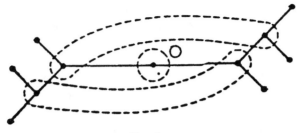

FIG. 1.

immediate, they are all straightforward and easily obtained. The important question is: What happens if we derive a general n-person game from a von Neumann game again? The answer is quite obvious from the above, namely, the set Ω has been reduced to include only those plays which are allowed by the rules of the game. More precisely, Ω has been reduced so that (1) all partitions of a set are such that the union of the elements of the partition is exactly the set and (2) plays $\pi \in C_\kappa \in \mathscr{C}_\kappa(0)$ such that $p_\kappa(C_\kappa) = 0$ have been dropped. The discussion may be summarized by:

THEOREM OF CATEGORIZATION: *The von Neumann games, excluding illegal or impossible plays, are the general n-person games in which a player's information at the occasion of a decision includes the number of choices preceding that decision.*

Figure 1 concludes this section with an example[9] of a general n-person game in which a player is not informed of the number of choices preceding his move. The dotted lines enclose information sets.]

§3. PURE AND MIXED STRATEGIES

DEFINITION 3. Let $\mathscr{U}_i = \{U | U \subset P_i\}$. A *pure strategy for player i* is a function π_i mapping $\mathscr{U}_i$ into the positive integers such that $U \subset A_j$ implies $\pi_i(U) \leq j$. We will say that π_i *chooses the alternative e* incident at $X \in U$ if $\pi_i(U) = \nu(e)$.

[The interpretation of a pure strategy is that it is a plan formulated before a play by a *strategist* for player i who then communicates his choices to the agents of player i. One may do this without violating the

[9] This example, as well as substantial contributions to the interpretations, were communicated by L. S. Shapley and J. C. C. McKinsey.

rules of the game regarding the information possessed by the agents by imagining that the strategist writes a book with one page for each of player i's information sets. If an information set has j alternatives, its page will contain a positive integer not greater than j and is given to the agent who is acting on that information set for player i. If his information set appears in a play then he is to choose the alternative indexed by the integer. This interpretation motivates the following definitions.]

Any n-tuple $\pi = (\pi_1, \ldots, \pi_n)$ of pure strategies for the n players defines a probability distribution on the alternatives in each of the information sets of K as follows:

If e is an alternative at a personal move for player i in the information set U:

$$p_\pi(e) = \begin{cases} 1 & \text{if } \pi_i(U) = \nu(e), \\ 0 & \text{otherwise.} \end{cases}$$

It e is an alternative at a chance move: $p_\pi(e)$ is the probability assigned to $\nu(e)$ by (III). This, in turn, defines a probability distribution on the plays of K by: $p_\pi(W) = \Pi_{e \in W} p_\pi(e)$ for all W.

[The interpretation is immediate; if the n strategists choose the pure strategies $\pi_1, \ldots, \pi_n$, then the probability that the play W will result is $p_\pi(W)$.]

DEFINITION 4. The *expected payoff* $H_i(\pi)$ to player i for pure strategies $\pi_1, \ldots, \pi_n$ is defined by $H_i(\pi) = \sum_W p_\pi(W) h_i(W)$ for $i = 1, \ldots, n$.

[Again redundancy is the price of a facile definition of pure strategies. Its nature is made clear once it is remarked that a pure strategy may make a choice at an early part of a game that makes many later moves impossible and hence renders the choices on those moves irrelevant. However, it can be viewed in another manner, namely, the efficacy of a strategy π_i is to be measured by the payoff H_i and hence two strategies should be considered equivalent if they produce the same payoff against all counter-strategies employed by the other players. Leaving the payoff function h arbitrary for the moment, this can be restated: two strategies should be considered equivalent if they yield the same probability for every play against all counter-strategies employed by the other players. The rest of this section shows that these two views of the redundancy are the same and revises the definition of pure strategy accordingly.]

If $\pi = (\pi_1, \ldots, \pi_i, \ldots, \pi_n)$ we shall write π/π_i' for $(\pi_1, \ldots, \pi_i', \ldots, \pi_n)$.

DEFINITION 5. The pure strategies π_i and π_i' are *equivalent*, written $\pi_i \equiv \pi_i'$, if and only if $p_\pi(W) = p_{\pi/\pi_i'}(W)$ for all plays W and all π containing π_i.

DEFINITION 6. A personal move X for player i is called *possible when playing* π_i if there exists a play W and π containing π_i such that $p_\pi(W) > 0$ and $X \in W$. An information set U for player i is called *relevant when playing* π_i if some $X \in U$ is possible when playing π_i. We will denote the set of moves which are possible when playing π_i by Poss π_i and the family of information sets which are relevant when playing π_i by Rel π_i.

PROPOSITION 1. *A move X for player i is possible when playing π_i if and only if π_i chooses all alternatives on the path W_X from 0 to X which are incident at moves of player i.*

PROOF. Let X be possible when playing π_i; then there exists a play W containing X and π containing π_i such that $p_\pi(W) = \Pi_{e \in W} p_\pi(e) > 0$. Hence it is clear that π_i chooses all of the alternatives for player i on W, thus certainly those of player i on W_X.

Now assume π_i chooses all the alternatives for player i on W_X. To demonstrate the possibility of X it is necessary to construct a play and strategies for the remaining players. Since no information set contains two moves on the same play, in constructing the strategies, choices on a unicursal path may be assigned independently. The choices for personal moves on W_X are assigned so as to conform to W_X. At X the choice is made by π_i; the construction of a play is continued by making choices arbitrarily *except when they are specified* by π_i. Call the resulting play W. The unspecified choices are made arbitrarily and the resulting pure strategies are called $\pi_1, \ldots, \pi_{i-1}, \pi_{i+1}, \ldots, \pi_n$. Then, since the probabilities of chance alternatives are positive, $p_\pi(W) > 0$ and X is possible when playing π_i.

COROLLARY. *Let the information set U contain the first move X for player i on a play W. Then U is relevant for all pure strategies π_i for player i.*

THEOREM 1. *The pure strategies π_i and π_i' are equivalent if and only if they define the same relevant information sets and coincide on these sets.*

PROOF. Suppose $\pi_i \equiv \pi_i'$ and U is relevant when playing π_i. Then there exists a move $X \in U$, a play W, and π containing π_i such that:

$$p_\pi(W) > 0 \quad \text{and} \quad X \in W.$$

Hence $p_{\pi/\pi_i'}(W) = p_\pi(W) > 0$, and U is relevant for π_i'. Moreover, by the definition of $p(W)$, $\pi_i(U) = \pi_i'(U) = v(e)$ where e is the alternative at X which lies on W and thus π_i and π_i' coincide on U.

To show that the condition is sufficient, assume that a play W and π containing π_i are given. If $p_\pi(W) = \prod_{e \in W} p_\pi(e) > 0$ then $\pi_i(U) = v(e)$ for all alternatives e lying on W and incident at personal moves of player i in U. All of these moves are possible when playing π_i and hence the same conclusion holds for π_i'. But then

$$\prod_{e \in W} p_{\pi/\pi_i'}(e) = \prod_{e \in W} p_\pi(e).$$

But, interchanging π_i and π_i' in this argument, $p_\pi(W) = 0$ implies $p_\pi(W) = 0$. Consequently, $\pi_i \equiv \pi_i'$.

[The interpretation of a pure strategy as a strategy book can be extended to the equivalence classes defined above by leaving blank those pages which correspond to irrelevant information sets.]

[Even the simplest games, e.g., Matching Pennies, reveal that a player is at a disadvantage if he uses the same pure strategy in each play. Instead he should randomize his choices. Two methods of randomization are described in this paper. In the first, the player uses a probability distribution on his pure strategies, choosing the particular pure strategy that he will employ in a given play of the game according to this distribution. Following von Neumann, it is called a mixed strategy. The second method is studied in Section 5.]

DEFINITION 7. A *mixed strategy for player* i, μ_i, is a probability distribution on the pure strategies for player i, which assigns probability q_{π_i} to π_i.

Any n-tuple $\mu = (\mu_1, \ldots, \mu_n)$ of mixed strategies for the n players defines a probability distribution on the plays of K by:

$$p_\mu(W) = \sum_\pi q_{\pi_1} \cdots q_{\pi_n} p_\pi(W) \quad \text{for all } W.$$

DEFINITION 8. The *expected payoff* $H_i(\mu)$ to player i for mixed strategies $\mu_1, \ldots, \mu_n$ is defined by $H_i(\mu) = \sum_W p_\mu(W) h_i(W)$ for $i = 1, \ldots, n$.

PROPOSITION 2. *For each move X, let $c(X)$ be the product of the chance probabilities at alternatives on W_X, the path from 0 to X. Then*

$$p_\mu(X) = c(X) \sum_{\substack{X \in \text{Poss } \pi_i \\ i=1,\ldots,n}} q_{\pi_1} \cdots q_{\pi_n} = c(X) \prod_{i=1}^{n} \left(\sum_{X \in \text{Poss } \pi_i} q_{\pi_i} \right)$$

gives the probability that the move X will occur when the players play μ.

PROOF. This is an immediate consequence of Proposition 1 and the interpretation of a mixed strategy.

DEFINITION 9. A personal move X for player i is called *possible when playing* μ_i if there exists an n-tuple μ of mixed strategies containing μ_i such that $p_\mu(X) > 0$. An information set U for player i is called *relevant when playing* μ_i if some $X \in U$ is possible when playing μ_i. Again, we shall write Poss μ_i and Rel μ_i for the sets of X and U which are possible and relevant when playing μ_i.

§4.   THE DECOMPOSITION OF GAMES

[It often occurs that the moves of a game which are subsequent to a fixed move X form a subgame in a natural manner. They constitute the vertices of a game tree with X as first move while the player partition, the probabilities at the chance moves, and the payoff on the plays in this tree carry over from the original game. This is also true of the information partition if, at every move of the original game, the player choosing is informed whether or not his move is in the subgame.

If this happens, the moves which do not lie in the subgame also form the moves of a game which is determined up to the payoff at the vertex X (which is a play in this game!). In this section, this decomposition of a game into a pair of games will be studied to show that the equilibrium points of the pair determine the equilibrium points of the original game.]

DEFINITION 10. Given a move X in a game Γ, let K_X be the component of K which contains X if we delete the unique edge (if any) at X which is not an alternative at X. We shall say that Γ *decomposes at X into Γ_X and $\Gamma_D(\mu_X)$* if every information set U either is contained in K_X or does not intersect K_X. The game Γ_X is called the *subgame* and is defined as follows:

The game tree is K_X; as such it is imbedded in the same oriented plane as K and has X as distinguished vertex.

(I_X) (II_X) The player and information partitions of the moves of Γ_X are the respective partitions of the moves of K restricted to K_X. The family of information sets for player i will be denoted by $\mathscr{X}_i$.

(III_X) For each chance move in Γ_X the probability distribution is that specified by (III) in Γ.

(IV_X) The payoff function h_X for Γ_X is the payoff function h restricted to the plays of K_X.

A game $\Gamma_D(\mu_X)$, called a *difference game*, in defined *for each n-tuple of mixed strategies*, μ_X, in Γ_X. Its game tree is $K - K_X$ completed by X and has 0 as its distinguished vertex. The specifications (I_D)–(IV_D) are made as above with the additional definition that $h_D(X) = H_X(\mu_X)$; to emphasize the dependence of the payoff in $\Gamma_D(\mu_X)$ on μ_X, we shall write this payoff as $H_D(\mu_D, \mu_X)$. The family of information sets for player i will be denoted by $\mathscr{D}_i$.

[Corresponding to this natural decomposition of Γ into the subgame Γ_X and the difference game Γ_D, there is a natural decomposition of the pure strategies for Γ into pairs of pure strategies for Γ_X and Γ_D. The main burden of our proofs in this section lies in analyzing the effect on the payoff of this decomposition and the analogous splitting of mixed strategies for Γ.]

DEFINITION 11. Let the game Γ decompose at X. Then we shall say that a pure strategy π_i for player i *decomposes at* X into pure strategies $\pi_{X|i}$ and $\pi_{D|i}$ for player i in Γ_X and Γ_D if

(a) $\pi_{X|i}$ is the restriction of π_i from $\mathscr{U}_i$ to $\mathscr{X}_i$ and

(b) $\pi_{D|i}$ is the restriction of π_i from $\mathscr{U}_i$ to $\mathscr{D}_i$. Since $\mathscr{U}_i$ is the disjoint union of $\mathscr{X}_i$ and $\mathscr{D}_i$, we can also *compose* a pure strategy π_i from pure strategies $\pi_{X|i}$ and $\pi_{D|i}$ which will be denoted by $\pi_i = (\pi_{X|i}, \pi_{D|i})$.

LEMMA 1. *If* π_i *decomposes into* $\pi_{X|i}$ *and* $\pi_{D|i}$ *for* $i = 1, \ldots, n$, *then*

$$p_\pi(Y) = p_{\pi_D}(Y) \quad \text{for all } Y \in K_D$$

and

$$p_\pi(Y) = p_{\pi_D}(X) p_{\pi_X}(Y) \quad \text{for all } Y \in K_X$$

where $\pi = (\pi_1, \ldots, \pi_n)$, $\pi_X = (\pi_{X|1}, \ldots, \pi_{X|n})$, *and* $\pi_D = (\pi_{D|1}, \ldots, \pi_{D|n})$.

PROOF. The verification is immediate, using the definition of $p_\pi(Y)$ and the fact that all paths from 0 to a Y in K_X pass through X.

DEFINITION 12. Let the game Γ decompose at X. Then we shall say that a mixed strategy μ_i for player i *decomposes at* X into mixed strategies $\mu_{X|i}$ and $\mu_{D|i}$ for player i in Γ_X and Γ_D if

(a) $q_{\pi_{D|i}} = \Sigma_{D(\pi_i) = \pi_{D|i}} q_{\pi_i}$ for all $\pi_{D|i}$, where $D(\pi_i)$ denotes the restriction of π_i from $\mathcal{U}_i$ to $\mathcal{D}_i$.

(b) When $X \in \operatorname{Poss} \mu_i$,

$$
q_{\pi_{X|i}} = \sum_{\substack{X(\pi_i) = \pi_{X|i} \\ X \in \operatorname{Poss} \pi_i}} q_{\pi_i} \Bigg/ \sum_{X \in \operatorname{Poss} \pi_i} q_{\pi_i} \quad \text{for all } \pi_{X|i},
$$

where $X(\pi_i)$ denotes the restriction of π_i from $\mathcal{U}_i$ to $\mathcal{X}_i$.

When $X \notin \operatorname{Poss} \mu_i$,

$$
q_{\pi_{X|i}} = \sum_{X(\pi_i) = \pi_{X|i}} q_{\pi_i} \quad \text{for all } \pi_{X|i}.
$$

LEMMA 2. *Every pair,* $\mu_{X|i}$ *and* $\mu_{D|i}$, *of mixed strategies for player i in* Γ_X *and* Γ_D *is obtained from the decomposition of some* μ_i *in* Γ.

PROOF. Let $(\pi_{D|i}, \pi_{X|i})$ denote the pure strategy for Γ obtained by composing $\pi_{D|i}$ and $\pi_{X|i}$. We then set

$$
q_{(\pi_{D|i}, \pi_{X|i})} = q_{\pi_{D|i}} q_{\pi_{X|i}} \quad \text{for all } \pi_{D|i} \text{ and } \pi_{X|i}.
$$

Then (a) and (b) are easily verified once it is remarked that $X \in \operatorname{Poss} \mu_i$ if and only if $X \in \operatorname{Poss} \mu_{D|i}$. It should be noticed that our composition and decomposition of mixed strategies is a consistent extension of the original definitions for pure strategies.

THEOREM 2. *If* Γ *decomposes at X then there is a mapping of the n-tuples μ of mixed strategies for Γ onto the pairs (μ_D, μ_X) of n-tuples of mixed strategies for Γ_D and Γ_X in such a way that*

$$
H(\mu) = H_D(\mu_D, \mu_X) \tag{1}
$$

if (μ_D, μ_X) *corresponds to* μ *under the mapping.*

PROOF. The mapping is the decomposition of Definition 10. Lemma 2 says that it is a mapping *onto* all pairs. To prove (1) we consider the members of the equation separately. First,

$$H(\mu) = \sum_W p_\mu(W)h(W)$$

$$= \sum_{W \in K - K_X} p_\mu(W)h(W) + \sum_{W \in K_X} p_\mu(W)h(W). \qquad (2)$$

Second,

$$H_D(\mu_D, \mu_X) = \sum_{W \in K_D} p_{\mu_D}(W)h_D(W)$$

$$= \sum_{W \in K - K_X} p_{\mu_D}(W)h_D(W) + p_{\mu_D}(X)H_X(\mu_X). \qquad (3)$$

Noting that, if $W \in K - K_X$,

$$p_\mu(W) = \sum_\pi q_{\pi_1} \cdots q_{\pi_n} p_\pi(W)$$

$$= \sum_{\pi_D} \left(\sum_{D(\pi) = \pi_D} q_{\pi_1} \cdots q_{\pi_n} \right) p_{\pi_D}(W)$$

$$= \sum_{\pi_D} q_{\pi_{D|1}} \cdots q_{\pi_{D|n}} p_{\pi_D}(W) = p_{\mu_D}(W)$$

and hence, comparing (2) and (3), we need only show

$$\sum_{W \in K_X} p_\mu(W)h(W) = p_{\mu_D}(X)H_X(\mu_X). \qquad (4)$$

However, since

$$H_X(\mu_X) = \sum_{W \in K_X} p_{\mu_X}(W)h_X(W) = \sum_{W \in K_X} p_{\mu_X}(W)h(W),$$

to prove (4) it is sufficient to show

$$p_\mu(W) = p_{\mu_D}(X)p_{\mu_X}(W) \quad \text{for all } W \in K_X. \qquad (5)$$

(It should be remarked that this is the analogue of the second half of Lemma 2, stated for mixed strategies, and that our definition of the decomposition of mixed strategies has been framed intentionally to preserve this property.)

Noting that

$$p_{\mu_D}(X) = \sum_{\pi_D} q_{\pi_{D|1}} \cdots q_{\pi_{D|n}} p_{\pi_D}(X)$$

$$= c(X) \prod_{i=1}^{n} \left(\sum_{X \in \text{Poss } \pi_{D|i}} q_{\pi_{D|i}} \right) = c(X) \prod_{i=1}^{n} \left(\sum_{X \in \text{Poss } \pi_i} q_{\pi_i} \right)$$

where $c(X)$ is the product of the probabilities of chance alternatives on the path from 0 to X (the void product is taken to be unity) and

$$p_{\mu_X}(W) = \sum_{\pi_X} q_{\pi_{X|1}} \cdots q_{\pi_{X|n}} p_{\pi_X}(W)$$

$$= \sum_{\pi_X} \left\{ \prod_{i=1}^{n} \left(\sum_{\substack{X(\pi_i) = \pi_{X|i} \\ X \in \text{Poss } \pi_i}} q_{\pi_i} \bigg/ \sum_{X \in \text{Poss } \pi_i} q_{\pi_i} \right) \right\} p_{\pi_X}(W)$$

we have

$$p_{\mu_D}(X) p_{\mu_X}(W) = c(X) \sum_{\pi_X} \left\{ \prod_{i=1}^{n} \left(\sum_{\substack{X(\pi_i) = \pi_{X|i} \\ X \in \text{Poss } \pi_i}} q_{\pi_i} \right) \right\} p_{\pi_X}(W)$$

$$= \sum_{\pi} q_{\pi_1} \cdots q_{\pi_n} p_{\pi_D}(X) p_{\pi_X}(W)$$

$$= \sum_{\pi} q_{\pi_1} \cdots q_{\pi_n} p_{\pi}(W) = p_{\mu}(W).$$

This completes the proof.

[The basic consequence of Theorem 2 is that solutions for Γ can be composed from solutions to Γ_X and Γ_D if we take equilibrium points as our definition of a solution to an n-person game.]

DEFINITION 13. An n-tuple $\bar{\mu} = (\bar{\mu}_1, \ldots, \bar{\mu}_n)$ of mixed strategies for a game Γ is called an *equilibrium point*[10] if

$$H_i(\bar{\mu}) \geqq H_i(\bar{\mu}/\mu_i) \quad \text{for } i = 1, \ldots, n$$

[10] See Nash, J. "Non-cooperative games," *Ann. of Math.*, *54* (1951), pp. 286–295.

for all μ_i, where $\bar{\mu}/\mu_i$ denotes the n-tuple of mixed strategies obtained by replacing $\bar{\mu}_i$ by μ_i in $\bar{\mu}$.

THEOREM 3. *Let Γ decompose at X, let $\bar{\mu}_X$ be an equilibrium point in the subgame Γ_X, and let $\bar{\mu}_D$ be an equilibrium point in $\Gamma_D(\bar{\mu}_X)$. If $\bar{\mu}$ is any n-tuple of mixed strategies for Γ which decomposes into $\bar{\mu}_X$ and $\bar{\mu}_D$, then $\bar{\mu}$ is an equilibrium point for Γ.*

PROOF. Let μ_i be any mixed strategy for player i which decomposes into $\mu_{X|i}$ and $\mu_{D|i}$. Then it is clear that $\bar{\mu}/\mu_i$ decomposes into $\bar{\mu}_X/\mu_{X|i}$ and $\bar{\mu}_D/\mu_{D|i}$, and

$$H_i(\bar{\mu}) = H_{D|i}(\bar{\mu}_D, \bar{\mu}_X) \geq H_{D|i}(\bar{\mu}_D/\mu_{D|i}, \bar{\mu}_X)$$

$$= \sum_{W \in K - K_X} p_{\bar{\mu}_D/\mu_{D|i}}(W) h_{D|i}(W) + p_{\bar{\mu}_D/\mu_{D|i}}(X) H_{X|i}(\bar{\mu}_X)$$

$$\geq \sum_{W \in K - K_X} p_{\bar{\mu}_D/\mu_{D|i}}(W) h_{D|i}(W) + p_{\bar{\mu}_D/\mu_{D|i}}(X) H_{X|i}(\bar{\mu}_X/\mu_{X|i})$$

$$= H_{D|i}(\bar{\mu}_D/\mu_{D|i}, \bar{\mu}_X/\mu_{X|i}) = H_i(\bar{\mu}/\mu_i).$$

[The computational consequences of Theorem 3 are clear; it is generally easier to solve two small games than one large game. We present two applications which derive directly or indirectly from this remark.]

(A) THE THEOREM OF ZERMELO–VON NEUMANN. It is well known that a zero-sum two-person game with *perfect information* always has a saddle-point in pure strategies.[11] In our formalization, a game with *perfect information* is one in which all information sets are one-element sets and a saddle-point is the zero-sum two-person specialization of the concept of an equilibrium point.

COROLLARY 1. *A general n-person game Γ with perfect information always has an equilibrium point in pure strategies.*

PROOF. The proof is an induction on the number of moves in Γ. For a game with no moves it is trivially true. For a game with one move it is immediate, since if that move is a personal move of player i then he should choose the alternative that maximizes his payoff, while if that move is a chance move then the theorem is again vacuously satisfied.

[11] Zermelo, E. "Über eine Anwendung der Mengenlehre auf die Theorie des Schachspiels," *Proc. Fifth Int. Cong. Math.*, Vol. II, Cambridge (1912), p. 501.

For a game with m moves, since it assumed to have perfect information, it can be decomposed into two games with less than m moves. By the induction hypothesis, these have pure strategy equilibrium points whose composition is again a pure strategy and, by Theorem 3, an equilibrium point for Γ.

(B) SIMULTANEOUS GAMES. A class of games, introduced by G. Thompson as a natural extension of games with perfect information and called *simultaneous games*, can easily be solved through the use of Theorem 3. These are zero-sum two-person games which may be described verbally as consisting of a sequence of simultaneous moves by the two players; following each such move both are informed of the choices. Since our formal system does not handle simultaneous moves (even Matching Pennies has two successive moves) we must describe the games as follows: Player 1 has a_{2k-1} alternatives at the moves of rank $2k-1$ and Player 2 has a_{2k} alternatives at the moves of rank $2k$ where $k = 1, \ldots, K$. Player 1 has perfect information at all of his moves, while Player 2 is informed at his moves of rank $2k$ of everything except the choices by Player 1 at the moves of rank $2k-1$. Clearly we can decompose a simultaneous game at any move of Player 1.

§5.   BEHAVIOR STRATEGIES

[In this section, another natural method of randomization is investigated. Using this method, a player chooses a probability distribution on the alternatives in each of his information sets, thus randomizing on the occasion of a choice as he knows it. It is explicitly assumed that the choices of alternatives at different information sets are made independently. Thus it might be reasonable to call them "uncorrelated" or "locally randomized" strategies; however, since these are the distributions that one would measure in attempting to observe the behavior of a player, we have called them *behavior strategies*.]

DEFINITION 14. To each $U \subset \mathscr{U}_i$ such that $U \subset A_j$, a *behavior strategy for player* i, β_i, assigns j non-negative numbers $b(U, \nu)$, $\nu = 1, \ldots, j$, such that $\sum_\nu b(U, \nu) = 1$.

Any n-tuple $\beta = (\beta_1, \ldots, \beta_n)$ of behavior strategies for the n players defines a probability distribution on the plays of K as follows:

If e is an alternative at a personal move $X \in U \in \mathscr{U}_i$ then $p_\beta(e) = b(U, \nu(e))$.

If e is an alternative at a chance move then $p_\beta(e)$ is the probability assigned to $\nu(e)$ by (III).

Finally, $p_\beta(W) = \prod_{e \in W} p_\beta(e)$.

DEFINITION 15. The *expected payoff* $H_i(\beta)$ to player i for behavior strategies $\beta_1, \ldots, \beta_n$ is defined by $H_i(\beta) = \Sigma_W p_\beta(W) h_i(W)$ for $i = 1, \ldots, n$.

[From our interpretation of behavior strategies, it is clear that each mixed strategy determines a behavior strategy. The next definition establishes this correspondence and the lemma following show that we can achieve every behavior by some mixed strategy.]

DEFINITION 16. The *behavior* β_i of a mixed strategy $\mu_i = (q_{\pi_i})$ for player i is a behavior strategy defined by:

If $U \in \mathrm{Rel}\ \mu_i$ then

$$b(U, \nu) = \sum_{\substack{U \in \mathrm{Rel}\ \pi_i \\ \pi_i(U) - \nu}} q_{\pi_i} \Bigg/ \sum_{U \in \mathrm{Rel}\ \pi_i} q_{\pi_i}.$$

If $U \notin \mathrm{Rel}\ \mu_i$ then

$$b(U, \nu) = \sum_{\pi_i(U) = \nu} q_{\pi_i}.$$

LEMMA 3. *Given a behavior strategy* β_i *for player i, define a mixed strategy* $\mu_i = (q_{\pi_i})$ *for player i by:*

$$q_{\pi_i} = \prod_{U \in \mathscr{U}_i} b(U, \pi_i(U)). \tag{6}$$

Then β_i *is the behavior of* μ_i.

PROOF. This lemma is a direct consequence of Definition 16 and (6).

[To apply these notions to a concrete example, consider the following game:

A PARTNER GAME. In this zero-sum, two-person game, player 1 consists of two agents, called the Dealer and Partner, respectively. Two cards, one marked "High," the other "Low," are dealt to the Dealer and player 2—the two possible deals occurring with equal probabilities. The agent with the High card then receives one dollar from the agent with the Low card *and* has the alternatives of terminating or continuing

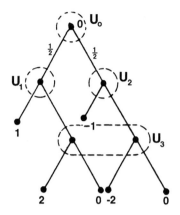

the play. If the play continues, the Partner, not knowing the nature of the deal, can instruct the Dealer to change cards with player 2 or to hold his card. Again, the holder of the High card receives a dollar from the holder of the Low card.

In our formalization, this game is described by the following diagram (remark that the possible plays are labeled with the payoff to player 1).

$$\mathcal{U}_0 = \{U_0\}, \mathcal{U}_1 = \{U_1, U_3\}, \mathcal{U}_2 = \{U_2\}.$$

For simplicity's sake, we will denote the pure strategies π_1 for player 1 by $(\pi_1(U_1), \pi_1(U_3))$ and the pure strategies π_2 for player 2 by $(\pi_2(U_2))$. Then the game matrix of expectations $H_1(\pi_1, \pi_2)$ is:

	(1)	(2)
(1, 1)	0	$-\frac{1}{2}$
(1, 2)	0	$\frac{1}{2}$
(2, 1)	$\frac{1}{2}$	0
(2, 2)	$-\frac{1}{2}$	0

and the "solution" $q_{(1,1)} = q_{(2,2)} = 0$, $q_{(1,2)} = q_{(2,1)} = \frac{1}{2}$, and $q_{(1)} = q_{(2)} = \frac{1}{2}$ insures player 1 the expectation $\frac{1}{4}$ while player 2 can expect to lose no more than $\frac{1}{4}$. On the other hand, if we let $x = b(U_1, 1)$, $1 - x = b(U_1, 2)$, and $y = b(U_3, 1)$, $1 - y = b(U_3, 2)$ be the behavior strategies for player

1, we have:

$$\text{Player 1's expectation against } \pi_2 = \begin{cases} (1) \\ (2) \end{cases} \text{ is } \begin{cases} -\tfrac{1}{2} + \tfrac{1}{2}x + y - xy \\ \tfrac{1}{2}x \qquad - xy \end{cases}$$

Hence the maximum amount that player 1 can assure himself is

$$\max_{\substack{0 \leq x \leq 1 \\ 0 \leq y \leq 1}} \min\{-\tfrac{1}{2} + \tfrac{1}{2}x + y - xy, \tfrac{1}{2}x - xy\} = 0.$$

Thus, behavior strategies may do a poorer job than mixed strategies. Remark that the mixed strategy $\mu_1 = (q_{(1,1)}, q_{(1,2)}, q_{(2,1)}, q_{(2,2)})$ has as its behavior $\beta_1 = (x, y) = (q_{(1,1)} + q_{(1,2)}, q_{(1,1)} + q_{(2,1)})$. Hence, if we consider the optimal mixed strategy $(0, \tfrac{1}{2}, \tfrac{1}{2}, 0)$ for player 1, the associated behavior is $x = y = \tfrac{1}{2}$ and, while the optimal mixed strategy assures 1 the amount $\tfrac{1}{4}$, even its associated behavior only assures 1 the amount 0. This discrepancy is due, of course, to the uncorrelated nature of the behavior strategy. To obtain positive results on the use of behavior strategies we must restrict the nature of the information partition.

DEFINITION 17. A game Γ is said to have *perfect recall* if $U \in \text{Rel } \pi_i$ and $X \in U$ implies $X \in \text{Poss } \pi_i$ for all U, X and π_i.

[The reader should verify that this condition is equivalent to the assertion that each player is allowed by the rules of the game to remember everything he knew at previous moves and all of his choices at those moves. This obviates the use of agents; indeed the only games which do not have perfect recall are those, such as Bridge, which include the description of the agents in their verbal rules.]

LEMMA 4. *Let Γ be a game with perfect recall for $i = 1, \ldots, n$. If i has a move on W, let the last alternative e for i on W be incident at $X \in U$ and set $T_i(W) = \{\pi_i | U \in \text{Rel } \pi_i \text{ and } \pi_i(U) = \nu(e)\}$; otherwise, $T_i(W)$ is the set of all π_i. Finally, let $c(W)$ be the product of the probabilities of the chance alternatives on W or 1 if there are none. Then, for all π and all W,*

$$p_\pi(W) = \begin{cases} c(W) & \text{if } \pi_i \in T_i(W) \text{ for } i = 1, \ldots, n. \\ 0 & \text{otherwise.} \end{cases}$$

PROOF. It is clear that we need only show that $\pi_i \in T_i(W)$ implies that π_i chooses all of the alternatives for i on W (if any such exist). But

$\pi_i \in T_i(W)$ implies $U \in \text{Rel } \pi_i$ and hence, since Γ has perfect recall, $X \in \text{Poss } \pi_i$. Therefore, by Proposition 1, π_i chooses all of the alternatives for i on W.

LEMMA 5. *Let e be an alternative on a play W incident at $X \in U \in \mathscr{U}_i$ and let the next move for player i, if any, be $Y \in V$. Further, let*

$$S = \{\pi_i | U \in \text{Rel } \pi_i \quad and \quad \pi_i(U) = \nu(e)\}$$

and

$$T = \{\pi_i | V \in \text{Rel } \pi_i\}.$$

Then $S = T$.

PROOF. Let $\pi_i \in S$. Then $U \in \text{Rel } \pi_i$ and hence, since Γ has perfect recall, $X \in \text{Poss } \pi_i$ and therefore, by Proposition 1, π_i chooses all alternatives for i on the path 0 to X. But $\pi_i(U) = \nu(e)$ and hence π_i chooses all alternatives for i on the path 0 to Y. Therefore $Y \in \text{Poss } \pi_i$, $V \in \text{Rel } \pi_i$, and $\pi_i \in T$.

Let $\pi_i \in T$. Then $V \in \text{Rel } \pi_i$ and hence, since Γ has perfect recall, $Y \in \text{Poss } \pi_i$, and therefore $X \in \text{Poss } \pi_i$ and $\pi_i(U) = \nu(e)$. That is, $\pi_i \in S$ and the lemma is proved.

THEOREM 4. *Let β be the behavior associated with an arbitrary n-tuple of mixed strategies μ in a game Γ (in which all moves possess at least two alternatives). Then a necessary and sufficient condition that*

$$H_i(\beta) = H_i(\mu) \quad for \, i = 1, \ldots, n$$

and all μ and for all assignments of the payoff function h is that Γ have perfect recall.

PROOF. Assume that Γ has perfect recall; then we need only show

$$p_\beta(W) = p_\mu(W) \quad \text{for all } W.$$

If there is an alternative e on W belonging to player i which is incident at a move in an irrelevant information set for μ_i then both sides are clearly zero. Hence we may assume that all such information sets are relevant for μ_i. Working with each side separately, first:

$$p_\beta(W) = \prod_{e \in W} p_\beta(e).$$

Considering those alternatives e on W which belong to i, their probabilities are given by the *fractions* of Definition 16. The first denominator is clearly 1 while each numerator is the denominator of the next fraction by Lemma 5. Hence

$$p_\beta(W) = c(W) \prod_{i=1}^{n} \left(\sum_{\pi_i \in T_i(W)} q_{\pi_i} \right)$$

where $c(W)$ and the $T_i(W)$ are defined as in Lemma 4. On the other hand,

$$p_\mu(W) = \sum_\pi q_{\pi_1} \cdots q_{\pi_n} p_\pi(W)$$

$$= \sum_{\text{all } \pi_i \in T_i(W)} q_{\pi_1} \cdots q_{\pi_n} c(W)$$

by Lemma 4. Comparing the two expressions for $p(W)$, it is seen that the sufficiency is proved.

For the necessity, if Γ does not have perfect recall then there must be a pure strategy π_i and two moves X and Y in an information set U with $X \in \text{Poss } \pi_i$ and $Y \notin \text{Poss } \pi_i$. Choose a π_i' for which Y is shown to lie in $\text{Poss } \pi_i'$ by the play W and the n-tuple π'. If one sets $\mu_i = \frac{1}{2}\pi_i + \frac{1}{2}\pi_i'$, then

$$p_{\pi'/\mu_i}(W) = \tfrac{1}{2}c(W).$$

However, there is an alternative e lying on the path from 0 to Y which is chosen by π_i' but not by π_i and thus is assigned probability $\frac{1}{2}$ by the behavior of μ_i. If we assume $\pi_i'(U) \neq \pi_i(U)$ then the behavior of μ_i assigns the probability $\frac{1}{2}$ to the alternative at Y on W. Hence

$$p_\beta(W) \leqq \tfrac{1}{4}c(W)$$

and the proof is completed.

EXAMPLES. To illustrate the efficacy of behavior strategies, it is possible to draw three examples from the literature. It is mere coincidence that they are all variants of Poker; the essential common property is that they all have perfect recall.

EXAMPLE 1. von Neumann and Morgenstern give a Poker example[12] in which the number of pure strategies for each player is 3^S, S being the number of possible "hands." Hence the dimension of the set of mixed strategies is $e^S - 1$. However, the dimension of the set of behavior strategies is $2S$; when S is a large number the difference is considerable.

EXAMPLE 2. In the example given by the author,[13] domination arguments reduce the number of pure strategies from 27 to 8 for player 1 and from 64 to 4 for player 2. Nevertheless, the computation of solutions is still a tedious matter. With behavior strategies the payoff function has 3 variables for player 1 and 2 variables for player 2; moreover it has no terms of degree higher than 2 and so the computation of solutions is a mere exercise in elementary calculus.

EXAMPLE 3. In the Simple 3-Person Poker of Nash and Shapley,[14] domination arguments result in a game in which the three players have, respectively, 17, 19, and 31 dimensions of mixed strategies. However, they each have 5 dimensions of behavior strategies and this reduction makes possible the discovery of the unique equilibrium point for their game.

H. W. Kuhn
Princeton University

---

[12] *Loc. cit.*, pp. 190–196.

[13] Kuhn, H., "A simplified two-person Poker," *Annals of Math. Study*, 24 (1950) pp. 97–103.

[14] Nash, J. and Shapley, L. S., "A simple three-person poker game," *Annals of Math. Study*, 24 (1950), pp. 105–116.

7

A VALUE FOR *n*-PERSON GAMES[1]

L. S. SHAPLEY

§1. INTRODUCTION

At the foundation of the theory of games is the assumption that the players of a game can evaluate, in their utility scales, every "prospect" that might arise as a result of a play. In attempting to apply the theory to any field, one would normally expect to be permitted to include in the class of "prospects," the prospect of having to play a game. The possibility of evaluating games is therefore of critical importance. So long as the theory is unable to assign values to the games typically found in application, only relatively simple situations—where games do not depend on other games—will be susceptible to analysis and solution.

In the finite theory of von Neumann and Morgenstern[2] difficulty in evaluation persists for the "essential" games, and for only those. In this note we deduce a value for the "essential" case and examine a number of its elementary properties. We proceed from a set of three axioms, having simple intuitive interpretations, which suffice to determine the value uniquely.

Our present work, though mathematically self-contained, is founded conceptually on the von Neumann-Morgenstern theory up to their introduction of characteristic functions. We thereby inherit certain important underlying assumptions: (a) that utility is objective and transferable; (b) that games are cooperative affairs; (c) that games, granting (a) and (b), are adequately represented by their characteristic functions. However, we are not committed to the assumptions regarding rational behavior embodied in the von Neumann-Morgenstern notion of "solution."

We shall think of a "game" as a set of rules with specified players in the playing positions. The rules alone describe what we shall call an "abstract game." Abstract games are played by *roles*—such as "dealer,"

[1] The preparation of this paper was sponsored (in part) by the RAND Corporation.
[2] Reference [3] at the end of this paper. Examples of infinite games without values may be found in [2], pp. 58–59, and in [1], p. 110. See also Karlin [2], pp. 152–153.

or "visiting team"—rather than by *players* external to the game. The theory of games deals mainly with abstract games.[3] The distinction will be useful in enabling us to state in a precise way that the value of a "game" depends only on its abstract properties. (Axiom 1 below).

§2. DEFINITIONS

Let U denote the universe of players, and define a *game* to be any superadditive set-function v from the subsets of U to the real numbers, thus:

$$v(0) = 0, \tag{1}$$

$$v(S) \geq v(S \cap T) + v(S - T) \qquad \text{(all } S, T \subseteq U). \tag{2}$$

A *carrier* of v is any set $N \subseteq U$ with

$$v(S) = v(N \cap S) \qquad \text{(all } S \subseteq U). \tag{3}$$

Any superset of a carrier of v is again a carrier of v. The use of carriers obviates the usual classification of games according to the number of players. The players outside any carrier have no direct influence on the play since they contribute nothing to any coalition. We shall restrict our attention to games which possess finite carriers.

The *sum* ("superposition") of two games is again a game. Intuitively it is the game obtained when two games, with independent rules but possibly overlapping sets of players, are regarded as one. If the games happen to possess disjunct carriers, then their sum is their "composition."[4]

Let $\Pi(U)$ denote the set of permutations of U—that is, the one to one mappings of U onto itself. If $\pi \in \Pi(U)$, then, writing πS for the image of S under π, we may define the function πv by

$$\pi v(\pi S) = v(S) \qquad \text{(all } S \subseteq U). \tag{4}$$

If v is a game, then the class of games πv, $\pi \in \Pi(U)$, may be regarded as the "abstract game" corresponding to v. Unlike composition, the operation of addition of games cannot be extended to abstract games.

---

[3] An exception is found in the matter of symmetrization (see for example [2], pp. 81–83), in which the players must be distinguished from their roles.

[4] See [3], §§26.7.2 and 41.3.

By the *value* $\phi[v]$ of the game v we shall mean a function which associates with each i in U a real number $\phi_i[v]$, and which satisfies the conditions of the following axioms. The value will thus provide an additive set-function (an *inessential game*) $\bar{v}$:

$$\bar{v}(S) = \sum_S \phi_i[v] \qquad (\text{all } S \subseteq U), \tag{5}$$

to take the place of the superadditive function v.

AXIOM 1. *For each π in $\Pi(U)$,*

$$\phi_{\pi i}[\pi v] = \phi_i[v].$$

AXIOM 2. *For each carrier N of v,*

$$\sum_N \phi_i[v] = v(N).$$

AXIOM 3. *For any two games v and w,*

$$\phi[v + w] = \phi[v] + \phi[w].$$

COMMENTS. The first axiom ("symmetry") states that the value is essentially a property of the abstract game. The second axiom ("efficiency") states that the value represents a distribution of the full yield of the game. This excludes, for example, the evaluation $\phi_i[v] = v((i))$, in which each player pessimistically assumes that the rest will all cooperate against him. The third axiom ("law of aggregation") states that when two independent games are combined, their values must be added player by player. This is a prime requisite for any evaluation scheme designed to be applied eventually to systems of *inter*dependent games.

It is remarkable that no further conditions are required to determine the value uniquely.[5]

§3. DETERMINATION OF THE VALUE FUNCTION

LEMMA 1. *If N is a finite carrier of v, then, for $i \notin N$,*

$$\phi_i[v] = 0.$$

---

[5] Three further properties of the value which might suggest themselves as suitable axioms will be proved as Lemma 1 and Corollaries 1 and 3 below.

PROOF. Take $i \notin N$. Both N and $N \cup (i)$ are carriers of v; and $v(N) = v(N \cup (i))$. Hence $\phi_i[v] = 0$ by Axiom 2, as was to be shown.

We first consider certain symmetric games. For any $R \subseteq U$, $R \neq 0$ define v_R:

$$v_R(S) = \begin{cases} 1 & \text{if } S \supseteq R, \\ 0 & \text{if } S \not\supseteq R. \end{cases} \tag{6}$$

The function cv_R is a game, for any non-negative c, and R is a carrier.

In what follows, we shall use $r, s, n, \ldots$ for the numbers of elements in $R, S, N, \ldots$ respectively.

LEMMA 2. For $c \geq 0$, $0 < r < \infty$, we have

$$\phi_i[cv_R] = \begin{cases} c/r & \text{if } i \in R, \\ 0 & \text{if } i \notin R. \end{cases}$$

PROOF. Take i and j in R, and choose $\pi \in \Pi(U)$ so that $\pi R = R$ and $\pi i = j$. Then we have $\pi v_R = v_R$, and hence, by Axiom 1,

$$\phi_j[cv_R] = \phi_i[cv_R].$$

By Axiom 2,

$$c = cv_R(R) = \sum_{j \in R} \phi_j[cv_R] = r\phi_i[cv_R],$$

for any $i \in R$. This, with Lemma 1, completes the proof.

LEMMA 3.[6] Any game with finite carrier is a linear combination of symmetric games v_R:

$$v = \sum_{\substack{R \subseteq N \\ R \neq 0}} c_R(v)v_R, \tag{7}$$

N being any finite carrier of v. The coefficients are independent of N, and are given by

$$c_R(v) = \sum_{T \subseteq R} (-1)^{r-t} v(T) \qquad (0 < r < \infty). \tag{8}$$

[6] The use of this lemma was suggested by H. Rogers.

PROOF. We must verify that

$$v(S) = \sum_{\substack{R \subseteq N \\ R \neq 0}} c_R(v) v_R(S) \tag{9}$$

holds for all $S \subseteq U$, and for any finite carrier N of v. If $S \subseteq N$, then (9) reduces, by (6) and (8), to

$$v(S) = \sum_{R \subseteq S} \sum_{T \subseteq R} (-1)^{r-t} v(T)$$

$$= \sum_{T \subseteq S} \left[\sum_{r=t}^{s} (-1)^{r-t} \binom{s-t}{r-t} \right] v(T).$$

The expression in brackets vanishes except for $s = t$, so we are left with the identity $v(S) = v(S)$. In general we have, by (3),

$$v(S) = v(N \cap S) = \sum_{R \subseteq N} c_R(v) v_R(N \cap S) = \sum_{R \subseteq N} c_R(v) v_R(S).$$

This completes the proof.

REMARK. It is easily shown that $c_R(v) = 0$ if R is not contained in every carrier of v.

An immediate corollary to Axiom 3 is that $\phi[v - w] = \phi[v] - \phi[w]$ if v, w, and $v - w$ are all games. We can therefore apply Lemma 2 to the representation of Lemma 3 and obtain the formula:

$$\phi_i[v] = \sum_{\substack{R \subseteq N \\ R \ni i}} c_R(v)/r \qquad \text{(all } i \in N). \tag{10}$$

Inserting (8) and simplifying the result gives us

$$\phi_i[v] = \sum_{\substack{S \subseteq N \\ S \ni i}} \frac{(s-1)!(n-s)!}{n!} v(S) - \sum_{\substack{S \subseteq N \\ S \not\ni i}} \frac{s!(n-s-1)!}{n!} v(S)$$

$$\text{(all } i \in N). \tag{11}$$

Introducing the quantities

$$\gamma_n(s) = (s - 1)!(n - s)!/n!, \tag{12}$$

we now assert:

THEOREM. *A unique value function ϕ exists satisfying Axioms 1–3, for games with finite carriers; it is given by the formula*

$$\phi_i[v] = \sum_{S \subseteq N} \gamma_n(s)[v(S) - v(S - (i))] \quad (all\ i \in U), \tag{13}$$

where N is any finite carrier of v.

PROOF. (13) follows from (11), (12), and Lemma 1. We note that (13), like (10), does not depend on the particular finite carrier N; the ϕ of the theorem is therefore well defined. By its derivation it is clearly the only value function which could satisfy the axioms. That it does in fact satisfy the axioms is easily verified with the aid of Lemma 3.

§4. ELEMENTARY PROPERTIES OF THE VALUE

COROLLARY 1. *We have*

$$\phi_i[v] \geq v((i)) \quad (all\ i \in U), \tag{14}$$

with equality if and only if i is a dummy—*i.e., if and only if*

$$v(S) = v(S - (i)) + v((i)) \quad (all\ S \ni i). \tag{15}$$

PROOF. For any $i \in U$ we may take $N \ni i$ and obtain, by (2),

$$\phi_i[v] \geq \sum_{\substack{S \subseteq N \\ S \ni i}} \gamma_n(s)v((i)),$$

with equality if and only if (15), since none of the $\gamma_n(s)$ vanishes. The proof is completed by noting that

$$\sum_{\substack{S \subseteq N \\ S \ni i}} \gamma_n(s) = \sum_{s=1}^{n} \binom{n-1}{s-1} \gamma_n(s) = \sum_{s=1}^{n} \frac{1}{n} = 1. \tag{16}$$

Only in this corollary have our results depended on the superadditive nature of the functions v.

COROLLARY 2. *If v is decomposable—i.e., if games $w^{(1)}, w^{(2)}, \ldots, w^{(p)}$ having pairwise disjunct carriers $N^{(1)}, N^{(2)}, \ldots, N^{(p)}$ exist such that*

$$v = \sum_{k=1}^{p} w^{(k)},$$

—then, for each $k = 1, 2, \ldots, p$,

$$\phi_i[v] = \phi_i[w^{(k)}] \qquad (all\ i \in N^{(k)}).$$

PROOF. By Axiom 3.

COROLLARY 3. *If v and w are strategically equivalent—i.e., if*

$$w = cv + \bar{a}, \tag{17}$$

where c is a positive constant and $\bar{a}$ an additive set-function on U with finite carrier[7]—then

$$\phi_i[w] = c\phi_i[v] + \bar{a}((i)) \qquad (all\ i \in U).$$

PROOF. By Axiom 3, Corollary 1 applied to the inessential game $\bar{a}$, and the fact that (13) is linear and homogeneous in v.

COROLLARY 4. *If v is constant-sum—i.e., if*

$$v(S) + v(U - S) = v(U) \qquad (all\ S \subseteq U), \tag{18}$$

—then its value is given by the formula:

$$\phi_i[v] = 2\left[\sum_{\substack{S \subseteq N \\ S \ni i}} \gamma_n(s)v(S) \right] - v(N) \qquad (all\ i \in N) \tag{19}$$

where N is any finite carrier of v.

PROOF. We have, for $i \in N$,

$$\phi_i[v] = \sum_{\substack{S \subseteq N \\ S \ni i}} \gamma_n(s)v(S) - \sum_{\substack{T \subseteq N \\ T \not\ni i}} \gamma_n(t+1)v(T)$$

$$= \sum_{\substack{S \subseteq N \\ S \ni i}} \gamma_n(s)v(S) - \sum_{\substack{S \subseteq N \\ S \ni i}} \gamma_n(n-s+1)[v(N) - v(S)].$$

But $\gamma_n(n - s + 1) = \gamma_n(s)$; hence (18) follows with the aid of (16).

---

[7] This is McKinsey's "S-equivalence" (see [2], p. 120), wider than the "strategic equivalence" of von Neumann and Morgenstern ([3], §27.1).

§5. EXAMPLES

If N is a finite carrier of v, let A denote the set of n-vectors (α_i) satisfying

$$
\begin{cases}
\sum_N \alpha_i = v(N), \\
\alpha_i \geq v((i)), \qquad \text{(all } i \in N).
\end{cases}
$$

If v is inessential A is a single point; otherwise A is a regular simplex of dimension $n - 1$. The value of v may be regarded as a point ϕ in A, by Axiom 2 and Corollary 1. Denote the centroid of A by θ:

$$
\theta_i = v((i)) + \frac{1}{n}\left[v(N) - \sum_{j \in N} v((j)) \right].
$$

EXAMPLE 1. For *two-person* games, *three-person constant-sum* games, and *inessential* games, we have

$$
\phi = \theta. \tag{20}
$$

The same holds for arbitrary symmetric games—i.e., games which are invariant under a transitive group of permutations of N—and, most generally, games strategically equivalent to them. These results are demanded by symmetry, and do not depend on Axiom 3.

EXAMPLE 2. For general *three-person* games the positions taken by ϕ in A cover a regular hexagon, touching the boundary at the midpoint of each 1-dimensional fact (see Figure 1). The latter cases are of course the decomposable games, with one player a dummy.

EXAMPLE 3. *The quota games*[8] are characterized by the existence of constants ω_i satisfying

$$
\begin{cases}
\omega_i + \omega_j = v((i, j)) \qquad \text{(all } i, j \in N, i \neq j) \\
\sum_N \omega_i = v(N).
\end{cases}
$$

---

[8] Discussed in [4].

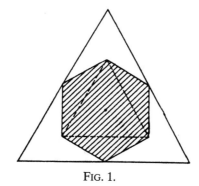

Fɪɢ. 1.

For $n = 3$, we have

$$\phi - \theta = \frac{\omega - \theta}{2}. \tag{21}$$

Since ω can assume any position in A, the range of ϕ is a triangle, inscribed in the hexagon of the preceding example (see Figure 1).

Eхамрle 4. All *four-person constant-sum* games are quota games. For them we have

$$\phi - \theta = \frac{\omega - \theta}{3}. \tag{22}$$

The quota ω ranges over a certain cube,[9] containing A. The value ϕ meanwhile ranges over a parallel, inscribed cube, touching the boundary of A at the midpoint of each 2-dimensional face. In higher quota games the points ϕ and ω are not so directly related.

Eхамрle 5. The *weighted majority games*[10] are characterized by the existence of "weights" w_i such that never $\Sigma_S w_i = \Sigma_{N-S} w_i$, and such that

$$\begin{cases} v(S) = n - s & \text{if } \displaystyle\sum_S w_i > \sum_{N-S} w_i, \\ v(S) = -s & \text{if } \displaystyle\sum_S w_i < \sum_{N-S} w_i. \end{cases}$$

[9] Illustrated in [4], figure 1.
[10] See [3], §50.1.

The game is then denoted by the symbol $[w_1, w_2, \ldots, w_n]$. It is easily shown that

$$\phi_i < \phi_j \quad \text{implies} \quad w_i < w_j \qquad (\text{all } i, j \in N) \qquad (23)$$

in any weighted majority game $[w_i, w_2, \ldots, w_n]$. Hence "weight" and "value" rank the players in the same order.

The exact values can be computed without difficulty for particular cases. We have

$$\phi = \frac{n-3}{n-1}(-1, -1, \ldots, -1, n-1)$$

for the game $[1, 1, \ldots, 1, n-2]$,[11] and

$$\phi = \tfrac{2}{5}(1, 1, 1, -1, -1, -1)$$

for the game $[2, 2, 2, 1, 1, 1]$,[12] etc.

§6. Derivation of the Value from a Bargaining Model

The deductive approach of the earlier sections has failed to suggest a bargaining procedure which would produce the value of the game as the (expected) outcome. We conclude this paper with a description of such a procedure. The form of our model, with its chance move, lends support to the view that the value is best regarded as an *a priori* assessment of the situation, based on either ignorance or disregard of the social organization of the players.

The players constituting a finite carrier N agree to play the game v in a grand coalition, formed in the following way: 1. Starting with a single member, the coalition adds one player at a time until everyone has been admitted. 2. The order in which the players are to join is determined by chance, with all arrangements equally probable. 3. Each player, on his admission, demands and is promised the amount which his adherence contributes to the value of the coalition (as determined by the function v). The grand coalition then plays the game "efficiently" so as to obtain the amount $v(N)$—exactly enough to meet all the promises.

---

[11] Discussed at length in [3], §55.
[12] Discussed in [3], §53.2.2.

The expectations under this scheme are easily worked out. Let $T^{(i)}$ be the set of players preceding i. For any $S \ni i$, the payment to i if $S - (i) = T^{(i)}$ is $v(S) - v(S - (i))$, and the probability of that contingency is $\gamma_n(s)$. The total expectation of i is therefore just his value, (13), as was to be shown.

BIBLIOGRAPHY

[1] Borel, E. and Ville, J. "Applications aux jeux de hasard," *Traité du Calcul des Probabilitiés et de ses Applications*, Vol. 4, Part 2. Paris: Gauthier-Villars, 1938.

[2] Kuhn, H. W. and Tucker, A. W. (eds.). "Contributions to the Theory of Games," *Annals of Mathematics Study*, No. 24, Princeton, 1950.

[3] von Neumann, J. and Morgenstern, O. *Theory of Games and Economic Behavior*. Princeton, 1944; 2nd ed., 1947.

[4] Shapley, L. S. "Quota solutions of n-person games," this study.

L. S. Shapley
Princeton University and
The RAND Corporation

8

STOCHASTIC GAMES*

L. S. SHAPLEY

PRINCETON UNIVERSITY

Communicated by J. von Neumann, July 17, 1953

Introduction—In a stochastic game the play proceeds by steps from position to position, according to transition probabilities controlled jointly by the two players. We shall assume a finite number, N, of positions, and finite numbers m_k, n_k of choices at each position, nevertheless, the game may not be bounded in length. If, when at position k, the players choose their ith and jth alternatives, respectively, then with probability $s_{ij}^k > 0$ the game stops, while with probability p_{ij}^{kl} the game moves to position l. Define

$$s = \min_{k,i,j} s_{ij}^k.$$

Since s is positive, the game ends with probability 1 after a finite number of steps, because, for any number t, the probability that it has *not* stopped after t steps is not more than $(1 - s)^t$.

Payments accumulate throughout the course of play: the first player takes a_{ij}^k from the second whenever the pair i, j is chosen at position k. If we define the bound M:

$$M = \max_{k,i,j} |a_{ij}^k|,$$

then we see that the expected total gain or loss is bounded by

$$M + (1 - s)M + (1 - s)^2 M + \cdots = M/s. \tag{1}$$

The process therefore depends on $N^2 + N$ matrices

$$P^{kl} = \left(p_{ij}^{kl} \,\middle|\, i = 1, 2, \ldots, m_k; j = 1, 2, \ldots, n_k \right)$$

$$A^k = \left(a_{ij}^k \,\middle|\, i = 1, 2, \ldots, m_k; j = 1, 2, \ldots, n_k \right),$$

*The preparation of this paper was sponsored (in part) by the Office of Naval Research.

with $k, l = 1, 2, \ldots, N$, with elements satisfying

$$p_{ij}^{kl} \geq 0, \quad \left| a_{ij}^k \right| \leq M, \quad \sum_{l=1}^{N} p_{ij}^{kl} = 1 - s_{ij}^k \leq 1 - s < 1.$$

By specifying a starting position we obtain a particular game Γ^k. The term "stochastic game" will refer to the collection $\Gamma = \{ \Gamma^k \mid k = 1, 2, \ldots, N \}$.

The full sets of pure and mixed strategies in these games are rather cumbersome, since they take account of much information that turns out to be irrelevant. However, we shall have to introduce a notation only for certain behavior strategies,[1] namely those which prescribe for a player the same probabilities for his choices every time the same position is reached, by whatever route. Such *stationary strategies*, as we shall call them, can be represented by N-tuples of probability distributions, thus:

$$\vec{x} = (x^1, x^2, \ldots, x^N), \quad \text{each } x^k = \left(x_1^k, x_2^k, \ldots, x_{m_k}^k \right),$$

for the first player, and similarly for the second player. This notation applies without change in all of the games belonging to Γ.

Note that a stationary strategy is not in general a mixture of *pure* stationary strategies (all x_i^k zero or one), since the probabilities in a behavior strategy must be uncorrelated.

Existence of a Solution—Given a matrix game B, let val$[B]$ denote its minimax value to the first player, and $X[B]$, $Y[B]$ the set of optimal mixed strategies for the first and second players, respectively.[2] If B and C are two matrices of the same size, then it is easily shown that

$$|\text{val}[B] - \text{val}[C]| \leq \max_{i, j} |b_{ij} - c_{ij}|. \tag{2}$$

Returning to the stochastic game Γ, define $A^k(\vec{\alpha})$ to be the matrix of elements

$$a_{ij}^k + \sum_{l} p_{ij}^{kl} \alpha^l,$$

$i = 1, 2, \ldots, m_k; \ j = 1, 2, \ldots, n_k$, where $\vec{\alpha}$ is any N-vector with numerical components. Pick $\vec{\alpha}_{(0)}$ arbitrarily, and define $\vec{\alpha}_{(t)}$ by the recursion:

$$\alpha_{(t)}^k = \text{val}\left[A^k\left(\vec{\alpha}_{(t-1)} \right) \right], \qquad t = 1, 2, \dots .$$

(If we had chosen $\alpha_{(0)}^k$ to be the value of A^k, for each k, then $\alpha_{(t)}^k$ would be the value of the truncated game $\Gamma_{(t)}^k$ which starts at position k, and which is cut off after t steps if it lasts that long.) We shall show that the limit of $\vec{\alpha}_{(t)}$ as $t \to \infty$ exists and is independent of $\vec{\alpha}_{(0)}$, and that its components are the values of the infinite games Γ^k.

Consider the transformation T:

$$T\vec{\alpha} = \vec{\beta}, \qquad \text{where } \beta^k = \text{val}[A^k(\vec{\alpha})].$$

Define the norm of $\vec{\alpha}$ to be

$$\|\vec{\alpha}\| = \max_k | \alpha^k |.$$

Then we have

$$\|T\vec{\beta} - T\vec{\alpha}\| = \max_k \left| \text{val}\left[A^k\left(\vec{\beta} \right) \right] - \text{val}[A^k(\vec{\alpha})] \right|$$

$$\le \max_{k, i, j} \left| \sum_l p_{ij}^{kl} \beta^l - \sum_l p_{ij}^{kl} \alpha^l \right|$$

$$\le \max_{k, i, j} \left| \sum_l p_{ij}^{kl} \right| \max_l | \beta^l - \alpha^l |$$

$$= (1 - s)\| \vec{\beta} - \vec{\alpha}\|, \qquad (3)$$

using (2). In particular, $\|T^2\vec{\alpha} - T\vec{\alpha}\| \le (1 - s)\|T\vec{\alpha} - \vec{\alpha}\|$. Hence the sequence $\vec{\alpha}_{(0)}, T\vec{\alpha}_{(0)}, T^2\vec{\alpha}_{(0)}, \dots$ is convergent. The limit vector $\vec{\phi}$ has the property $\vec{\phi} = T\vec{\phi}$. But there is only one such vector, for $\vec{\psi} = T\vec{\psi}$ implies

$$\|\vec{\psi} - \vec{\phi}\| = \|T\vec{\psi} - T\vec{\phi}\| \le (1 - s)\|\vec{\psi} - \vec{\phi}\|,$$

by (3), whence $\|\vec{\psi} - \vec{\phi}\| = 0$. Hence $\vec{\phi}$ is the unique fixed point of T and is independent of $\vec{\alpha}_{(0)}$.

To show that ϕ^k is the value of the game Γ^k, we observe that by following an optimal strategy of the finite game $\Gamma_{(t)}^k$ for the first t steps and playing arbitrarily thereafter, the first player can assure himself an amount within $\epsilon_t = (1-s)^t M/s$ of the value of $\Gamma_{(t)}^k$; likewise for the other player. Since $\epsilon_t \to 0$ and the value of $\Gamma_{(t)}^k$ converges to ϕ^k, we conclude that ϕ^k is indeed the value of Γ^k. Summing up:

THEOREM 1. *The value of the stochastic game Γ is the unique solution $\vec{\phi}$ of the system*

$$\phi^k = \mathrm{val}\left[A^k(\vec{\phi}) \right], \qquad k = 1, 2, \ldots, N.$$

Our next objective is to prove the existence of optimal strategies.

THEOREM 2. *The stationary strategies $\vec{x}^*, \vec{y}^*$, where $x^l \in X[A^l(\vec{\phi})]$, $y^l \in Y[A^l(\vec{\phi})]$, $l = 1, 2, \ldots, N$, are optimal for the first and second players respectively in every game Γ^k belonging to Γ.*

PROOF. Let a finite version of Γ^k be defined by agreeing that on the tth step the play shall stop, with the first player receiving the amount $a_{ij}^h + \Sigma_l p_{ij}^{hl} \phi^l$ instead of just a_{ij}^h. Clearly, the stationary strategy $\vec{x}^*$ assures the first player the amount ϕ^k in this finite version. In the original game Γ^k, if the first player uses $\vec{x}^*$, his expected winnings after t steps will be at least

$$\phi^k - (1-s)^{t-1} \max_{h,i,j} \sum_l p_{ij}^{hl} \phi^l,$$

and hence at least

$$\phi^k - (1-s)^t \max_l \phi^l.$$

His total expected winnings are therefore at least

$$\phi^k - (1-s)^t \max_l \phi^l - (1-s)^t M/s.$$

Since this is true for arbitrarily large values of t, it follows that $\vec{x}^*$ is optimal in Γ^k for the first player. Similarly, $\vec{y}^*$ is optimal for the second player.

Reduction to a Finite-Dimensional Game—The non-linearity of the "val" operator often makes it difficult to obtain exact solutions by means of Theorems 1 and 2. It therefore becomes desirable to express the payoff directly in terms of stationary strategies. Let $\bar{\Gamma} = \{\bar{\Gamma}^k\}$ denote the collection of games whose *pure* strategies are the stationary strategies of Γ. Their payoff functions $\mathfrak{H}^k(\vec{x}, \vec{y})$ must satisfy

$$\mathfrak{H}^k(\vec{x}, \vec{y}) = x^k A^k y^k + \sum_l x^k P^{kl} y^k \mathfrak{H}^l(\vec{x}, \vec{y}),$$

for $k = 1, 2, \ldots, N$. This system has a unique solution; indeed, for the linear transformation $T_{\vec{x}\vec{y}}$:

$$T_{\vec{x}\vec{y}} \vec{\alpha} = \vec{\beta}, \quad \text{where } \beta^k = x^k A^k y^k + \sum_l x^k P^{kl} y^k \alpha^l$$

we have at once

$$\|T_{\vec{x}\vec{y}} \vec{\beta} - T_{\vec{x}\vec{y}} \vec{\alpha}\| = \max_k \left| \sum_l x^k P^{kl} y^k (\beta^l - \alpha^l) \right| \leq (1 - s) \|\vec{\beta} - \vec{\alpha}\|,$$

corresponding to (3) above. Hence, by Cramer's rule,

$$\mathfrak{H}^k(\vec{x}, \vec{y})$$

$$= \frac{\begin{vmatrix} x^1 P^{11} y^1 - 1 & x^1 P^{12} y^1 & \cdots & -x^1 A^1 y^1 & \cdots & x^1 P^{1N} y^1 \\ x^2 P^{21} y^2 & x^2 P^{22} y^2 - 1 & & & & \\ \cdots & & & \cdots & & \cdots \\ x^N P^{N1} y^N & \cdots & & \cdots & -x^N A^N y^N & \cdots & x^N P^{NN} y^N - 1 \end{vmatrix}}{\begin{vmatrix} x^1 P^{11} y^1 - 1 & x^1 P^{12} y^1 & \cdots & x^1 P^{1k} y^1 & \cdots & x^1 P^{1N} y^1 \\ x^2 P^{21} y^2 & x^2 P^{22} y^2 - 1 & & \cdots & & \\ \cdots & & & x^k P^{kk} y^k - 1 & & \cdots \\ & & & \cdots & & \\ x^N P^{N1} y^N & \cdots & & x^N P^{Nk} y^N & \cdots & x^N P^{NN} y^N - 1 \end{vmatrix}}.$$

THEOREM 3. *The games $\bar{\Gamma}^k$ possess saddle points:*

$$\min_{\vec{y}} \max_{\vec{x}} \mathfrak{H}^k(\vec{x}, \vec{y}) = \max_{\vec{x}} \min_{\vec{y}} \mathfrak{H}^k(\vec{x}, \vec{y}), \tag{4}$$

for $k = 1, 2, \ldots, N$. Any stationary strategy which is optimal for all $\Gamma^k \in \Gamma$ is an optimal pure strategy for all $\bar{\Gamma}^k \in \bar{\Gamma}$, and conversely. The value vectors of Γ and $\bar{\Gamma}$ are the same.

The proof is a simple argument based on Theorem 2. It should be pointed out that a strategy $\vec{x}$ may be optimal for one game Γ^k (or $\bar{\Gamma}^k$) and not optimal for other games belonging to Γ (or $\bar{\Gamma}$). This is due to the possibility that Γ might be "disconnected"; however if none of the p_{ij}^{kl} are zero this possibility does not arise.

It can be shown that the sets of optimal stationary strategies for Γ are closed, convex polyhedra. A stochastic game with rational coefficients does not necessarily have a rational value. Thus, unlike the minimax theorem for bilinear forms, the equation (4) is not valid in an arbitrary ordered field.

Examples and Applications—1. When $N = 1$, Γ may be described as a simple matrix game A which is to be replayed according to probabilities that depend on the players' choice. The payoff function of Γ is

$$\mathfrak{H}(x, y) = \frac{xAy}{xSy},$$

where S is the matrix of (non-zero) stop probabilities. The minimax theorem (4) for rational forms of this sort was established by von Neumann;[3] an elementary proof was subsequently given by Loomis.[4]

2. By setting all the stop probabilities s_{ij}^k equal to $s > 0$, we obtain a model of an indefinitely continuing game in which future payments are discounted by a factor $(1 - s)^t$. In this interpretation the actual transition probabilities are $q_{ij}^{kl} = p_{ij}^{kl}/(1 - s)$. By holding the q_{ij}^{kl} fixed and varying s, we can study the influence of interest rate on the optimal strategies.

3. A stochastic game does not have perfect information, but is rather a "simultaneous game," in the sense of Kuhn and Thompson.[1] However, perfect information can be simulated within our framework by putting either m_k or n_k equal to 1, for all values of k. Such a stochastic game of perfect information will of course have a solution in stationary *pure* strategies.

4. If we set $n_k = 1$ for all k, effectively eliminating the second player, the result is a "dynamic programming" model.[5] Its solution is given by

any set of integers $\vec{i} = \{i_1, i_1, \ldots, i_N | 1 \le i_k \le m_k\}$ which maximizes the expression

$$\mathfrak{H}^k(\vec{i}) = \frac{\begin{vmatrix} p_{i_1}^{11} - 1 & p_{i_1}^{12} & \cdots & -a_{i_1}^{1} & \cdots & p_{i_1}^{1N} \\ p_{i_2}^{21} & p_{i_2}^{22} - 1 & & & & \\ \cdots & & & \cdots & & \cdots \\ p_{i_N}^{N1} & & \cdots & -a_{i_N}^{N} & \cdots & p_{i_N}^{NN} - 1 \end{vmatrix}}{\begin{vmatrix} p_{i_1}^{11} - 1 & p_{i_1}^{12} & \cdots & p_{i_1}^{1k} & \cdots & p_{i_1}^{1N} \\ p_{i_2}^{21} & p_{i_2}^{22} - 1 & & \cdots & & \\ \cdots & & & p_{i_k}^{kk} - 1 & & \cdots \\ & & & \cdots & & \\ p_{i_N}^{N1} & & \cdots & p_{i_N}^{Nk} & \cdots & p_{i_N}^{NN} - 1 \end{vmatrix}} .$$

For example (taking $N = 1$), let there be alternative procedures $i = 1, \ldots, m$ costing $c_i = -a_i$ to apply and having probability s_i of success. The above then gives us the rule: adopt that procedure i^* which maximizes the ratio a_{i*}/s_{i*}, or equivalently, the ratio s_{i*}/c_{i*}.

5. Generalizations of the foregoing theory to infinite sets of alternatives, or to an infinite number of states, readily suggest themselves (see for example ref. 6). We shall discuss them in another place.

NOTES

[1] Kuhn, H. W., *Contributions to the Theory of Games II*, Annals of Mathematics Studies No. 28, Princeton, 1953, pp. 209–210.

[2] von Neumann, J., and Morgenstern, O., *Theory of Games and Economic Behavior*, Princeton, 1944 and 1947, p. 158.

[3] von Neumann, J., *Ergebnisse eines Math. Kolloquiums*, **8**, 73–83 (1937).

[4] Loomis, L. H., these PROCEEDINGS, **32**, 213–215 (1946).

[5] Bellman, R., these PROCEEDINGS, **38**, 716–719 (1952).

[6] Isbell, J. R., *Bull. A. M. S.*, **59**, 234–235 (1953).

9

RECURSIVE GAMES

H. EVERETT[1]

INTRODUCTION

A recursive game is a finite set of "game elements," which are games for which the outcome of a single play (payoff) is either a real number, or another game of the set, but not both. By assigning real numbers to game payoffs, each element of the recursive game becomes an ordinary game, whose value and optimal strategies (if they exist) of course depend upon the particular assignment. It is shown that if every game element possesses a solution for arbitrary assignments, then the recursive game possesses a solution. In particular, if the game elements possess minimax solutions for all assignments of real numbers to game payoffs, then the recursive game possesses a supinf solution in stationary strategies, while if the game elements possess only supinf solutions, then the recursive game possesses a supinf solution which may, however, require nonstationary strategies. No restrictions are placed upon the type of game elements, other than the condition that they possess solutions for arbitrary assignments of real numbers to game payoffs. Some extensions to more general games are given.

§1. DEFINITIONS

A recursive game, $\vec{\Gamma}$, is a finite set of n "game elements," denoted by $\Gamma^1, \Gamma^2, \ldots, \Gamma^n$, each of which possesses a pair of strategy spaces, denoted by S_1^k and S_2^k corresponding to Γ^k for Players 1 and 2 respectively. To every pair of strategies $X^k \in S_1^k$, $Y^k \in S_2^k$, there is associated an expression (generalized payoff):

$$H^k(X^k, Y^k; \vec{\Gamma}) = p^k e^k + \sum_{j=1}^{n} q^{kj} \Gamma^j, \qquad (1.1)$$

where

$$p^k, q^{kj} \geqq 0 \quad \text{and} \quad p^k + \sum_j q^{kj} = 1.$$

[1] National Science Foundation Predoctoral Fellow 1953–56.

The interpretation of this generalized payoff is that if Player 1 and Player 2 play Γ^k with strategies X^k and Y^k respectively, the possible outcomes of the single round are either to terminate play with Player 1 receiving an amount e^k from Player 2, or to have no payoff and proceed to play another game of the set, where p^k and the q^{kj} are the probabilities of these events.

A strategy $\chi \in \mathscr{S}_1$ for P_1 is an infinite sequence of vectors, $\chi = \{\vec{X}_t\}$ $= \vec{X}_1, \vec{X}_2, \ldots, \vec{X}_t, \ldots$ where $\vec{X}_t = (X_t^1, X_t^2, \ldots, X_t^n)$ and $X_t^i \in S_1^i$ for all $t_k\ t$ and all i, with the interpretation that if P_1 finds himself in Γ^k for the t-th round of play, he will use strategy X_t^k. A strategy χ is stationary in component i if $X_t^i = X_1^i$ for all t. A strategy χ is stationary if it is stationary in all components. Similar definitions hold for a strategy $\Psi \in \mathscr{S}_2$ for P_2.

A pair of strategies χ, Ψ and a starting position Γ^j define a random walk with absorbing barriers among the game elements. Since absorption in Γ^k in the t-th round carries the payoff e_t^k, and expectation, $\mathrm{Ex}^j[\chi, \Psi]$ is defined. Thus to each strategy pair there corresponds an expectation *vector*, whose components correspond to the starting positions. If we define the $n \times n$ matrices P_t and Q_t and the column vector $\vec{E}_t$ for the strategy pair (χ, Ψ) by:

$$[P_t]^{ij} = \delta^{ij} p^i \qquad [Q_t]^{ij} = q^{ij}$$

$$[Q_0]^{ij} = \delta^{ij} \qquad\qquad E_t^i = e^i \qquad\qquad (1.2)$$

where p^i, q^{ij}, and e^i are given by $H^i(X_t^i, Y_t^i; \vec{\Gamma})$ through (1.0), then straightforward calculation gives the expectation vector for n rounds of play as:

$$\overrightarrow{\mathrm{Ex}}_n(\chi, \Psi) = \sum_{k=1}^{n} \left(\prod_{t=0}^{k-1} Q_t \right) P_k \vec{E}_k, \qquad\qquad (1.3)$$

where

$$\prod_{t=0}^{k-1} Q_t = Q_0 Q_1 Q_2 Q_3 \cdots Q_{k-1}$$

and hence that the ultimate expectation is

$$\overrightarrow{\mathrm{Ex}}(\chi,\Psi) = \lim_{n\to\infty} \overrightarrow{\mathrm{Ex}}_n(\chi,\Psi) = \sum_{k=1}^{\infty} \left(\prod_{t=0}^{k-1} Q_t\right) P_k \vec{E}_k \qquad (1.4)$$

which for bounded payoffs always converges, and which assigns zero expectation to a nonterminating play.

A recursive game will be said to possess a solution if there exists a vector $\vec{V}$, and if for all $\epsilon > 0$ there exist strategies $\chi^\epsilon \in \mathscr{S}_1$, $\Psi^\epsilon \in \mathscr{S}_2$, such that:

$$\overrightarrow{\mathrm{Ex}}(\chi^\epsilon,\Psi) \geqq \vec{V} - \epsilon\vec{1} \quad \text{for all } \Psi \in \mathscr{S}_2 \qquad (1.5)$$

and

$$\overrightarrow{\mathrm{Ex}}(\chi,\Psi^\epsilon) \leqq \vec{V} + \epsilon\vec{1} \quad \text{for all } \chi \in \mathscr{S}_1$$

where $\vec{U} \geq \vec{W} \rightleftarrows U^i \geq W^i$ for all i, and $\vec{1} = (1,1,\ldots,1)$. Then χ^ϵ and Ψ^ϵ are called ϵ-best strategies, and $\vec{V}$ is the value of the recursive game. (Our definition thus corresponds to a solution of *all* of the games which can arise from the different possible starting positions Γ^i.)

§2. THE VALUE MAPPING, M

For an arbitrary vector $\vec{W} = (W^1, W^2, \ldots, W^n)$ we can reduce a game element Γ^k to an ordinary (nonrecursive) game $\Gamma^k(\vec{W})$ by defining the (numerical valued) payoff function for $\Gamma^k(\vec{W})$ to be:

$$H^k(X^k, Y^k; \vec{W}) = p^k e^k + \sum_j q^{kj} W^j, \qquad (X^k, Y^k) \in S_1^k \times S_2^k \quad (2.1)$$

which results from $H(X^k, Y^k; \vec{\Gamma})$ by replacing the symbols Γ^j by the real numbers W^j in (1.0). In effect we are arbitrarily assigning a "value," W^j, to the command to play Γ^j.

(2.2) DEFINITION 1. A game element Γ^i satsifies the *supinf condition* if the ordinary game $\Gamma^i(\vec{W})$ possesses a supinf solution in the usual sense for all $\vec{W}$.

(2.3) DEFINITION 2. A game element Γ^i satisfies the *minimax condition* if $\Gamma^i(\vec{W})$ possesses a minimax solution for all $\vec{W}$.

Of course, if a game element satisfies the minimax condition, it also satisfies the supinf condition. We shall henceforth deal only with recursive games, all of whose elements satisfy at least the supinf condition.

If each of the n game elements of a recursive game $\vec{\Gamma}$ satisfies the supinf condition, then for any n-vector $\vec{U}$, we define the n-vector $\vec{U'} = M(\vec{U})$ through:

$$U'^i = \text{Val } \Gamma^i(\vec{U}). \tag{2.4}$$

The mapping, M, of n-vectors into n-vectors is then called the *value mapping* for the game $\vec{\Gamma}$.

We now define the relations $\geq^{\cdot}$ and $\leq$ for vectors (or numbers) to mean:

$$\vec{U} \geq^{\cdot} \vec{V} \rightleftarrows \left\{ \begin{array}{ll} U^i > V^i & \text{if } V^i > 0 \\ U^i \geq V^i & \text{if } V^i \leq 0 \end{array} \right\} \quad \text{for all } i$$

$$\vec{U} \leq \vec{V} \rightleftarrows \left\{ \begin{array}{ll} U^i < V^i & \text{if } V^i < 0 \\ U^i \leq V^i & \text{if } V^i \geq 0 \end{array} \right\} \quad \text{for all } i \tag{2.5}$$

and we further define, for $\vec{\Gamma}$, the classes $C_1(\vec{\Gamma})$, $C_2(\vec{\Gamma})$ of n-vectors by:

$$\vec{W} \in C_1(\vec{\Gamma}) \rightleftarrows M(\vec{W}) \geq^{\cdot} \vec{W}$$

$$\vec{W} \in C_2(\vec{\Gamma}) \rightleftarrows M(\vec{W}) \leq \vec{W} \tag{2.6}$$

and we note that $C_1(\vec{\Gamma})$ and $C_2(\vec{\Gamma})$ are always disjoint except possibly for the zero vector.

THEOREM 1. (*a*) $\vec{W} \in C_1(\vec{\Gamma}) \Rightarrow$ *for every $\epsilon > 0$ there exists a strategy* $\chi^\epsilon \in \mathscr{S}_1$ *such that*

$$\overrightarrow{\text{Ex}}(\chi^\epsilon, \Psi) \geq \vec{W} - \epsilon \vec{1} \quad (\text{all } \Psi \in \mathscr{S}_2)$$

(*b*) $\vec{W} \in C_2(\vec{\Gamma}) \Rightarrow$ *for every $\epsilon > 0$ there exists a strategy* $\Psi^\epsilon \in \mathscr{S}_2$ *such that*

$$\overrightarrow{\text{Ex}}(\chi, \Psi) \leq \vec{W} + \epsilon 1 \quad (\text{all } \chi \in \mathscr{S}_1).$$

PROOF. We shall prove (a) by supposing that we are given a $\vec{W} \in C_1(\vec{\Gamma})$ and an $\epsilon > 0$, and then using $\vec{W}$ to construct a strategy $\chi^\epsilon \in \mathscr{S}_1$, which we subsequently prove gives the desired result.

Let $\vec{W}' = M(\vec{W})$. Because $\vec{W} \in C_1(\vec{\Gamma})$ all components W^i which are positive increase under the value mapping, and since there are only a finite number, there exists a $\gamma > 0$ such that $W^i > 0 \Rightarrow W'^i - W^i \geq \gamma$ for all i. Choose δ such that $0 < \delta < \min(\gamma, \epsilon)$, and then let strategy $\chi^\epsilon \in \mathscr{S}_1$ for P_1 have components X_t^i as follows:

1) If $\Gamma^i(\vec{W})$ possesses an optimal strategy, $\bar{X}^i \in S_1^i$, for P_1, then let $X_t^i = \bar{X}^i$ for all t. (stationary in comp. i)

2) If $\Gamma^i(\vec{W})$ fails to possess an optimal strategy for P_1, but $W^i > 0$, then let $X_t^i = \tilde{X}^i$ for all t, where $\tilde{X}^i \in S_1^i$ is δ-best in $\Gamma^i(\vec{W})$. (stationary in comp. i) $\qquad$ (2.7)

3) If $\Gamma^i(\vec{W})$ fails to possess an optimal strategy for P_1, and $W^i \leq 0$, then let $X_t^i \in S_1^i$ be a strategy which is δ_t-best in $\Gamma^i(\vec{W})$, where $\delta_t = (\frac{1}{2})^t \delta$. (nonstationary in comp. i).

Then by (2.1), (2.4), and (2.6), for $\vec{X}_t$ so defined and for all $\vec{Y}_t$:

$$H^i\left(X_t^i, Y_t^i; \vec{W}\right) = p_t^i e_t^i + \sum_j q_t^{ij} W^j$$

$$\geq \begin{cases} W^i + \gamma - \delta \text{ in 1) and 2)} & (w^i > 0) \\ W^i - \delta_t \text{ in 3)} & (W^i \leq 0) \end{cases} \quad \text{all } i,$$

$$(2.8)$$

so that if we define the nonnegative vectors $\vec{\mu}$ and $\vec{\delta}_t$ by:

$$\mu^i = \begin{cases} \gamma - \delta & \text{if } W^i > 0 \\ 0 & \text{if } W^i \leq 0 \end{cases} \qquad \delta_t^i = \begin{cases} 0 & \text{if } W^i > 0 \\ \delta_t & \text{if } W^i \leq 0 \end{cases} \quad (2.9)$$

then we can summarize (2.8) in matrix notation as

$$P_t \vec{E_t} + Q_t \vec{W} \geqq \vec{W} + \vec{\mu} - \vec{\delta_t}$$ (2.10)

for all t, under χ^ϵ, and for all Ψ.

Using the facts that the addition of a constant vector to each side, and the multiplication of each side of a matrix with nonnegative elements will preserve the inequality, we calculate the expectation in the recursive game for n rounds under χ^ϵ and for an arbitrary Ψ:

$$\vec{Ex}_n = \sum_{k=1}^{n} \left(\prod_{t=0}^{k-1} Q_t \right) P_k \vec{E_k}$$

$$\geqq \sum_{k=1}^{n} \left(\prod_{t=0}^{k-1} Q_t \right) \left[(I - Q_k) \vec{W} + \vec{\mu} - \vec{\delta_k} \right]$$ (2.11)

by (1.3) and (2.10), where I is the identity matrix. This may be rewritten, by collapsing the terms involving $\vec{W}$, as

$$\vec{Ex}_n \geqq \vec{W} - \left(\prod_{t=0}^{n} Q_t \right) \vec{W}$$

$$+ \sum_{k=1}^{n} \left(\prod_{t=0}^{k-1} Q_t \right) \vec{\mu} - \sum_{k=1}^{n} \left(\prod_{t=0}^{k-1} Q_t \right) \vec{\delta_k}.$$ (2.12)

Now let

$$\tau = \operatorname*{Max}_{i} W^i / (\gamma - \delta)$$

if this is positive, and zero otherwise. (In case all $W^i < 0$.) Then, clearly, by definition of $\vec{\mu}$ we have $\tau \vec{\mu} \geqq \vec{W}$, and hence

$$\sum_{k=1}^{n} \left(\prod_{t=0}^{k-1} Q_t \right) \vec{\mu} - \left(\prod_{t=0}^{n} Q_t \right) \vec{W} \geqq \sum_{k=1}^{n} \left(\prod_{t=0}^{k-1} Q_t \right) \vec{\mu} - \tau \left(\prod_{t=0}^{n} Q_t \right) \vec{\mu}$$

$$= \sum_{k=1}^{n} \vec{\mu}_{k-1} - \tau \vec{\mu}_n,$$ (2.13)

where we have defined

$$\vec{\mu}_k = \left(\prod_{t=0}^{k} Q_t\right)\vec{\mu}.$$

Now, because $\vec{0} \le \vec{\mu}_k \le (\gamma - \delta)\vec{1}$ and $\tau \ge 0$, it is clear that there exists an m such that for all $n > m$,

$$\sum_{k=1}^{n} \vec{\mu}_{k-1} \ge \tau\vec{\mu}_n$$

since, if a component of the sum diverges, the boundedness of $\tau\vec{\mu}_n$ insures the result, and if a component converges, it means the corresponding component of $\vec{\mu}_n \to 0$, which also insures the result. This result, with (2.12) and (2.13), implies that there exists an m such that for all $n > m$:

$$\vec{\mathrm{Ex}}_n \ge \vec{W} - \sum_{k=1}^{n}\left(\prod_{t=0}^{k-1} Q_t\right)\vec{\delta}_k \tag{2.14}$$

and hence that

$$\vec{\mathrm{Ex}}(\chi^\epsilon, \Psi) = \lim_{n \to \infty} \vec{\mathrm{Ex}}_n$$

$$\ge \vec{W} - \lim_{n \to \infty} \sum_{k=1}^{n}\left(\prod_{t=0}^{k-1} Q_t\right)\vec{\delta}_k \quad \text{(all } \Psi) \tag{2.15}$$

but $\vec{\delta}_k \le \delta_k\vec{1}$ by definition of $\vec{\delta}_k$, so that

$$\left(\sum_{t=0}^{k-1} Q_t\right)\vec{\delta}_k \le \delta_k\vec{1}$$

because all components of the matrix are nonnegative and ≤ 1. This implies that

$$\lim_{n \to \infty} \sum_{k=1}^{n}\left(\prod_{t=0}^{k-1} Q_t\right)\vec{\delta}_k \le \left(\sum_{k=1}^{\infty} \delta_k\right)\vec{1}$$

$$= \left(\sum_{k=1}^{\infty}\left(\frac{1}{2}\right)^k \delta\right)\vec{1} = \delta\vec{1}, \tag{2.16}$$

by definition of $\delta_k = (\frac{1}{2})^k \delta$. Moreover, since δ was chosen $< \epsilon$, we have finally, from (2.16) and (2.15) that

$$\overrightarrow{\mathrm{Ex}}(\chi^\epsilon, \Psi) \geq \vec{W} - \epsilon\vec{1} \quad \text{for all } \Psi \in \mathscr{S}_2 \qquad (2.17)$$

and the proof of (a) of Theorem 1 is completed. Identical treatment reversing the roles of the players proves (b).

———————

Since the games are zero-sum, an immediate consequence of Theorem 1 is:

$$\vec{W}_1 \in C_1(\vec{\Gamma}), \qquad \vec{W}_2 \in C_2(\vec{\Gamma}) \Rightarrow \vec{W}_1 \leq \vec{W}_2. \qquad (2.18)$$

§3. THE CRITICAL VECTOR

(3.1) DEFINITION 3. $\vec{V} = \vec{V}(\vec{\Gamma})$ is a critical vector for $\vec{\Gamma} \rightleftarrows$ for every $\epsilon > 0$ there exists a pair of vectors, $\vec{W}_1$ and $\vec{W}_2$, lying componentwise within an ϵ-neighborhood of $\vec{V}$, $(\epsilon N_\epsilon(\vec{V}))$, such that $\vec{W}_1 \in C_1(\vec{\Gamma})$ and $\vec{W}_2 \in C_2(\vec{\Gamma})$. ($\vec{V}$ is in the intersection of the closures of $C_1(\vec{\Gamma})$ and $C_2(\vec{\Gamma})$.)

THEOREM 2. $\vec{V}$ *is a critical vector in* $\vec{\Gamma} \Rightarrow \vec{\Gamma}$ *possesses a solution, with value* $\vec{V}$. (*Hence* $\vec{V}$ *is unique.*)

PROOF. Follows immediately from definition of critical vector, Theorem 1, and the definition of a solution.

COROLLARY. *If* $\vec{\Gamma}$ *possesses a critical vector,* $\vec{V}$, *then there exist for all* $\epsilon > 0$, ϵ-*best strategies* $\chi^\epsilon, \Psi^\epsilon$, *for the players which are stationary in all components i for which either* Γ^i *satisfies the minimax condition, or* V^i *is favorable.* ($V^i > 0$ *is favorable for* P_1, $V^i < 0$ *is favorable for* P_2.)

PROOF. Follows from construction of ϵ-best strategies (2.7) in proof of Theorem 1.

REMARK 1. The value of an ordinary (nonrecursive) game is obviously a critical 1-vector in that game.

§4. REDUCTIONS OF RECURSIVE GAMES

For any recursive games $\vec{\Gamma} = (\Gamma^1, \Gamma^2, \ldots, \Gamma^n)$, we can form a reduced recursive game $\vec{\Gamma}^s$ ($\vec{W}^s$), from any subset s of the game elements of $\vec{\Gamma}$, by assigning real numbers W^i to the game payoffs Γ^i for the remaining set $\bar{s}$ of game elements. That is, the generalized payoff function for the game element Γ^i, $i \in s$, in the reduced game, is defined to be:

$$H^i\left(X^i, Y^i; \vec{\Gamma}^s(\vec{W}^{\bar{s}})\right) = p^i e^i + \sum_{j \in \bar{s}} q^{ij} W^j + \sum_{j \in s} q^{ij} \Gamma^j. \quad (4.1)$$

We shall say that a game element Γ^i has bounded payoff if there exist finite numbers α, β such that $\beta \le e^i \le \alpha$ for all $(X^i, Y^i) \in S_1^i \times S_2^i$.

We shall now investigate the behavior of the value, if it exists, of a reduced game $\vec{\Gamma}^s(\nu \vec{1}^{\bar{s}})$ formed by assigning the single real number ν to the set $\bar{s}$. We will abbreviate the k-th component of

$$\vec{\mathrm{Val}}^s\left\{\vec{\Gamma}^s(\nu\vec{1}^{\bar{s}})\right\}$$

by $V^k(\nu)$, and the game element $\Gamma^k(\nu\vec{1}^{\bar{s}})$ by $\Gamma^k(\nu)$. We then have:

LEMMA 1. (a) $\alpha > 0$, $\beta < 0$ *are any payoff bounds for all* Γ^k, $k \in s$.
(b) $V^k(\nu)$ *exists for all* ν.

These together imply
(c) $\beta \le V^k(\nu) \le \alpha$ *for* $\beta \le \nu \le \alpha$.
(d) *For every* $\delta > 0$, *and for all* ν, $V^k(\nu) - \delta^* \le V^k(\nu - \delta) \le V^k(\nu)$ $\le V^k(\nu + \delta)^* \le V^k(\nu) + \delta$, *where* " $*\le$ " *means* " $<$ " *unless* $V^k(\nu) = \nu$.

That is, that if the reduced game possesses a solution for all ν, *its value components change monotonically with* ν *at a rate less than or equal to the rate of change of* ν, *with strict inequality holding whenever* $V^k(\nu) \ne \nu$.

PROOF. We shall consider a single game element $\Gamma^k(\nu)$, with value $V(\nu)$, and shall, for convenience, drop the superscript. For any strategy pair χ, Ψ, in the reduced game, the expectation for the starting position $\Gamma(\nu)$ can always be written in the form:

$$\mathrm{Ex}(\chi, \Psi) = (1 - S)E + S\nu, \qquad 0 \le S \le 1 \quad (4.2)$$

where S is the probability of ultimately receiving a game payoff to one of the game elements not in s, and $(1 - S)E$ is the rest of the expectation, where E satisfies the relation $\beta \le E \le \alpha$ for all χ, Ψ, and α, β are payoff bounds for the game elements of s. Then (c) is proved immediately, since (4.2) implies that:

$$(1 - S)\beta + S\beta \leq (1 - S)E + Sv \leq (1 - S)\alpha + S\alpha \quad \text{all } \chi, \Psi$$

$$\Rightarrow \beta \leq \text{Ex}(\chi, \Psi) \leq \alpha \qquad\qquad\qquad \text{all } \chi, \Psi$$

$$\Rightarrow \beta \leq V(v) \leq \alpha. \tag{4.3}$$

Now (b) implies that for every $\epsilon > 0$ there exist strategies $\chi^\epsilon, \Psi^\epsilon$, which are ϵ-best in $\Gamma(v)$, so that

$$(1 - S)E + Sv \geq V(v) - \epsilon \quad \text{under } \chi^\epsilon \quad \text{for all } \Psi, \tag{4.4}$$

$$(1 - S)E + Sv \leq V(v) + \epsilon \quad \text{under } \Psi^\epsilon \quad \text{for all } \chi. \tag{4.5}$$

We shall now prove (d) by considering the effects of *these* strategies in $\Gamma(v + \delta)$. Consider first $\Gamma(v - \delta)$. In this case the expectation for χ and Ψ may be written as

$$(1 - S)E + S(v - \delta) \tag{4.6}$$

where S and E are the same as in $\Gamma(v)$, for χ and Ψ. Now, because of (4.5), P_2 possesses a strategy Ψ^ϵ, for all $\epsilon > 0$, which yields, when applied to $\Gamma(v - \delta)$, an expectation such that (from (4.6)):

$$\text{Ex} = (1 - S)E + S(v - \delta) \leq V(v) + \epsilon - S\delta \leq V(v) + \epsilon \tag{4.7}$$

and since such a strategy exists for all $\epsilon > 0$ we can conclude that

$$V(v - \delta) \leq V(v). \tag{4.8}$$

We now prove that

$$V(v - \delta) \geq^* V(v) - \delta. \tag{4.9}$$

From (4.4) there exists, for every $\epsilon > 0$, a χ^ϵ, such that, when applied to $\Gamma(v - \delta)$, by (4.6):

$$\text{Ex} = (1 - S)E + S(v - \delta) \geq V(v) - \epsilon - S\delta \quad \text{for all } \Psi \tag{4.10}$$

and, since P_1 possesses such a strategy for all $\epsilon > 0$, and because $S \leq 1$, we can conclude that

$$V(\nu - \delta) \geq V(\nu) - \delta. \tag{4.11}$$

We shall now show that the equality can hold only if $V(\nu) = \nu$.

$$\text{Assume } V(\nu - \delta) = V(\nu) - \delta. \tag{4.12}$$

Then, by (b), P_2 has for every $\epsilon > 0$ a strategy $\overline{\Psi}^\epsilon$ which is ϵ-best in $\Gamma(\nu - \delta)$, so that for such a $\overline{\Psi}^\epsilon$, and by (4.12); and (4.6):

$$(1 - S)E + S(\nu - \delta) \leq V(\nu - \delta) + \epsilon = V(\nu) - \delta + \epsilon \tag{4.13}$$

for all $\chi \in \mathscr{S}_1$.

We now find some bounds on S for the two mentioned strategies χ^ϵ and $\overline{\Psi}^\epsilon$ by making use of the payoff bounds α and β, where we require that our lower bound β is $< \nu - \delta$. This requirement can always be met, for every ν, δ, since we are always free to replace a lower bound by a still lower bound, if necessary. (The reader is cautioned that the strategies χ^ϵ and $\overline{\Psi}^\epsilon$ are not ϵ-best in the same game; χ^ϵ refers to $\Gamma(\nu)$ while $\overline{\Psi}^\epsilon$ is ϵ-best in $\Gamma(\nu - \delta)$.) From (4.13) we conclude that

$$(1 - S)\beta + S(\nu - \delta) \leq V(\nu) - \delta + \epsilon \tag{4.14}$$

$$\Rightarrow S \leq \frac{V(\nu) - \delta + \epsilon - \beta}{\nu - \delta - \beta} \quad \text{for } \chi^\epsilon, \overline{\Psi}^\epsilon \tag{4.15}$$

while from (4.4) we get

$$(1 - S)\alpha + S\nu \geq V(\nu) - \epsilon \tag{4.16}$$

$$\Rightarrow S \leq \frac{\alpha - V(\nu) + \epsilon}{\alpha - \nu} \quad \text{for } \chi^\epsilon, \overline{\Psi}^\epsilon \tag{4.17}$$

and applying (4.15) and (4.17) to (4.10) we obtain the two relations for $\Gamma(\nu - \delta)$ which must both hold:

$$\text{Ex}(\chi^\epsilon, \overline{\Psi}^\epsilon) \geq V(\nu) - \epsilon - \delta\left(\frac{V(\nu) - \delta + \epsilon - \beta}{\nu - \delta - \beta}\right), \tag{4.18}$$

$$\text{Ex}(\chi^\epsilon, \tilde{\Psi}^\epsilon) \geq V(\nu) - \epsilon - \delta\left(\frac{\alpha - V(\nu) + \epsilon}{\alpha - \nu}\right). \tag{4.19}$$

But since $\overline{\Psi}^\epsilon$ is ϵ-best in this game ($\Gamma(\nu - \delta)$) and since for every $\epsilon > 0$ the corresponding $\overline{\Psi}^\epsilon$ can be defended against by a χ^ϵ, with the results (4.18) and (4.19), we can conclude that the value of $\Gamma(\nu - \delta), V(\nu - \delta)$, must satisfy both of the following relations:

$$V(\nu - \delta) \geq V(\nu) - \delta\left(\frac{V(\nu) - \delta - \beta}{\nu - \delta - \beta}\right), \qquad (4.20)$$

$$V(\nu - \delta) \geq V(\nu) - \delta\left(\frac{\alpha - V(\nu)}{\alpha - \nu}\right). \qquad (4.21)$$

However, one or the other multipliers of δ in (4.20) and (4.21) is less than 1 whenever $V(\nu) \neq \nu$, which would contradict assumption (4.12), so that we can conclude that

$$V(\nu - \delta) = V(\nu) - \delta \Rightarrow V(\nu) = \nu, \qquad (4.22)$$

which, together with (4.11), establishes the truth of

$$V(\nu - \delta) \overset{*}{\geq} V(\nu) - \delta, \quad \text{(all } \nu, \delta). \qquad (4.23)$$

Finally, reversal of the roles of the players suffices to establish the analogues of (4.9) and (4.23) for the game $\Gamma(\nu + \delta)$ and the proof is completed.

———————

We shall now state an analogous result for the ordinary game $\Gamma^i(\vec{W})$ for arbitrary $\vec{W}$:

LEMMA 2. (a) $\alpha > 0$, $\beta < 0$ are payoff bounds for Γ^i,
(b) $\vec{\delta} = (\delta^1, \delta^2, \ldots, \delta^n)$, $\delta^i \geq 0$ all i,
(c) $\gamma = \max_i \delta^i$,
(d) $\beta\vec{1} \leq \vec{W} \leq \alpha\vec{1}$,

imply that:
(e) $\beta \leq \text{Val } \Gamma^i(\vec{W}) \leq \alpha$,
(f) $\text{Val } \Gamma^i(\vec{W}) - \gamma \leq \text{Val } \Gamma^i(\vec{W} - \vec{\delta}) \leq \text{Val } \Gamma^i(\vec{W}) \leq \text{Val } \Gamma^i(\vec{W} + \vec{\delta}) \leq \text{Val } \Gamma^i(\vec{W}) + \gamma$.

This is simply a statement of the well-known fact that the value of an ordinary game is a continuous, monotonic function of its pay-offs, obeying the Lipschitz condition of order 1. This may, however, be proved easily, if desired, by suitable modification of the proof of Lemma 1.

REMARK 2. Since critical vectors are values, Lemmas 1 and 2 apply to critical vectors.

Since Lemma 2 establishes the continuity of the value mapping M we are in a position to draw some useful conclusions about critical vectors.

THEOREM 3. *If $\vec{V}$ is the critical vector for $\vec{\Gamma}$, then $\vec{V}$ is a fixed point of the value mapping, and, furthermore, $\vec{W}_1 \in C_1(\vec{\Gamma}) \Rightarrow \vec{W}_1 \leq \vec{V}$ and $\vec{W}_2 \in C_2(\vec{\Gamma}) \Rightarrow \vec{W}_2 \geq \vec{V}$.*

PROOF. Follows from the definition of a critical vector, the continuity of the value mapping, and (2.18).

THEOREM 4. *If $\vec{\Gamma}$ possesses a critical vector, $\vec{V}$, then for any subset s of the game elements of $\vec{\Gamma}$ the reduced game, $\vec{\Gamma}^s(\vec{V}^{\bar{s}})$ (which is formed by assigning V^i to payoff Γ^i for $i \in \bar{s}$), possesses a critical vector, $V(\vec{\Gamma}^s(\vec{V}^{\bar{s}}))$, whose components are the same as the components of $\vec{V}$ restricted to the subset s. Symbolically: $V(\vec{\Gamma}^s(\vec{V}^{\bar{s}})) = \vec{V}^s$.*

PROOF. $\vec{V}$ critical in $\vec{\Gamma}$ implies that for every $\epsilon > 0$ there exists a $\vec{W}_1 \in N_\epsilon(\vec{V})$ such that

$$M(\vec{W}_1) \gtrdot \vec{W}_1 \quad \text{i.e.,} \quad \left(\vec{W}_1 \in C_1(\vec{\Gamma})\right). \tag{4.24}$$

Now, by Theorem 3 we know that $\vec{W}_1 \leq \vec{V}$ which implies that

$$\vec{W}_1^{\bar{s}} \leq \vec{V}^{\bar{s}} \tag{4.25}$$

so that using (4.25), Lemma 2, (4.24), and the definition of the value map:

$$\text{Val } \Gamma^i\left(\vec{V}^{\bar{s}}, \vec{W}_1^s\right) \geq \text{Val } \Gamma^i\left(\vec{W}_1^{\bar{s}}, \vec{W}_1^s\right)$$

$$= \text{Val } \Gamma^i\left(\vec{W}_1\right) \gtrdot W_1^i \quad (i \in s) \tag{4.26}$$

which is simply a statement that for the value map $\tilde{M}$ for the reduced game $\vec{\Gamma}^s(\vec{V}^{\tilde{s}})$:

$$\tilde{M}\left(\vec{W}_1^s\right) \geqq \vec{W}_1^s \quad \left(\text{so that } \vec{W}_1^s \in C_1\left(\vec{\Gamma}^s(\vec{V}^{\tilde{s}})\right)\right). \tag{4.27}$$

Similar treatment holds for $\vec{W}_2^s$ and we conclude that $\vec{V}^s$ is critical in $\vec{\Gamma}^s(\vec{V}^{\tilde{s}})$, and the proof is completed.

§5. Existence of the Critical Vector—Main Theorem

THEOREM 5. *Every recursive game whose game elements have bounded payoffs and satisfy the supinf condition possesses a critical vector.*

PROOF. Induction on the number of game elements using:

HYPOTHESIS (k). Every recursive game consisting of k or fewer game elements, all of which have bounded payoffs, and satisfy the supinf condition, possesses a critical vector.

Now consider any recursive game $\vec{\Gamma}$ which consists of $k + 1$ game elements with the above properties. Remove one element, say Γ^q, and consider the remaining set, $\vec{\Gamma}^r$, as a reduced game $\vec{\Gamma}^r(\nu)$ which is a function of the "value" ν assigned to Γ^q. This is then a recursive game with k elements and hence by hypothesis possesses a critical vector $\vec{V}^r(\nu)$ for all ν. Moreover, since Lemma 1 applies to critical vectors, as we have seen, we conclude that $\vec{V}^r(\nu)$ is a continuous monotonic function of ν in all components.

Now consider the ordinary game $\Gamma^q(\vec{V}^r(\nu), \nu)$, which possesses a value for all ν by virtue of its satisfying the supinf condition. Define:

$$\tilde{V}(\nu) = \text{Val } \Gamma^q\left(\vec{V}^r(\nu), \nu\right), \tag{5.1}$$

then applying Lemmas 1 and 2 we obtain the conditions on $\tilde{V}(\nu)$:

$$\tilde{V}(\nu) - \delta \leq \tilde{V}(\nu - \delta) \leq \tilde{V}(\nu) \leq \tilde{V}(\nu + \delta)$$

$$\leq \tilde{V}(\nu) + \delta, \quad \text{all } \delta \geq 0. \tag{5.2}$$

$$\beta \leq \tilde{V}(\nu) \leq \alpha \quad \text{for all } \nu \quad \text{such that } \beta \leq \nu \leq \alpha, \tag{5.3}$$

where α, β are the upper and lower payoff bounds for all of the game elements of $\vec{\Gamma}$. Therefore, $\tilde{V}(\nu)$ is a continuous mapping of the closed line segment $[\beta, \alpha]$ into itself, so that there exists a closed, nonempty set of fixed points, and hence in particular there exists a fixed point of minimum absolute value which we shall designate as ν^*. That is, there always exists a ν^* such that:

$$V(\nu^*) = \nu^*, \quad \text{and for all } \nu, \qquad V(\nu) = \nu \Rightarrow |\nu| \geq |\nu^*|. \quad (5.4)$$

We shall now show that the $(k + 1)$-vector $\vec{V}$ defined by $\vec{V} = [\vec{V}^r(\nu^*), \nu^*]$, and which always exists, is critical in $\vec{\Gamma}$. To do this we proceed to show that for any $\epsilon > 0$ there is a $\vec{W}_1 \in N_\epsilon(\vec{V})$ such that $\vec{W}_1 \in C_1(\vec{\Gamma})$:

CASE 1. $\nu^* > 0$.
We first remark that

$$\tilde{V}(\nu) > \nu \quad \text{for } 0 \leq \nu < \nu^* \qquad (5.5)$$

since by (5.2) and (5.4) if we let $\nu = \nu^* - \delta$, $\delta > 0$, then $\tilde{V}(\nu) = \tilde{V}(\nu^* - \delta) \geq \tilde{V}(\nu^*) - \delta = \nu^* - \delta = \nu$; but the equality cannot hold, or it would contradict the minimum absolute value property of the fixed point ν^*. Now, (5.5) implies that for every $\epsilon > 0$ there is a $\nu^\epsilon \in N_\epsilon(\nu^*)$ for which $V(\nu^\epsilon) > \nu^\epsilon$, and hence there exists a $\delta, 0 < \delta < \epsilon$, for which:

$$\tilde{V}(\nu^\epsilon) > \nu^\epsilon + \delta. \qquad (5.6)$$

We now turn to the reduced game $\vec{\Gamma}^r(\nu^\epsilon)$, for which the induction hypothesis guarantees the existence of a vector $\vec{V}^r(\nu^\epsilon)$ which is critical in $\vec{\Gamma}^r(\nu^\epsilon)$, so that there is a vector $\vec{W}_1^r \in N_\delta(\vec{V}^r(\nu^\epsilon))$ with the property that:

$$\text{Val } \Gamma^k\left(\vec{W}_1^r, \nu^\epsilon\right) \gtrdot W_1^k \quad \text{for all } k \in r. \qquad (5.7)$$

Now, (5.6) implies that in Γ^q, Val $\Gamma^q(\vec{V}^r(\nu^\epsilon), \nu^\epsilon) > \nu^\epsilon + \delta$, and since $\vec{W}_1^r \in N_\delta(\vec{V}^r(\nu^\epsilon))$, applying Lemma 2 we get

$$\text{Val } \Gamma^q\left(\vec{W}_1^r, \nu^\epsilon\right) > \epsilon. \qquad (5.8)$$

But (5.7) and (5.8) are simply the statement that the $(k + 1)$-vector

$\vec{W}_1 = [\vec{W}_1^r, \nu^\epsilon]$ has the property that $M(\vec{W}_1) \geqq \vec{W}_1$, so that

$$\vec{W}_1 \in C_1(\vec{\Gamma}). \tag{5.9}$$

But by choice of ν^ϵ:

$$\nu^\epsilon \in N_\epsilon(\nu^*) \tag{5.10}$$

which implies, using Lemma 1 for critical vectors, that $\vec{V}^r(\nu^\epsilon) \in N_\epsilon(\vec{V}^r(\nu^*))$, which, because $\vec{W}_1^r \in N_\delta\{\vec{V}^r(\nu^\epsilon)\}$, and $\delta < \epsilon$, implies that

$$\vec{W}_1^r \in N_{2\epsilon}(\vec{V}^r(\nu^*)) \tag{5.11}$$

and combining (5.10) and (5.11) we see that

$$\vec{W}_1 = \left[\vec{W}_1^r, \nu^\epsilon\right] \in N_{2\epsilon}\left\{\left[\vec{V}^r(\nu^*), \nu^*\right]\right\} \tag{5.12}$$

and hence that $\vec{W}_1 \in N_{2\epsilon}(\vec{V})$, and Case 1 is completed.

CASE 2. $\nu^* \leqq 0$.

Now, $\vec{V} = [\vec{V}^r(\nu^*), \nu^*]$ may have some other components besides the q-th component equal to ν^*. Let $\bar{p}$ denote the set of all those indices k for which $V^k = \nu^*$. Then, since by Theorem 3 $\vec{V}^r$ is a fixed point of $\tilde{M}$, the value mapping for the reduced game $\vec{\Gamma}^r(\nu^*)$, we can conclude that:

$$\text{Val } \Gamma^k(\vec{V}^r(\nu^*), \nu^*) = \nu^* \quad \text{all } k \in \bar{p}. \tag{5.13}$$

We now turn to the reduced game $\vec{\Gamma}^p(\nu \vec{1}^{\bar{p}})$ consisting of the elements $\Gamma^i, i \in p$, with "value" ν assigned to all of the game elements in $\bar{p}$. By applying Theorem 4 to $\vec{\Gamma}^r(\nu^*)$, which has critical vector $\vec{V}^r(v^*)$, we have:

$$V\left(\vec{\Gamma}^p\left(\nu^* \vec{1}^{\bar{p}}\right)\right) = \vec{V}^p = \vec{V}$$

restricted to p, is critical in $\vec{\Gamma}^p\left(\nu^* \vec{1}^{\bar{p}}\right)$, \tag{5.14}

since $V^k = \nu^*$ for $k \in \bar{p}$. Now, if we denote $V(\vec{\Gamma}^p(\nu \vec{1}^{\bar{p}}))$ by $\tilde{V}^p(\nu)$, we get, by Lemma 1, that for any $\epsilon > 0$, and for $\nu = \nu^* - \epsilon$:

$$\tilde{V}^p(\nu^*) - \tilde{V}^p(\nu) < \nu^* - \nu = \epsilon, \tag{5.15}$$

in all components since $\tilde{\vec{V}}^p(\nu^*) = \vec{V}^p$ is not equal to ν^* in any component. Hence there exists a δ, $0 < \delta < \epsilon$, such that:

$$\tilde{\vec{V}}^p(\nu^*) - \tilde{\vec{V}}^p(\nu) \leq \epsilon\vec{1}^p - \delta\vec{1}^p \qquad (5.16)$$

$$\Rightarrow \tilde{\vec{V}}^p(\nu) - \delta\vec{1}^p \geq \tilde{\vec{V}}^p(\nu^*) - \epsilon\vec{1}^p. \qquad (5.17)$$

Now choose $\vec{W}_1^p \in N_\delta\{\tilde{\vec{V}}^p(\nu)\}$ for which:

$$\text{Val } \Gamma^i\left(\nu\vec{1}^p, \vec{W}_1^p\right) \overset{\cdot}{\geq} W_1^i \quad \text{all } i \in p. \qquad (5.18)$$

The existence of such a $\vec{W}_1^p$ is assured, since $\tilde{\vec{V}}^p(\nu)$ is critical in $\vec{\Gamma}^p(\nu\vec{1}^p)$. Now, returning to (5.13) and rewriting it as

$$\text{Val } \Gamma^k\left(\nu^*\vec{1}^p, \tilde{\vec{V}}^p(\nu^*)\right) = \nu^* \quad \text{all } k \in \bar{p}, \qquad (5.19)$$

we apply Lemma 2 to get:

$$\text{Val } \Gamma^k\left((\nu^* - \epsilon)\vec{1}^p, \tilde{\vec{V}}^p(\nu^*) - \epsilon\vec{1}^p\right) \geq \nu^* - \epsilon \quad \text{all } k \in \bar{p} \qquad (5.20)$$

which means, since $\nu = \nu^* - \epsilon$, and by (5.17) and Lemma 2, that

$$\text{Val } \Gamma^k\left(\nu\vec{1}^p, \tilde{\vec{V}}^p(\nu) - \delta\vec{1}^p\right) \geq \nu \quad \text{all } k \in \bar{p}. \qquad (5.21)$$

But

$$\vec{W}_1^p \in N_\delta\{\tilde{\vec{V}}^p(\nu)\},$$

and Lemma 2 applied to (5.21) yield:

$$\text{Val } \Gamma^k\left(\nu\vec{1}^p, \vec{W}_1^p\right) \geq \nu \quad \text{all } k \in \bar{p}, \qquad (5.22)$$

Now we see that, because $\nu \leq 0$, (5.18) and (5.22) are precisely the statement that the $(k + 1)$-vector $\vec{W}_1 = [\nu\vec{1}^p, \vec{W}_1^p]$ is in $C_1(\vec{\Gamma})$.
    Finally, since

$$\vec{W}_1^p \in N_\delta\{\tilde{\vec{V}}^p(\nu)\}$$

and $\vec{V}^p(v) \in N_\epsilon(\vec{V}^p)$ (by (5.14) and (5.15)) and $\delta < \epsilon$, we have that $\vec{W}_1^p \in N_{2\epsilon}(\vec{V}^p)$, and because

$$v\vec{1}^{\bar{p}} \in N_\epsilon\{v*\vec{1}^{\bar{p}} = \vec{V}^{\bar{p}}\}$$

we conclude that $\vec{W}_1 \in N_{2\epsilon}(\vec{V})$ and Case 2 is completed.

We have seen that in each case, the $\vec{V}$ previously constructed possesses for every $\epsilon > 0$ a $\vec{W}_1 \in N_\epsilon(\vec{V})$ such that $\vec{W}_1 \in C_1(\vec{\Gamma})$. Similar treatment with use of alternate relations and reversal of all inequalities (effectively reversing the roles of the players) yields the existence of a $\vec{W}_2 \in C_2(\vec{\Gamma})$, $\vec{W}_2 \in N_\epsilon(\vec{V})$ and we have proved that the vector $\vec{V}$, which we have constructed, is critical in $\vec{\Gamma}$.

We have shown that $\mathrm{Hyp}(k) \Rightarrow \mathrm{Hyp}(k + 1)$. Furthermore, for the empty game, which has zero payoff for all strategies, and hence value zero, the hypothesis is obviously satisfied, since zero itself is critical in this game, and the proof of Theorem 5 is completed. (An independent proof of the hypothesis for recursive games with one element is contained in Theorem 8.) We can now summarize our results, using Theorems 2, 5, and Corollary 1:

THEOREM 6. MAIN THEOREM. *Every recursive game whose elements have bounded payoffs and satisfy the supinf condition possesses a solution, $\vec{V}$, and ϵ-best strategies χ^ϵ and Ψ^ϵ for the players which are stationary in all components i for which either Γ^i satisfies the minimax condition of V^i is favorable.*

An important consequence of Theorem 6 is that any recursive game whose game elements are matrix games (matrices with generalized payoff elements of the form $a_{ij} = p_{ij}e_{ij} + \Sigma_k q_{ij}^k \Gamma^k$) possesses a solution and ϵ-best *stationary* strategies for the players, since all such matrix game elements satisfy the minimax condition.

§6.   GENERALIZATIONS

For our purposes we shall define a *stochastic game*, $\vec{\Gamma}$, to be a collection of game elements (Γ^i), each with strategy spaces S_1^i and S_2^i, and generalized payoff function of the form:

$$H^i(X^i, Y^i; \vec{\Gamma}) = e^i + p^i S + \sum_j q^{ij}\Gamma^j; \qquad (X^i, Y^i) \in S_1^i \times S_2^i$$

$$p^i, q^{ij} \geq 0; \qquad p^i + \sum_j q^{ij} = 1, \tag{6.1}$$

where now e^i is a payoff which takes place whether or not the play stops, p^i is the stop probability, and the q^{ij} are the transition probabilities to other game elements, as before. With such games the payoffs are allowed to accumulate throughout the course of the play, in distinction to recursive games, where payoff can take place only when the play stops.

If we now extend all of our definitions and formulas in the obvious manner (which amounts to replacing $\overrightarrow{PE}$ by $\overrightarrow{E}$ in the expectation formulas) to stochastic games, we notice that Theorems 1, 2, 3, and 4 remain true for stochastic games. Lemma 1, however, fails ((c) is no longer true, and the crucial $\leq^*$ of (d) must be replaced by the milder $\leq$) and (e) of Lemma 2 is no longer true, so that the extension of the main theorem to arbitrary stochastic games is prevented, and we must be content to examine a few special cases.

a. Pseudo-Recursive Games

A pseudo-recursive game[2] is a stochastic game for which e^i/p^i is bounded for all $X^i, Y^i \in S_1^i \times S_2^i$ in all game elements. Such a stochastic game can always be reduced to an equivalent recursive game, by simply rewriting the payoff function in the form:

$$H^i(X^i, Y^i; \overrightarrow{\Gamma}) = p^i(e^i/p^i) + \sum_j q^{ij}\Gamma^j \qquad (6.2)$$

which is formally the same as (1.0), so that:

THEOREM 7. *Theorem 6 holds for pseudo-recursive games.*

b. Simple Stochastic Games

A simple stochastic game is a stochastic game which consists of only one element, which can at most repeat itself, with payoff function of the form:

$$H(X, Y; \Gamma) = e + q\Gamma; \qquad (X, Y) \in S_1 \times S_2 \quad 0 \leq q \leq 1. \quad (6.3)$$

---

[2] The stochastic games treated by Shapley [1], which are generalized matrix games with the condition that the stop probability is bounded away from zero for all strategies in all game elements, are a subclass of pseudo-recursive games.

If we allow for the possibility of infinite values, which we define to mean:

$$\text{Val } \Gamma = +\infty \quad \Rightarrow \quad \text{for every } \xi \text{ there exists a } \chi^\xi \in \mathscr{S}_1$$

$$\text{such that } \text{Ex}(\chi^\xi, \Psi) \geqq \xi \quad \text{for all } \Psi \in \mathscr{S}_2, \quad (6.4)$$

and similarly for Val $= -\infty$, and make a similar extension of the notion of a critical vector (in this case simply a number), then we can give a *complete* answer to the existence of a solution for simple stochastic games, with no restrictions on e or q.

THEOREM 8. *Every simple stochastic game which satisfies the supinf condition possesses a solution.*

PROOF. We simply remark that every point of the extended real line is in either $C_1(\Gamma)$ or $C_2(\Gamma)$, and that neither is empty since $-\infty$ is always in C_1 and $+\infty$ is always in C_2, so that the intersection of their closures is nonempty. But a point in the closure of both is critical for Γ, and the theorem is proved.

c.   Univalent Stochastic Games

A univalent stochastic game is one for which the payoffs are always nonnegative (or nonpositive) for all strategies, and in all game elements. Such games are useful for describing certain pursuit games, in which Player 1, the player being pursued, receives some positive payoff from the pursuer (P_2) for every move that takes place for which he successfully avoids capture, the play ending with no payoff when capture takes place.

It is useful now to introduce the notion of a "trap" in a stochastic game, which is an element or set of elements such that once the play reaches an element of the trap one of the players can force the play to remain in trap indefinitely in such a way as to accumulate payoffs from the other player, and hence achieve an arbitrarily high expectation. Traps are, then, sets of elements which have infinite values in the sense of (6.4). A game contains no traps when each player can prevent infinite adverse expectations in all elements. We shall see, however, that even trap-free stochastic games do not always possess solutions, at least in the sense of our previous definition of a solution.

THEOREM 9. *Every univalent stochastic game, whose game elements satisfy the supinf condition, and which contains no traps, possesses a solution.*

PROOF. Consider the sequence $\{\vec{W}_k\}$:

$$\vec{W}_0 = \vec{0}$$

$$\vec{W}_{k+1} = M\left(\vec{W}_k\right)$$

which is generated by iterating the value mapping. Since all payoffs are nonnegative we know that $\vec{W}_1 \geq 0 = \vec{W}_0$. Now, assume that $\vec{W}_{k+1} \geq \vec{W}_k$. Then, by Lemma 2(f), Val $\Gamma^i(\vec{W}_{k+1}) \geq$ Val $\Gamma^i(\vec{W}_k)$ for all i, which implies that $M(\vec{W}_{k+1}) \geq M(\vec{W}_k)$ which means that $\vec{W}_{k+2} \geq \vec{W}_{k+1}$ so that by induction we have proved:

$$\left\{\vec{W}_n\right\} \text{ is monotone increasing in all components.} \qquad (6.5)$$

Let $\tilde{\vec{\Gamma}}(n)$ be the truncated game which results from $\vec{\Gamma}$ by introducing a compulsory stop with zero payoff after n rounds of play, whose value is obviously given by iterating the value mapping n times on the zero vector and hence equal to $\vec{W}_n$. We are assured that P_1 can come arbitrarily close to this value by simply playing a strategy which is ϵ-best in the truncated game and arbitrarily thereafter, since, because all of the payoffs are ≥ 0, he can lose nothing after n moves. So that for every $\vec{W}_n$ in the sequence $\{\vec{W}_k\}$, and for every $\epsilon > 0$, there exists a $\chi^\epsilon \in \mathscr{S}_1$ such that:

$$\overrightarrow{\text{Ex}}(\chi^\epsilon, \Psi) \geq \vec{W}_n - \epsilon\vec{1} \quad \text{for all } \Psi \in \zeta_2.$$

But, since the game is presumed to be trap-free, P_2 can prevent any infinite positive expectations, so that the sequence $\{\vec{W}_k\}$ is bounded above, and hence converges to some finite limit, $\vec{W}^*$, which P_1 can approach arbitrarily closely, and for which $M(\vec{W}^*) = \vec{W}^*$. However, this means that $\vec{W}^* \in C_2(\vec{\Gamma})$, so that P_2 can also, by Theorem 1, come arbitrary close to $\vec{W}^*$, so that the game has a solution with value $\vec{W}^*$. (It should be noted that in general the limit of the iterated value mapping is not the value of the game. See §9, Ex. 6.)

It can be shown that if the elements of a univalent stochastic game satisfy the extended minimax condition ($\Gamma^i(\vec{W})$ possesses a minimax solution for all $\vec{W}$ including the possibility of infinite components), then the game possesses a solution even if traps are present, where infinite values are, of course, allowed. The difficulty in extending this result to the supinf case lies in the fact that the limit of the iterated mapping, $\vec{W}^*$, which may now have infinite components, does not necessarily satisfy the relation that $M(\vec{W}^*) = \vec{W}^*$ in the supinf case as it does in the minimax case.

§7. EXTENSION TO THE CASE OF CONTINUOUS TIME

We shall now present a further generalization of the theory of recursive games developed so far, to include the case of a continuous, rather than discrete, time parameter. We wish to show that the theory of continuous time recursive games can be reduced in a simple manner to the earlier theory.

A *continuous time recursive game* $\vec{\Gamma}$ is a collection of game elements $\{\Gamma^i\}$, with payoff functions of the form:

$$H^i(X^i, Y^i; \vec{\Gamma}) = p^i e^i + \sum_j' q^{ij} r^j \quad (\Sigma' \text{ omits } j = i) \qquad (7.1)$$

where the interpretation is that if the players are playing strategies X^i, Y^i in Γ^i, then in the (infinitesimal) time interval dt the play stops with payoff e^i with probability $p^i\, dt$, while with probability $q^{ij}\, dt$ the players move on and play Γ^j. The p^i and q^{ij} are referred to as *transition rates*. They are nonnegative, but do not necessarily sum to unity.

In such games the players are at each instant playing some strategy, but they are free to change at any time. However, we assume that with all admissible time dependent strategies the transition rates are integrable, i.e., $\int p^i\, dt$ and the $\int q^{ij}\, dt$ always exist. (In any actual game it is simply impossible that the players could change strategies so fast that this condition would not be met.) We furthermore assume that the transition rates p^i and q^{ij}, as well as the payoffs e^i, are bounded for all strategies, in all elements.

We shall show that we can, in a simple manner, associate with $\vec{\Gamma}$ a *discrete* time recursive game $\vec{\Gamma}(\Delta)$, which, if it has a critical vector, supplies all the information necessary for optimal (or ϵ-best) play in $\vec{\Gamma}$ —i.e., which has the same value, and whose ϵ-best strategies furnish

ϵ-best strategies for $\vec{\Gamma}$. Thus the problem of continuous time recursive games will be reduced to that of discrete time games which we have already discussed.

The reduction to a discrete time game is accomplished as follows: Let Δ be a positive number such that $\Delta(p^i + \epsilon'_j q^{ij})$ is ≤ 1 for all strategies in all elements. (The existence of such a Δ is guaranteed by the boundedness of the transition rates.) Then let $\vec{\Gamma}(\Delta)$ be the discrete time recursive game whose payoff function for the i-th element is:

$$H^i\left(X^i, Y^i; \vec{\Gamma}(\Delta)\right) = p^{*i} e^{*i} + \sum_j q^{*ij} \Gamma^j(\Delta) \qquad (7.2)$$

where the numbers are defined from the payoff of $\vec{\Gamma}$ for the same strategies, given by (7.1) as follows:

$$p^{*i} = \Delta p^i, \qquad q^{*ij} = \Delta q^{ij} \quad (i \neq j),$$

$$q^{*ii} = 1 - \Delta\left(p^i + \sum_j{}' q^{ij}\right), \qquad e^{*i} = e^i. \qquad (7.3)$$

If the discrete recursive game $\vec{\Gamma}(\Delta)$ so constructed possesses a critical vector, then for every $\epsilon > 0$ there exists a strategy $\chi^\epsilon = \{\vec{X}_t\}$ for P_1 (constructed according to the method of (2.7)), which satisfies the inequalities (2.8). We wish to assert that this strategy χ^ϵ is also ϵ-best in the continuous time game $\vec{\Gamma}$, from which $\vec{\Gamma}(\Delta)$ was derived, but we must first supply a rule for the unambiguous application of χ^ϵ to $\vec{\Gamma}$ in case it is not a stationary strategy.

First, we define an *event* to be any time the play stops or there is a transition to another element. We define the k-th round to be the time between the occurrence of the $k - 1$-st event and the k-th event. We then state the rule:

RULE 1. If $\chi^\epsilon = \{\vec{X}_t\}$ is an ϵ-best strategy for $\vec{\Gamma}(\Delta)$, constructed according to (2.7), then in $\vec{\Gamma}$ play at the instant T the strategy $\vec{X}_t$ where $t = k + 1 + [T/\Delta]$, with k the number of the current round, and $[T/\Delta]$ the greatest integer $\leq T/\Delta$ (T measured from commencement of play.)

Thus according to Rule 1 one is always playing an element of the sequence $\{\vec{X}_t\}$, and changing to the next succeeding element each time that an event occurs and each time that an interval of time of duration

Δ elapses. Similar considerations hold for P_2, of course. With this understanding of how to play in $\vec{\Gamma}$ the strategies χ and Ψ which are constructed for $\vec{\Gamma}(\Delta)$, we can state:

THEOREM 10. $\vec{\Gamma}(\Delta)$ *possesses a critical vector* $\vec{V}$, *and* ϵ-*best strategies* $\chi^\epsilon, \Psi^\epsilon$ (*constructed according to* (2.7)) $\Rightarrow \chi^\epsilon$ *and* Ψ^ϵ *are also* ϵ-*best in* $\vec{\Gamma}$, *which has a solution with value* $\vec{V}$.

PROOF. Let us assume that it is the k-th round and that P_1 is playing χ^ϵ, and let t measure the time elapsed since the beginning of the round ($k - 1$-st event). P_1 is therefore playing $\vec{X}_{k+1+[T/\Delta]}$, which changes only at times $[T/\Delta]$, and for which, according to (2.8) and (2.9):

$$p^{*i}e^{*i} + \sum_j q^{*ij}W^j \geq W^i + \mu^i - \delta^i_{k+1+[T/\Delta]} \qquad (7.4)$$

for all $Y^i \in S_2^i$, and for all i. This implies, according to (7.3), that

$$p^i e^i + \sum_j {}'q^{ij}W^j \geq \left(p^i + \sum_j {}'q^{ij}\right)W^i + \frac{1}{\Delta}\mu^i - \frac{1}{\Delta}\delta^i_{k+1+[T/\Delta]} \quad (7.5)$$

for all Y^i and all i. Since (7.5) holds for all Y^i and all i, it holds at each instant of play of $\vec{\Gamma}$.

We are now interested in the ultimate outcome of the k-th round, regardless of the time involved, and wish to compute the probabilities $\tilde{p}_k^i, \tilde{q}_k^{ij}$ ($i \neq j$) for the various possible ultimate outcomes of the k-th round. We can then view the course of play as a discrete stochastic process which takes place only with each event, in which time is eliminated.

Whatever strategy $\Psi = \vec{Y}(t)$ P_2 is playing, the transition rates p^i, q^{ij}, as well as the payoffs e^i are functions of the time subject to (7.5). Let us restrict our attention to the i-th element, and let $n(t)\,dt$ be the probability of an event in the time interval dt, so that the transition rate $n(t)$ is:

$$n(t) = p^i(t) + \sum_j {}'q^{ij}(t). \qquad (7.6)$$

Furthermore, let $R(t)$ be the probability that the k-th event has not yet occurred at time t (Note: t measured from beginning of k-th round). Then clearly $R(t)$ is monotone decreasing, bounded between 0 and 1, and satisfies the relation:

$$\int_0^t R(\tau)n(\tau)\,d\tau = 1 - R(t). \tag{7.7}$$

The probability that by time t the k-th round will have resulted in a stop, $\bar{p}^i(t)$, is

$$\bar{p}^i(t) = \int_0^t R(\tau)p^i(\tau)\,d\tau \tag{7.8}$$

while the probability that it will have resulted in a transition to Γ^j, $\bar{q}^{ij}(t)$, is

$$\bar{q}^{ij}(t) = \int_0^t R(\tau)q^{ij}(\tau)\,d\tau. \tag{7.9}$$

Finally, if

$$\bar{e}^i(t) = \left(\int_0^t R(\tau)p^i(\tau)e^i(\tau)\,d\tau\right) \Big/ \left(\int_0^t R(\tau)p^i(\tau)\,d\tau\right)$$

denotes the mean payoff (which is, of course, bounded by any bounds for e^i), then we can write the total expected payoff as:

$$\bar{p}^i(t)\bar{e}^i(t) = \int_0^t R(\tau)p^i(\tau)e^i(\tau)\,d\tau. \tag{7.10}$$

However, making use of (7.5), we have that for the k-th round, in the i-th element, under χ^ϵ and for all $\vec{Y}(t)$:

$$\bar{p}^i(t)\bar{e}^i(t) + \sum_j{}' \bar{q}^{ij}(t)W^j$$

$$= \int_0^t R(\tau)p^i(\tau)e^i(\tau)\,d\tau + \sum_j{}' W^j \int_0^t R(\tau)q^{ij}(\tau)\,d\tau$$

$$= \int_0^t R(\tau)\left[p^i(\tau)e^i(\tau) + \sum_j{}' q^{ij}(\tau)W^j\right] d\tau$$

$$\geq \int_0^t R(\tau)\left[n(\tau)W^i + \frac{1}{\Delta}\mu^i - \frac{1}{\Delta}\delta^i_{k+1+[T/\Delta]}\right] d\tau \tag{7.11}$$

so that, using (7.7)

$$\bar{p}^i(t)\bar{e}^i(t) + \sum_j{}' \bar{q}^{ij}(t)W^j$$

$$\geq [1 - R(t)]W^i + \frac{1}{\Delta}\mu^i\left(\int_0^t R(\tau)\,d\tau\right)$$

$$-\frac{1}{\Delta}\int_0^t R(\tau)\delta_{k+1+[T/\Delta)}^i\,d\tau. \qquad (7.12)$$

Now by the construction of (2.7)

$$\delta_{k+1+[T/\Delta]}^i \leq \left(\tfrac{1}{2}\right)^{k+1+[T/\Delta]}\delta$$

so that, since $R(\tau)$ is bounded by 1, and certainly $\tau \leq T$, we have that

$$\int_0^\infty R(\tau)\delta_{k+1+[T/\Delta]}^i\,d\tau \leq \delta\int_0^\infty \left(\tfrac{1}{2}\right)^{k+1+[\tau/\Delta]}\,d\tau$$

$$= \delta\left(\tfrac{1}{2}\right)^{k+1}\Delta\sum_{n=0}^\infty \left(\tfrac{1}{2}\right)^n = \Delta\left(\tfrac{1}{2}\right)^k\delta,$$

and therefore the ultimate transition probabilities $\tilde{p}_k^i$ and $\tilde{q}_k^{ij}$ for the k-th round, which are given by the limit of (7.12) as $t \to \infty$, satisfy:

$$\tilde{p}_k^i\tilde{e}_k^i + \sum_j{}' \tilde{q}_k^{ij}W^j \geq [1 - R(\infty)]W^i + \frac{1}{\Delta}\mu^i\left(\int_0^\infty R(\tau)\,d\tau\right) - \left(\frac{1}{2}\right)^k\delta.$$

$$(7.13)$$

We now observe that if $W^i > 0$ (which implies $\mu^i > 0$) that $R(\infty)$ must be zero, since otherwise

$$\int_0^\infty R(\tau)\,d\tau$$

would be infinite ($R \in \downarrow$) and the left side of (7.13) would be infinite, an impossibility for bounded e^i and finite W^j. Therefore if W^i is positive $[1 - R(\infty)]W^i = W^i$, while if $W^i \leq 0$ then $[1 - R(\infty)]W^i \geq W^i$.

Hence (7.13) implies that

$$\tilde{p}_k^i \tilde{e}_k^i + {\sum_j}' \tilde{q}_k^{ij} W^j \geqq W^i + \frac{1}{\Delta} \mu^i \left(\int_0^\infty R(\tau)\, d\tau \right) - (\tfrac{1}{2})^k \delta. \quad (7.14)$$

Finally, since Δ was chosen so that $\Delta(p^i + \epsilon_j' q^{ij}) \leqq 1$, for all strategies in all elements, we have that $\Delta n(\tau) \leqq 1$ for all τ, so that

$$\int_0^t R(\tau) n(\tau)\, d\tau = 1 - R(t) \leq \int_0^t R(\tau) \frac{1}{\Delta}\, d\tau$$

$$= \frac{1}{\Delta} \int_0^t R(\tau)\, d\tau.$$

Therefore,

$$\frac{1}{\Delta} \int_0^\infty R(\tau)\, d\tau \geqq 1 - R(\infty).$$

But since $\mu^i = 0$ unless $W^i > 0$, and because $W^i > 0$ implies $R(\infty) = 0$, we can conclude that

$$\frac{1}{\Delta} \mu^i \left(\int_0^\infty R(\tau)\, d\tau \right) \geqq \mu^i.$$

It then follows from (7.14) that under χ^ϵ and for any Ψ:

$$\tilde{p}_k^i \tilde{e}_k^i + {\sum_j}' \bar{q}_k^{ij} W^j \geqq W^i + \mu^i - (\tfrac{1}{2})^k \delta. \quad (7.15)$$

Similar analysis holds for each element, so that (7.15) holds for all i.

This expression (7.15) involving the ultimate transition probabilities and expected payoffs for the k-th round is formally equivalent to the expressions (2.8), (2.9). But if we form matrices P_k, Q_k, and vectors $\vec{E}_k$ from $\tilde{p}_k^i$, $\tilde{q}_k^{ij}$, and $\tilde{e}_k^i$ by the formulas (1.2) then the formulas (1.3), (1.4) for the expectation are applicable to our case. Therefore the proof of Theorem 1 is also applicable, and we can conclude that the ultimate expectation for χ^ϵ satisfies

$$\overrightarrow{\text{Ex}}(\chi^\epsilon, \Psi) \geqq \vec{W} - \epsilon\vec{1} \quad \text{for all } \Psi. \quad (7.16)$$

Since $\vec{W}$ is $\in N_\epsilon(\vec{V})$, the strategy χ^ϵ is 2ϵ-best for P_1. Reversal of the roles of the players shows that same for P_2, and the theorem is proved.

Theorem 10 is easily generalized to the case of continuous time *stochastic games*, which are games $\vec{\Gamma}$ whose elements Γ^i have payoffs of the form

$$H^i(X^i, Y^i; \vec{\Gamma}) = e^i + p^i S + \sum_j{}' q^{ij} \Gamma^j$$

where the interpretation is that if the players are playing X^i, Y^i in Γ^i, then in time dt a payoff $e^i \, dt$ takes place, and with probability $p^i \, dt$ play stops while with probability $q^{ij} \, dt$ there is a transition to Γ^j. e^i is in this case a rate of payoff which is going on at all times (accumulating throughout the course of play) until play stops. Theorem 10 then goes through directly with substitution of e^i for $p^i e^i$ in all formulas ($\vec{E}$ for $\vec{PE}$), and we have

THEOREM 11. *Theorem* 10 *holds for continuous time stochastic games.*

Finally, we remark that there is no difficulty in handling recursive (or stochastic) games in which some elements are discrete time games and the others continuous. One simply reduces the continuous time game elements to discrete time elements in the manner presented here, leaving the discrete time elements unaltered.

§8. SUMMARY AND COMMENTS

Our main tools have been the concepts of the *value mapping* and the *critical vector*. Theorems 1 and 2 establish that the critical vector is a solution of discrete time recursive (and stochastic) games, while Theorems 10 and 11 extend this result to the continuous time case. We can therefore state with full generality:

If a recursive or stochastic game, either discrete or continuous (or mixed) time, possesses time, possesses a critical vector then that critical vector is unique and is the solution of the game.

Therefore 5 establishes the existence of a critical vector for all discrete time recursive games whose elements have bounded payoffs and satisfy the supinf condition. This result, together with the above result, implies that:

Every discrete time recursive game whose elements have bounded payoffs and satisfy the supinf condition, as well as every continuous time game for which a derived discrete time game is such a recursive game, possesses a solution.

This latter result cannot be extended to stochastic games since the existence of a critical vector is no longer guaranteed, as shown by Examples 3 and 4 of §9.

We should like now to emphasize several points. We have addressed ourselves solely to the combinatorial problem of what can be expected when a number of games are "hooked together" with various feedback paths (by allowing some outcomes to feed into other games instead of numerical payoffs), under the assumption that the individual games (elements) are "inherently soluble" (i.e., that when the loops are opened, by replacing game payoffs by numerical payoffs, the resulting ordinary games have solutions).

The situation is fully analogous to servomechanism analysis, where the complex behavior of a closed loop servomechanism is analyzed in terms of the (open loop) behavior of its parts. The theory of servomechanisms is concerned solely with the problem of predicting this closed loop behavior from known behavior of the components. An appropriate alternate name for recursive games would be "games with feedback."

Since it was not necessary to place any restrictions on the type of game elements to achieve our results, they are valid whether the elements be matrix games, games on the square, infinite games in extensive form, some type as yet undiscovered, or for that matter other recursive games. It is therefore improper to regard recursive games as a particular class of games. Rather, the concept is one which can be applied to any game (every game is trivially a one element recursive game), but which is useful only if the game is such that there are a number of different situations which can confront the players (the game elements) and the behavior of these elementary situations is completely understood.

§9. EXAMPLES, COUNTER-EXAMPLES, APPLICATIONS

In order to illustrate the results, and to motivate some of the restrictions imposed on the theorems, we list some simple examples of discrete time games. A supply of examples for the continuous time case may be easily obtained from these by suitable reinterpretation of probabilities as transition rates.

EXAMPLE 1.

$$\Gamma : \begin{pmatrix} \Gamma & 1 \\ 1 & 0 \end{pmatrix}$$

Value 1, ϵ-best strategy
$[1 - \epsilon, \epsilon]$ for P_1, all
strategies optimal for P_2.

This is an example of a recursive game which satisfies the minimax condition, yet which possesses no *optimal* strategy for P_1. Notice that 1 is the *unique* fixed point of the value map for this game.

EXAMPLE 2.

$$\Gamma : \begin{pmatrix} \Gamma & 5 \\ 1 & 0 \end{pmatrix}$$

Where P_2 is restricted to
strategies of the type $[1 - \alpha, \alpha]$
with $0 < \alpha \leq 1$.

This is a recursive game which satisfies the supinf condition, for which P_2 possesses no *stationary* ϵ-best strategy. The value is 1, $[4/5, 1/5]$ is optimal for P_1, and P_2 can also approach 1 by playing a nonstationary strategy $[1 - \alpha_t, \alpha_t]$ for which

$$\sum_{t=1}^{\infty} \alpha_t < \epsilon.$$

EXAMPLE 3.

$$\Gamma^1 : \begin{pmatrix} \Gamma^2 & 10 & \Gamma^3 \\ -10 & 0 & -10 \\ \Gamma^3 & 10 & \Gamma^2 \end{pmatrix};$$

$$\Gamma^2 : (1 + \Gamma^2); \qquad \Gamma^3 : (-2 + \Gamma^3).$$

Example of a stochastic game with traps, for which no solution exists.

EXAMPLE 4.

$$\Gamma^1 : \begin{pmatrix} 1 + \Gamma^2 & 5 \\ -5 & 0 \end{pmatrix} \qquad \Gamma^2 : \begin{pmatrix} -1 + \Gamma^1 & 5 \\ -5 & 0 \end{pmatrix}.$$

This is an example of a stochastic game which contains no traps, but which still does not possess a solution according to our definition, due to the fact that under the "best" strategies the expectation oscillates. This game does not possess a critical vector.

EXAMPLE 5.

$$\Gamma^1 : \begin{pmatrix} \Gamma^1 & \Gamma^1 \\ \Gamma^2 & 20 \\ 20 & \Gamma^2 \end{pmatrix} \qquad \Gamma^2 : (-10).$$

This is a recursive game satisfying the minimax condition, for which the value Γ^1 is 5, with optimal strategies $[0, 1/2, 1/2]$ for P_1 and $[1/2, 1/2]$ for P_2 in Γ^1. However, for *any* truncation (compulsory stop after n-rounds) the value of Γ^1 is 10 instead of 5. This shows that the solution of a recursive (or stochastic) game cannot in general be obtained as a limit of solutions of truncated games. Note also that for this example the iterated value mapping starting with $\vec{0}$ does not converge to the value of the game.

EXAMPLE 6. "Colonel Blotto commands a desert outpost staffed by three military units, and is charged with the task of capturing the encampment of two units of enemy tribesmen, which is located ten miles away. Blotto scores $+1$ if he successfully captures the enemy base without losing his own base, and -1 if he loses his own base under any circumstances. Daylight raids are impractical, and for night raids an attacking force needs one more unit than the defending force to effect capture. If an attacking force arrives with insufficient strength to effect capture, then it retreats to its own base without engaging."

In this game a strategy for a player for a single night's operation consists simply of a partition of units into attacking and defending forces. Letting A stand for attack, D for defend, the matrix for this recursive game is:

$$
\begin{array}{c}
\\
\text{ENEMY}
\end{array}
$$

				A	0	1	2
				D	2	1	0

$$
\begin{array}{cc}
& \text{A} \quad \text{D} \\[2pt]
& \underline{0} \quad \underline{3} \\
\text{BLOTTO} & 1 \quad 2 \\
& 2 \quad 1 \\
& 3 \quad 0
\end{array}
\qquad
\begin{bmatrix}
\Gamma & \Gamma & \Gamma \\
\Gamma & \Gamma & 1 \\
\Gamma & 1 & -1 \\
1 & -1 & -1
\end{bmatrix}
$$

The value of this game is easily seen to be $+1$, with strategies of the form $[0, 1 - \epsilon - \epsilon^2, \epsilon, \epsilon^2]$ ϵ-best for Blotto, and all strategies optimal for enemy.

It is tactical situations such as this which appear to be the main application of zero-sum two-person recursive games. It is amusing to note that Blotto's patience could be measured by the reciprocal of the ϵ he chooses for his ϵ-best strategy, since the smaller he chooses ϵ the surer he is of winning, while at the same time the possible duration of play is increased. This is always the case in a recursive game when a player is forced to resort to ϵ-best strategies even though the elements satisfy the minimax condition, since such a strategy becomes *necessary* only in the event that optimal play in $\Gamma^i(\vec{V})$ for all game elements would fail to make all of the favorable game states transient, to use the terminology of stochastic processes.

Although not zero-sum, many bargaining situations can be conveniently formalized as recursive games by allowing one or several players to submit offers or bits, the play ending with payoff if agreement is reached, and continuing with no payoff and subsequent attempts otherwise.

BIBLIOGRAPHY

[1] Shapley, L. S. "Stochastic games," *Proceedings of the National Academy of Sciences, U.S.A.*, 39 (1953), pp. 1095–1100.

H. Everett
Princeton University

10

VON NEUMANN-MORGENSTERN SOLUTIONS TO COOPERATIVE GAMES WITHOUT SIDE PAYMENTS

R. J. Aumann and B. Peleg

Communicated by A. W. Tucker, January 7, 1960

THE USE OF side payments in the classical[1] theory of n-person games involves three restrictive assumptions. First, there must be a common medium of exchange (such as money) in which the side payments may be effected; next, the side payments must be physically and legally feasible; and finally, it is assumed that utility is "unrestrictedly transferable," i.e., that each player's utility for money[2] is a linear function of the amount of money.[3] These assumptions severely limit the applicability of the classical theory; in particular, the last assumption has been characterized by Luce and Raiffa [2, p. 233] as being "exceedingly restrictive—for many purposes it renders n-person theory next to useless." It is the purpose of this paper to present the outline of a theory that parallels the classical theory, but *makes no use of side payments*.[4] Our definitions are related to those given in [2, p. 234] and in [3], but whereas the previous work went no further than proposing definitions, the theory outlined here contains results which generalize a considerable portion of the classical theory. It thus demonstrates that the restrictive side payment assumption is not necessary for the development of a theory based on the ideas of von Neumann and Morgenstern. Only a general description of the theory and statements of the more important theorems will be included here; details and proofs will be published elsewhere.

---

[1] We will use the word "classical" to denote the von Neumann-Morgenstern theory as described in [1] and in [4].

[2] Or any other medium of exchange.

[3] See [2, p. 168]. It can be proved that when $n \geq 3$, linearity of the utilities in money is necessary and sufficient for the existence of an unrestrictedly transferable utility.

[4] In particular, our theory is of course also applicable to the case in which side payments are permitted. When, in addition, utility is unrestrictedly transferable, then our theory reduces to the classical theory.

1. Effectiveness

Let us fix attention on a given finite n-person game, and let N denote the set of players. Let E^N denote an n-dimensional euclidean space, and let us index the coordinates of points in E^N by the members of N. The points of E^N will be called *payoff vectors*; if $x \in E^N$ and $i \in N$, then x_i will denote the coordinate of x corresponding to player i, and will be called the *payoff* to i.

Intuitively, a coalition B is *effective* for a payoff vector x if the members of B, by joining forces, can play so that each player i in B receives at least x_i. This intuitive definition is open to a number of interpretations. The rather conservative one adopted by von Neumann and Morgenstern assumes that the most the members of B can count on is what they can get if the players of $N - B$ form a coalition whose purpose it is to minimize the payoff to B. There are at least two generalizations of this notion of effectiveness to the case in which there are no side payments:

(i) *A coalition B is said to be α-effective for the payoff vector x if there is a strategy[5] for B, such that for each strategy used by $N - B$, each member i of B receives at least x_i.*

(ii) *A coalition B is said to be β-effective for the payoff vector x, if for each strategy used by $N - B$, there is a strategy for B such that each member i of B receives at least x_i.*

Roughly, α-effectiveness means that B can assure itself of its portion of x independently of the actions of $N - B$, whereas β-effectiveness means that $N - B$ cannot prevent B from obtaining its (B's) portion of x. In the classical theory the two notions are equivalent, but this is not the case when side payments are forbidden. Which of the two definitions is preferable is a matter of taste; both have appeared, in more or less disguised form, in the previous literature [2, p. 175; 3; 5]. There seems to be a tendency to consider α-effectiveness as intuitively more appealing; on the other hand, there is evidence that β-effectiveness may eventually turn out to be the more significant concept.[6] The present theory applies equally well to both notions.

---

[5] The word "strategy" as used in this paper means what has been variously called "correlated mixed strategy" [2, p. 116], "joint randomized strategy" [2, p. 116], "cooperative strategy" [2, p. 175], and "correlated strategy B-vector" [5].

[6] See §6.

2. Axiomatic Treatment

It is possible to define many of the basic notions of n-person theory—domination, solution, core, etc.—in terms of effectiveness. Of course the objects we will get will usually depend on what kind of effectiveness we started with. Thus we will define the α-core and the β-core, but for a given game they usually differ; similarly with α-solutions and β-solutions, etc. Nevertheless, it is possible to prove a considerable number of general theorems which hold for either kind of effectiveness; the proofs of these theorems make use only of certain basic properties common to both kinds. The situation invites axiomatic treatment.

An n-person "characteristic function" is a set N with n members, together with a function v that carries each subset B of N into a subset $v(B)$ of E^N so that

(1) $v(B)$ *is convex;*

(2) $v(B)$ *is closed;*

(3) $v(\varnothing) = E^N;$[7]

(4) *if $x \in v(B)$, $y \in E^N$, and for all $i \in B$, $y_i \leqq x_i$, then $y \in v(B)$; and*

(5) *if B_1 and B_2 are disjoint, then $v(B_1 \cup B_2) \supset v(B_1) \cap v(B_2)$.*

An n-person "game" is an n-person characteristic function (N, v) together with a convex compact polyhedral subset H of $v(N)$.

An n-person game as just defined actually represents more than a game in the usual sense; it is a game together with a concept of effectiveness. The set H is the set of all "feasible" or "attainable" payoff vectors, i.e., the set of all payoff vectors which can be attained by a joint strategy of N. $v(B)$ represents the set of all payoff vectors for which B is effective. Conditions (1), (2), and (3) are self-explanatory. Condition (4) says that if a coalition B is effective for a payoff vector x, then it is also effective for any payoff vector with smaller (or equal) payoffs to its members. Condition (5) is the natural generalization of super-additivity of the characteristic function in the classical theory.[8]

In order to justify these definitions, it is necessary to show that an arbitrary finite game, when combined with the concept either of α-ef-

---

[7] $\varnothing$ denotes the empty set.

[8] Condition (5) is not needed for any of the results stated in this paper. It was included in order to underscore the parallelism with the classical theory, and with the hope that the stronger axioms will eventually yield a richer theory.

fectiveness or of β-effectiveness, satisfies our definition of a game.[9] For the most part this is straightforward; the only deep part occurs in verifying condition (5) in the case of β-effectiveness, where use is made of Kakutani's fixed point theorem.

One of the chief advantages of the axiomatic approach is its flexibility: it can be used not only with the notions of effectiveness described in §1, which are based on the conservative approach that characterizes the classical theory, but also with many other notions of effectiveness. For example, we may prefer an effectiveness notion based on the ideas of ψ-stability [2, p. 163–168, 174–176, 220–236]. Such notions can be constructed in a number of ways; the general idea would be that in order for a coalition B to be effective for a pair consisting of a payoff vector x and a coalition structure τ, the coalition B must be attainable from the given coalition structure τ, and no attainable combination of coalitions in $N - B$ should be able to prevent B from obtaining its portion of the payoff vector x. Specifically, we would say that B is effective for (x, τ) if $B \in \psi(\tau)$ and there is a strategy for B such that for each partition $(B_1, \cdots, B_k)$ of $N - B$ into members of $\psi(\tau)$, and each k-tuple of strategies used by the B_j, each member i of B receives at least x_i. A related but different notion can be obtained if we reverse the quantifiers. For fixed τ these two notions of effectiveness satisfy all the axioms except (5), and therefore the results stated in this paper hold for them as well (cf. footnote 8).

3. Domination and Solution

Fix an n-person game $G = (N, v, H)$. A payoff vector x is said to *dominate* a payoff vector y via B if $x \in v(B)$ and $x_i > y_i$ for all $i \in B$; x is said to *dominate* y if there is a B such that x dominates y via B. If K is an arbitrary set of payoff vectors, we define dom K to be the set of all payoff vectors dominated by at least one member of K. If P is an arbitrary set of payoff vectors, then a subset K of P is said to be *P-stable* if $K \cap$ dom K is empty and $K \cup$ dom $K \supset P$. The set $P -$ dom P is called the *P-core*. A payoff vector x is said to *majorize* a payoff vector y if x dominates y and all z that dominate x also dominate y. All the lemmas and theorems of §1 of [4] concerning domination, P-stability, the P-core and majorization remain true in this context; the proofs go through essentially unchanged.

---

[9] Where $v(B)$ and H have the meanings described in the previous paragraph.

It is easy to show that for each $i \in N$, there is an extended real number[10] v_i such that $v(\{i\}) = \{x \in E^N : x_i \leq v_i\}$. A payoff vector x is called *individually rational* if $x_i \geq v_i$ for each $i \in N$. x is called *group rational* if there is no $y \in H$ such that $y_i > x_i$ for each $i \in N$. Let us denote by $\overline{A}$ the set of individually rational members of H, and by A the set of members of $\overline{A}$ that are also group rational. Then it can be proved that a subset of H is A-stable if and only if it is $\overline{A}$-stable. This justifies us in defining a *solution* of G to be an A-stable set.[11]

The next step is to investigate which games are solvable and what their solutions are. First of all, it is easy to show that all 2-person games have a unique solution,[12] namely all of A. We next investigate 3-person zero-sum[13] games. Unlike the situation in the classical theory, we now find a large number of essentially different games, whose solutions exhibit the greatest variety and complexity. The basic theorem is

THEOREM 1. *Every 3-person zero-sum game is solvable.*

The proof, which is rather involved, proceeds by dividing A into regions, solving each region separately, and then combining the regional solutions into a solution for all of A. The shapes of the regions depend on the $v(B)$ and on their interrelationships. The question of the solvability of 3-person general-sum or 4-person zero-sum games remains open.

Incidentally, Theorem 1 is the only one of our results for which the assumption that the $v(B)$ be convex (condition (1)) is required.[14]

4. COMPOSITION

Let $G_1 = (N_1, v_1, H_1)$ and $G_2 = (N_2, v_2, H_2)$ be games whose player sets N_1 and N_2 are disjoint. Intuitively, the *composition* G of G_1 and G_2 is the game each play of which consists of a play of G_1 and a play of G_2, played without any interconnection. Formally, we define[15] $G = (N, v, H)$, where $N = N_1 \cup N_2$, $H = H_1 \times H_2$, and for each $B \subset N$, $v(B) = v_1(B \cap N_1) \times v_2(B \cap N_2)$.

---

[10] A real number or $+\infty$ or $-\infty$.

[11] If K is a solution of G, we will also say that K *solves* G, and that G is *solvable*.

[12] This solution is closely related to the *negotiation set* of a 2-person cooperative game [2, p. 118], but is *not* the same thing.

[13] A game is said to be *zero-sum* if H is contained in the plane $\sum_{i \in N} x_i = 0$.

[14] The question as to whether the theorem holds without this assumption remains open; certainly the proof does not go through.

[15] $\times$ denotes cartesian product.

THEOREM 2. *A necessary and sufficient condition that a subset K of H solve G is that it be of the form $K_1 \times K_2$, where K_1 solves G_1 and K_2 solves G_2.*

The simplicity of this result is somewhat surprising, in view of the fact that the corresponding result in the classical theory is much more complicated. The complexity of the classical result is explained by the fact that it permits side payments between members of N_1 and members of N_2, whereas no such intercourse can be possible in our framework; thus although the classical theory is a special case of our theory, the composition of two games in the classical sense yields a game which is in general not the composition of the games in our sense. All of our solutions appear in the classical theory, but the converse is not true. Our solutions are precisely those in which no "tribute" is paid by either group of players (cf. [1, §46.11.2, p. 401]).

5.   THE CORE

THEOREM 3. *The A-core and the $\bar{A}$-core coincide.*

This theorem justifies us in defining the *core* of a game to be its A-core. The proof makes essential use of the fact that H is polyhedral (this is the only theorem for which this assumption is needed); indeed, if this assumption is dropped, the theorem becomes false.

Shapley [6] has conjectured that in the classical theory, the intersection of all solutions is the core. This is not true in our theory; indeed, there is a 3-person zero-sum game with a unique solution, which strictly includes the core.[16]

6.   THE β-CORE AND THE SUPERGAME

The *supergame* of a game[17] Γ is the game each play of which consists of an infinite sequence of plays of Γ. A *strong equilibrium point* in an n-person game [5] is, roughly speaking, an n-tuple $\{\xi_i\}_{i \in N}$ of strategies with the property that if the members j of any coalition B use strategies different from the ξ_j, while the players not in B keep using

---

[16] This solution is disconnected, so that it also provides a counter-example to another conjecture of Shapley, namely that the union of all solutions is connected. Of course this example does not yet settle these questions for the classical theory.

[17] We are now referring to "game" in the ordinary sense of the word, not that defined in §2.

the ξ_i, then at least one player in B will not profit from the change, i.e., will get no more than he would have gotten had all the players used the ξ_i. Strong equilibrium points are strengthened forms of the Nash equilibrium points [7]; at a Nash equilibrium point there is no direct incentive for any *individual* to change his strategy, whereas at a strong equilibrium point there is no direct incentive for any *coalition* to change its strategy.

In [5] the concept of a c-acceptable payoff vector is defined, and it is shown that a payoff vector is c-acceptable in a given finite game if and only if it is the vector of payoffs to a strong equilibrium point in the corresponding supergame. It can be shown that the set of c-acceptable payoff vectors coincides with the β-core.[18] Hence *the β-core of a finite game is precisely the set of payoff vectors to strong equilibrium points in the corresponding supergame.*

7. The "Extended" Theory

The definition of an *extended game* is similar to that of a game, with the single exception that H is not assumed to be a subset of $v(N)$ but merely of E^N. In the classical theory extended games are important as a theoretical tool in composition theory; they were first considered by von Neumann and Morgenstern [1, Chapter X].

Theorems 1 and 2 remain true as they stand for extended games. Theorem 3 must be adjusted to read "The A-core is the intersection of the $\overline{A}$-core with A." §1 of [4] generalizes as before; however, we have not succeeded in obtaining a relation between the A-stable sets and the $\overline{A}$-stable sets in extended games.

References

1. J. von Neumann and O. Morgenstern, *Theory of games and economic behavior*, Princeton University Press, 1943, 2nd ed., 1947.
2. R. D. Luce and H. Raiffa, *Games and decisions*, John Wiley, 1957.
3. L. S. Shapley and M. Shubik, *Solutions of n-person games with ordinal utilities* (abstract), Econometrica vol. 21 (1953) p. 348.
4. D. B. Gillies, *Solutions to general non-zero-sum games*, Contributions to the Theory of Games IV, Princeton University Press, 1959, pp. 47–85.

---

[18] I.e., the core if we use β-effectiveness as our definition of effectiveness.

5. R. J. Aumann, *Acceptable points in general cooperative n-person games*, ibid., pp. 287–324.
6. L. S. Shapley, *Open questions* (dittoed), Report of an Informal Conference on the Theory of n-Person Games held at Princeton University, March 20–21, 1953, p. 15.
7. J. F. Nash, *Non-cooperative games*, Ann. of Math. vol. 54 (1951) pp. 286–295.

The Hebrew University,
Jerusalem, Israel

11

A LIMIT THEOREM ON THE CORE OF AN ECONOMY*

GERARD DEBREU AND HERBERT SCARF[1]

1. INTRODUCTION

In his *Mathematical Psychics* [5], Edgeworth presented a remarkable study of the exchanges of two commodities that might arise in an economy with two types of consumers. The first case that he considers concerns two individuals each of whom initially possesses certain quantities of each commodity. The result of trading consists of a reallocation of the total amounts of the two commodities and may, therefore, be described geometrically by a point in the Edgeworth box corresponding to that economy.

Edgeworth confines his attention to those exchanges which are Pareto optimal, i.e., those which cannot yield greater satisfaction for one consumer without impairing that of the other by means of additional trade. He further restricts the admissible final allocations to those which are at least as desired by *both* consumers as the allocation prevailing before trading. Those allocations which are not ruled out by either of these considerations constitute the "contract curve."

As Edgeworth remarks, a competitive allocation is on the contract curve (under assumptions listed in Section 2). But so are many other allocations, and nothing in the analysis of the case of two consumers indicates that the competitive solutions play a privileged role. In order to single out the competitive allocations Edgeworth introduces an expanded economy which consists of $2n$ consumers divided into two

* Manuscript received November 6, 1962.

[1] The work of Gerard Debreu was supported by the Office of Naval Research first under Task NR 047-006 with the Cowles Commission for Research in Economics, and then under Contract ONR 222 (77) with the University of California. The work of Herbert Scarf was supported by an Office of Naval Research Contract ONR-225 (28) with Stanford University. Reproduction in whole or in part is permitted for any purpose of the United States Government.

types; everyone of the same type having identical preferences and identical resources before trading takes place. The object is to demonstrate that as n becomes large, more and more allocations are ruled out, and eventually only the competitive allocations remain. This statement can be paraphrased by saying that the contract curve shrinks to the set of competitive equilibria as the number of consumers becomes infinite.

It is clear that the two principles mentioned above for ruling out allocations must be supplemented by some additional principles if this result is to be correct. The general principle which Edgeworth formulated is that of "recontracting." Consider an allocation of the total resources of the $2n$ consumers and consider any collection of consumers (which need not include the same number of each type). This collection "recontracts out," if it is possible for its members to redistribute their initial resources among themselves in such a way that some member of the collection prefers the new outcome to the allocation previously given while no member desires it less. The presumption is that an allocation is not made if it can be recontracted out by some group of consumers.

Edgeworth shows that the set of allocations which are not recontracted out decreases as n increases, and has the set of competitive equilibria as a limit. The proof given in *Mathematical Psychics* could easily be rewritten in the style of contemporary mathematical economics. It is, however, based on the geometrical picture of the Edgeworth box and does not seem to be applicable to the general case involving more than two commodities and more than two types of consumers.

As Martin Shubik pointed out, the question can be studied from the point of view of n-person game theory. In a very stimulating paper [12] he analyzed the Edgeworth problem, using the von Neumann-Morgenstern concept of a solution, and also Gillies' [6] concept of the "core." Other discussions of markets as n-person games may be found in von Neumann and Morgenstern [7] and in several papers by Shapley [9, 10].

In all these contributions, extensive use is made of a transferable utility. While this concept has been readily accepted in game theory, it has remained foreign to the mainstream of economic thought. Some recent work has been done, however, on a version of n-person game theory which avoids the assumption of transferable utility [1, 2] and which includes a definition of the core. It is this concept which corresponds to the Edgeworth notion of recontracting.

In [8] Scarf analyzed the core in the latter sense in an economy with an arbitrary number of types of consumers and an arbitrary number of commodities. Economies consisting of r consumers of each type were considered and it was proved that an allocation which assigns the same commodity bundle to all consumers of the same type and which is in the core for all r must be competitive. An economy consisting of an infinite sequence of consumers of each type was also studied and it was demonstrated that an allocation in the core of this economy is competitive. A suggestion for a simplification of the proofs of these theorems and for a weakening of their assumptions was given by Debreu [4].

Our main purpose is to show that the first of the two theorems mentioned in the last paragraph is very widely applicable and, thereby, to obtain a further considerable simplification of the study of the core and to discard an awkward assumption used in both papers [8, (Section 4, A.2)] and [4, (A.4)]. Our second purpose is to cover a case in which production is possible.

In the traditional Walrasian analysis of equilibrium the resources of the consumers and their shares in the producers' profits are specified. All the agents of the economy are assumed to adapt themselves to a price system which one then tries to choose as as to equate total demand and total supply. In the Paretian study of optimality, prices are seen from a second and very different point of view. The problem of efficient organization of an economy with an unspecified distribution of resources is considered, and it is essentially shown that a state of the economy is an optimum one if and only if there exists a price system to which every consumer and every producer is adapted. In Edgeworth's theorem, and in the generalization that we present here, prices appear in a third and again very different light. Given an economy with a specified distribution of resources composed of a certain number of types of consumers which is small relative to the numbers of consumers of each type, an outcome is viable, i.e., no coalition can block it, if and only if there exists a price system to which consumers and producers are adapted. That is to say, competitive equilibria, and only they, are viable. As in the study of Pareto optima, prices emerge from the analysis in a situation in which they were not introduced *a priori*.

2. THE CORE IN A PURE EXCHANGE ECONOMY

At first we study an economy in which no production can take place. We consider m consumers each with specific preferences for commodity bundles consisting of nonnegative quantities of a finite number of commodities. Such a commodity bundle is represented by a vector in the nonnegative orthant of the commodity space, and the preferences of the i-th consumer by a complete preordering, $\succeq_i$. The interpretation of $x' \succeq_i x$ is, of course, that the i-th consumer either prefers x' to x or is indifferent between them. If x' is strictly preferred to x, then we write $x' \succ_i x$.

Three assumptions will be made on the preferences:

1. *Insatiability.* Let x be an arbitrary nonnegative commodity bundle. We assume that there is a commodity bundle x' such that $x' \succ_i x$.

2. *Strong-convexity.* Let x' and x be an arbitrary different commodity bundles with $x' \succeq_i x$, and let α be an arbitrary number such that $0 < \alpha < 1$. We assume that $\alpha x' + (1 - \alpha)x \succ_i x$.

3. *Continuity.* We assume that for any nonnegative x', the two sets

$$\left\{x \mid x \underset{i}{\succeq} x'\right\} \quad \text{and} \quad \left\{x \mid x \underset{i}{\preceq} x'\right\}$$

are closed.

Each consumer owns a commodity bundle which he is interested in exchanging for preferred commodity bundles. The vector ω_i will represent the resources of the i-th consumer. We find it convenient to make the following assumption:

4. *Strict positivity of the individual resources.* We assume that every consumer owns a strictly positive quantity of every commodity.

The core can now be defined. Since production is not considered in the present section, the result of trading consists of an allocation of the total supply $\sum_{i=1}^{m} \omega_i$, and is therefore described by a collection of m nonnegative commodity bundles $(x_1, \ldots, x_m)$ such that

$$\sum_{i=1}^{m} (x_i - \omega_i) = 0.$$

An allocation is in the core if it cannot be recontracted out by any set of consumers S, i.e., if no set of consumers S can redistribute their own initial supply among themselves so as to improve the position of any one member of S without deterioration of that of any other. We emphasize here that it is permissible for an arbitrary set of consumers to combine and reallocate their own assets independently of the remaining consumers in the economy.

To give a formal definition of the core we introduce the notion of set of consumers blocking an allocation. Let $(x_1, \ldots, x_m)$ with $\sum_{i=1}^{m}(x_i - \omega_i)$

= 0 be an assignment of the total supply to the various consumers, and let S be an arbitrary set of consumers. We say that the allocation is blocked by S if it is possible to find commodity bundles x_i' for all i in S such that

$$\sum_{i \in S} (x_i' - \omega_i) = 0, \tag{1}$$

and

$$x_i' \succsim_i x_i, \tag{2}$$

for all i in S, with strict preference for at least one member of S.

The core of the economy is defined as the collection of all allocations of the total supply which cannot be blocked by any set S. *One immediate consequence of this definition is that an allocation in the core is Pareto optimal.* We prove this by taking for the set S all consumers. On the other hand, if we take the possible blocking set to consist of the i-th consumer himself, then we see that an allocation in the core must satisfy the condition $x_i \succsim_i \omega_i$; i.e., the i-th consumer does not prefer his initial holding to the commodity bundle that he receives on the basis of an allocation in the core. Many other conditions will, of course, be obtained as more general sets S are considered.

It is not clear that there always will be some allocations in the core. One can easily construct examples in n-person game theory in which every imputation is blocked by some coalition so that the core is empty. Economies with an empty core may also be found if the usual assumptions on preferences are relaxed. The following example due to Scarf, Shapley, and Shubik is typical.

Consider an economy with two commodities and three consumers, each of whom has preferences described by the indifference curves in Figure 1. It is shown in [11] that it the initial resources of each consumer consist of one unit of each commodity, then the core of the resulting economy is empty. This conclusion does not depend on the lack of smoothness of the indifference curves.

In this paper we make the customary assumptions listed above, in which case it may be shown that the core is not empty. The procedure for doing this is to observe that a competitive allocation exists, and then to demonstrate that every competitive allocation is in the core.

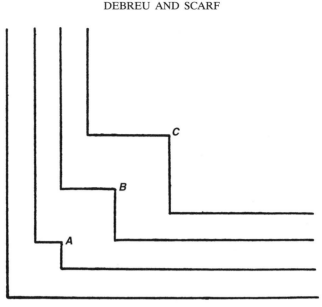

FIG. 1.

It is known that, given our four assumptions on preferences and initial holdings, there is a competitive equilibrium [3]. That is to say, there are nonnegative commodity bundles $x_1, \ldots, x_m$ with $\sum_{i=1}^{m}(x_i - \omega_i) = 0$ and a price vector p, such that x_i satisfies the preferences of the i-th consumer subject to the budget constraint $p \cdot x_i \leq p \cdot \omega_i$. The familiar argument of welfare economies by which a competitive allocation is proved to be Pareto optimal has been extended, as follows, by Shapley to prove:

THEOREM 1. *A competitive allocation is in the core.*

First notice that $x_i' \succ_i x_i$ obviously implies $p \cdot x_i' > p \cdot \omega_i$. For, otherwise, x_i does not satisfy the preferences of the i-th consumer under his budget constraint. Notice also that $x_i' \succsim_i x_i$ implies $p \cdot x_i' \geq p \cdot \omega_i$. For, if $p \cdot x_i' < p \cdot \omega_i$, there is, according to our assumptions 1 and 2, a consumption vector in a neighborhood of x_i' that satisfies the budget constraint and that is preferred to x_i.

Let S be a possible blocking set, so that $\sum_{i \in S}(x_i' - \omega_i) = 0$ with $x_i' \succsim_i x_i$ for all i in S, and with strict preferences for at least one i. From the two remarks we have just made, $p \cdot x_i' \geq p \cdot \omega_i$ for all i in S, with strict inequality for at least one i. Therefore

$$\sum_{i \in S} p \cdot x_i' > \sum_{i \in S} p \cdot \omega_i,$$

a contradiction on $\sum_{i \in S}(x_i' - \omega_i) = 0$.

3. THE CORE AS THE NUMBER OF CONSUMERS BECOMES INFINITE

We shall now follow the procedure first used by Edgeworth for enlarging the market. We imagine the economy to be composed of m types of consumers, with r consumers of each type. For two consumers to be of the same type, we require them to have precisely the same preferences and precisely the same vector of initial resources. The economy therefore consists of mr consumers, whom we index by the pair of numbers (i, q), with $i = 1, 2, \ldots, m$ and $q = 1, 2, \ldots, r$. The first index refers to the type of the individual and the second index distinguishes different individuals of the same type.

An allocation is described by a collection of mr nonnegative commodity bundles x_{iq} such that

$$\sum_{i=1}^{m} \sum_{q=1}^{r} x_{iq} - r \sum_{i=1}^{m} \omega_i = 0.$$

The following theorem makes for the simplicity of our study:

THEOREM 2. *An allocation in the core assigns the same consumption to all consumers of the same type.*

For any particular type i, let x_i represent the least desired of the consumption vectors x_{iq} according to the common preferences for consumers of this type and assume that for some type i' two consumers have been assigned different commodity bundles. Then

$$\frac{1}{r} \sum_{q=1}^{r} x_{iq} \succsim_i x_i, \quad \text{for all } i,$$

with strict preference holding for i'. However,

$$\sum_{i=1}^{m} \left(\frac{1}{r} \sum_{q=1}^{r} x_{iq} - \omega_i \right) = 0,$$

and therefore the set consisting of one consumer of each type, each of whom receives a least preferred consumption, would block.

The theorem we have just proved implies that an allocation in the core for the repeated economies considered here may be described by a collection of m nonnegative commodity bundles $(x_1, \ldots, x_m)$ with $\sum_{i=1}^{m}(x_i - \omega_i) = 0$. The particular collections of commodity bundles in the core will, of course, depend on r. It is easy to see that the core for $r + 1$ is contained in the core for r, for a coalition which blocks in the economy with r repetitions will certainly be available for blocking in the economy with $(r + 1)$ repetitions.

If we consider a competitive allocation in the economy consisting of one participant of each type and repeat the allocation when we enlarge the economy to r participant of each type, the resulting allocation is competitive for the larger economy and consequently is in the core. We see, therefore, that as a function of r, the cores form a nonincreasing sequence of sets, each of which contains the collection of competitive allocations for the economy consisting of one consumer of each type. Our main result is that no other allocation is in the core for all r.

THEOREM 3. If $(x_1, \ldots, x_m)$ is in the core for all r, then it is a competitive allocation.

Let Γ_i be the set of all z in the commodity space such that $z + \omega_i \succ_i x_i$, and let Γ be the convex hull of the union of the sets Γ_i. Since, for every i, Γ_i is convex (and nonempty), Γ consists of the set of all vectors z which may be written as $\sum_{i=1}^{m} \alpha_i z_i$, with $\alpha_i \geq 0$, $\sum_{i}^{m}\alpha_i = 1$, and $z_i + \omega_i \succ_i x_i$. Figure 2 describes this set in the case of two commodities and two types of consumers.

We first form the set of commodity bundles which are preferred to x_1 and from each of these subtract the vector ω_1, obtaining the set Γ_1. We do the same with x_2 and ω_2 in order to obtain Γ_2. We then take the union of Γ_1 and Γ_2 and form the convex hull obtaining Γ. Verifying that the origin does not belong to the set Γ in the general case of an arbitrary number of commodities and an arbitrary number of types of consumers is the key step in the proof of Theorem 3.

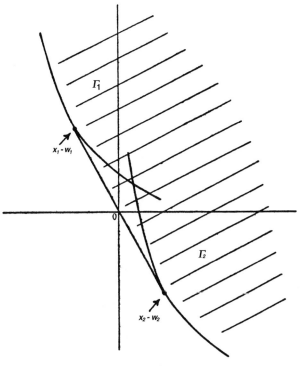

FIG. 2.

Let us suppose that the origin belongs to Γ. Then $\sum_{i=1}^{m} \alpha_i z_i = 0$ with $\alpha_i \geq 0$, $\sum_{i=1}^{m} \alpha_i = 1$, and $z_i + \omega_i \succ_i x_i$. Select an integer k, which will eventually tend to $+\infty$, and let α_i^k be the smallest integer greater than or equal to $k\alpha_i$. Also let I be the set of i for which $\alpha_i > 0$.

For each i in I we define z_i^k to be $[(k\alpha_i)/(\alpha_i^k)]z_i$ and observe that $z_i^k + \omega_i$ belongs to the segment $[\omega_i z_i + \omega_i]$ and tends to $z_i + \omega_i$ as k tends to infinity. The continuity assumption on preferences implies that $z_i^k + \omega_i \succ_i x_i$ for sufficiently large k. Moreover,

$$\sum_{i \in I} a_i^k z_i^k = k \sum_{i \in I} \alpha_i z_i = 0.$$

Consider the coalition composed of a_i^k members of type i to each one of whom we assign $\omega_i + z_i^k$, where i runs over the set I. Such a coalition blocks the allocation $(x_1, \ldots, x_m)$ repeated a number of times equal to $\max_{i \in I} a_i^k$. This contradicts the assumption that $(x_1, \ldots, x_m)$ is in the core for all r.

We have, therefore, established that the origin does not belong to the convex set Γ. Consequently, there is a hyperplane through it with normal p such that $p \cdot z \geq 0$ for all points z in Γ.

If $x' \succ_i x_i$, then $x' - \omega_i$ is in Γ_i, hence in Γ, and we obtain $p \cdot x' \geq p \cdot \omega_i$. Since in every neighborhood of x_i there are consumptions strictly preferred to x_i, we also obtain $p \cdot x_i \geq p \cdot \omega_i$. But

$$\sum_{i=1}^{m} (x_i - \omega_i) = 0.$$

Therefore $p \cdot x_i = p \cdot \omega_i$ for every i.

The argument is virtually complete at this stage. We have demonstrated the existence of prices p such that for every i, (1) $x' \succ_i x_i$ implies $p \cdot x' \geq p \cdot \omega_i$, and (2) $p \cdot x_i = p \cdot \omega_i$. As is customary in equilibrium analysis, there remains to show that x_i actually satisfies the preferences of the i-th consumer subject to his budget constraint, i.e., that $x' \succ_i x_i$ actually implies $p \cdot x' > p \cdot \omega_i$. Since ω_i has all of its components strictly positive, there is a nonnegative x^0 strictly below the budget hyperplane. If for some x'', both $x'' \succ_i x_i$ and $p \cdot x'' = p \cdot \omega_i$, the points of the segment $[x^0, x'']$ close enough to x'' would be strictly preferred to x_i and strictly below the budget hyperplane, a contradiction of (1). This completes the demonstration of Theorem 3.

4. THE CORE IN A PRODUCTIVE ECONOMY

An entirely straightforward extension of our results on the core to an economy in which production is possible can be given. We assume that all coalitions of consumers have access to the same production possibilities described by a subset Y of the commodity space. A point y in Y represents a production plan which can be carried out. Inputs into production appear as negative components of y and outputs as positive components. From now on, in addition to the four conditions given in Section 2 (insatiability, strong-convexity and continuity of preferences, and strict positivity of the individual resources), we impose on the economy the following condition:

5. *Y is a convex cone with vertex at the origin.*

Thus Sections 2 and 3 dealt with the particular case where the cone Y is degenerate to the set having the origin as its only element.

In the new contest, an allocation for an economy with m consumer is a collection of nonnegative commodity bundles $(x_1, \dots, x_m)$ such that there is in Y a production plan y satisfying the equality of demand and supply $\sum_{i=1}^{m} x_i = y + \sum_{i=1}^{m} \omega_i$, i.e., such that $\sum_{i=1}^{m}(x_i - \omega_i)$ belongs to Y. This allocation is blocked by the set S of consumers if it is possible to find commodity bundles x_i' for all i in S such that (1) $\sum_{i \in S}(x_i' - \omega_i)$ belongs to Y, and (2) $x_i' \succeq_i x_i$ for all i in S, with strict preference for at least one member of S. The core of the economy is defined as the collection of all allocations which cannot be blocked.

An allocation is competitive if there exists a price system p such that the profit is maximized on Y (since Y is a cone with vertex at the origin, the maximum profit is zero) and that x_i satisfies the preferences of the i-th consumer under the constraint $p \cdot x \leq p \cdot \omega_i$. Assumptions 1–5 are no longer sufficient to insure the existence of a competitive allocation, but Theorem 1 remains true: A competitive allocation is in the core.

The proof hardly differs from the one we have given. The two opening remarks are unchanged. Let S be a possible blocking set, so that $\sum_{i \in S}(x_i' - \omega_i) = y$ in Y with $x_i' \succeq_i x_i$ for all i in S, with strict preference for at least one i, and with $p \cdot y \leq 0$. Since $p \cdot x_i' \geq p \cdot \omega_i$ for all i in S, with strict inequality for at least one i, we have $\sum_{i \in S} p \cdot x_i' > \sum_{i \in S} p \cdot \omega_i$, or $p \cdot y > 0$, a contradiction.

As before we consider an economy composed of m types of consumers with r consumers of each type. An allocation is described by a collection of mr commodity bundles x_{iq} such that $\sum_{i=1}^{m} \sum_{r=1}^{q} x_{iq} - r\sum_{i=1}^{m} \omega_i$ belongs to Y. It is a simple matter to verify the analogue of Theorem 2: *An allocation in the core assigns the same consumption to all consumers of the same type.* The only modification in the previous proof involves the fact that $y \in Y$ implies $(1/r)y \in Y$.

The allocations in the cores may therefore be described by a collection of m commodity bundles $(x_1, \ldots, x_m)$ with $\sum_{i=1}^{m}(x_i - \omega_i)$ in Y. Again it is clear that the cores form a nonincreasing sequence of sets as r increases. We now indicate the proof of the analogue of Theorem 3: *If $(x_1, \ldots, x_m)$ is in the core for all r, then it is a competitive allocation.* The set Γ is defined, as before, to be the convex hull of the union of the m sets

$$\Gamma_i = \left\{ z \mid z + \omega_i \underset{i}{\succ} x_i \right\}.$$

We then show that Γ and Y are disjoint. Suppose, to the contrary, that $\sum_{i=1}^{m} \alpha_i z_i = y$ in Y with $\alpha_i \geq 0$, $\sum_{i=1}^{m} \alpha_i = 1$, and $z_i + \omega_i \succ_i x_i$. Using the same definitions of k, a_i^k, I, and z_i^k as in the proof of Theorem 3, we see that $z_i^k + \omega_i \succ_i x_i$ for sufficiently large k. Moreover,

$$\sum_{i \in I} a_i^k z_i^k = k \sum_{i \in I} \alpha_i z_i = ky.$$

Since $ky \in Y$, the allocation is blocked by the coalition we have described in proving Theorem 3. Thus a contradiction has been obtained.

The two convex sets Γ and Y may, therefore, be separated by a hyperplane with normal p such that $p \cdot z \geq 0$ for all points z in Γ and $p \cdot y \leq 0$ for all points y in Y. The demonstration then proceeds as before to verify that we indeed have a competitive allocation.

5. GENERALIZATIONS

Until now we have constrained the consumption bundles of consumers to belong to the nonnegative orthant of the commodity space. This restriction, which was made only keep the exposition as simple as possible, is not essential. Instead we can require the consumptions of all the consumers of the i-th type ($i + 1, \ldots, m$) to belong to a given subset X_i of the commodity space. We impose on these consumption sets the condition:

0. X_i is convex.

We make the appropriate modification on Assumptions 1–4; in particular, in 3, the two sets $\{x \mid x \succsim_i x'\}$ and $\{x \mid x \precsim_i x'\}$ are now assumed to be closed in X_i, and in 4, ω_i is now assumed to be interior to X_i. Then the three theorems are established without alteration of their proofs.

A second generalization consists in replacing Assumption 2 (strong-convexity of preferences) by

2'. Convexity. Let x' and x be arbitrary commodity bundles with $x' \succ_i x$, and let α be an arbitrary number such that $0 < \alpha < 1$. We assume that $\alpha x' + (1 - \alpha)x \succ_i x$.

This substitution affects neither the statement nor the proof of Theorem 1. In order to establish the analogue of Theorem 2, we consider an economy with r consumers of each one of m types. Given an allocation (x_{iq}) in its core, we define $\bar{x}_i$ to be $(1/r)\sum_{q=1}^r x_{iq}$ and, as before, we denote by x_i the least desired of the consumption bundles x_{iq} according to the common preferences for consumers of the i-th type. Since $\sum_{i=1}^m (\bar{x}_i - \omega_i)$ belongs to Y, the coalition consisting of one consumer of each type who receives a least preferred consumption blocks, unless $\bar{x}_i \sim_i x_i$ for every i. Therefore, by 2', *an allocation in the core assigns to all consumers of the same type consumptions indifferent to the average of the consumptions for that type.* This suggests defining the *strict core* of the economy as the collection of all unblocked allocations with the same consumption bundles assigned to all consumers of the same type. As we have just seen, with an allocation in the core is associated an allocation (consisting of the *m* average consumptions repeated *r* times) in the strict core which is indifferent to the first allocation for every consumer. Thus the distinction between the core and the strict core is not essential. However, we can treat the strict core under 2′ exactly as we treated the core under 2. As a function of *r*, the strict cores form a nonincreasing sequence of sets and *if* $(x_1, \ldots, x_m)$ *is in the strict core for all r, then it is a competitive allocation.*

Cowles Foundation, Yale University, and University of California, Berkeley, U.S.A. and Stanford University, U.S.A.

REFERENCES

[1] Aumann, R. J. "The Core of a Cooperative Game without Side Payments," *Transactions of the American Mathematical Society*, XCVIII (March 1961), 539–52.
[2] —— and Peleg, B. "Von Neumann-Morgenstern Solutions to Cooperative Games without Side Payments," *Bulletin of the American Mathematical Society*, LXVI (May 1960), 173–79.
[3] Debreu, Gerard. "New Concepts and Techniques for Equilibrium Analysis," *International Economic Review*, III (September 1962), 257–73.
[4] ——. "On a Theorem of Scarf," *The Review of Economic Studies*, XXX (1963).
[5] Edgeworth, F. Y. *Mathematical Psychics*. London: Kegan Paul, 1881.
[6] Gillies, D. B. *Some Theorems on n-Person Games*, Ph.D. Thesis. Princeton University, 1953.
[7] von Neumann, John and Oskar Morgenstern. *Theory of Games and Economic Behavior*. Princeton: Princeton University Press, 1947.
[8] Scarf, Herbert. "An Analysis of Markets with a Large Number of Participants," *Recent Advances in Game Theory*. The Princeton University Conference, 1962.
[9] Shapley, L. S. *Markets as Cooperative Games*. The Rand Corporation, 1955, p. 629.
[10] ——. *The Solutions of a Symmetric Market Game*. The Rand Corporation, 1958, p. 1392.
[11] —— and Shubik, Martin. *Example of a Distribution Game Having no Core*. Unpublished manuscript prepared at the Rand Corporation, July 1961.
[12] Shubik, Martin. "Edgeworth Market Games," in A. W. Tucker and R. D. Luce (eds.), *Contributions to the Theory of Games IV*. Princeton: Princeton University Press, 1959.

12

THE BARGAINING SET FOR
COOPERATIVE GAMES*

ROBERT J. AUMANN AND MICHAEL MASCHLER

§1. INTRODUCTION

This paper grew out of an attempt to translate into mathematical formulas what people may argue when faced with a cooperative n-person game described by a characteristic function.

The basic difficulty in n-person game theory is due to the lack of a clear meaning as to what is the purpose of the game. Certainly, the purpose is not just to get the maximum amount of profits, because if every player will demand the maximum he can get in a coalition, no agreement will be reached. Thus, one decides that the purpose of the game is to reach some kind of stability, to which the players would or should agree if they want any agreement to be enforced. This stability should reflect in some sense the power of each player, which results from the rules of the game.

In this paper, we assume that all the players can "bargain" together, with perfect communication, and settle at a "stable" outcome which is based on the "threats" and "counterthreats" that they possess. The set of all the stable outcomes, called the *bargaining set*,[1] is defined in Section 2 and some of its properties are discussed. In particular, this set can always be determined by solving systems of algebraic linear inequalities.

The bargaining sets for the 2- and 3-person games are fully described (Sections 3, 4, 5) and some cases of 4-person games are treated, in which not all the coalitions are permissible (Sections 6, 8).

Some counterexamples for various conjectures, as well as existence theorems, are treated in Section 7, and possible modifications are suggested in Sections 9 and 10. In Section 11 we discuss some similarities and deviations between our theory and other solution concepts,

* This research was sponsored in part by the Office of Naval Research under Nonr-1858(16), and in part by the Carnegie Corporation of New York.
[1] The definition of the bargaining set appeared in [1].

such as Vickrey's concept of self-policing patterns [8] and Luce's concept of ψ-stability [3]. We also outline how our theory can be modified if the game is given by Thrall's characteristic function [7] or by the Aumann-Peleg characteristic function for cooperative games without side payments.

We conclude by pointing out that "central parts" of the von Neumann-Morgenstern solution for some games appear also in the bargaining set. The reason for this phenomenon is obscure to us.

§2. THE BARGAINING SET

We shall consider an n-person cooperative game Γ, described by its characteristic function. More precisely, a set $N = \{1, 2, \ldots, n\}$ of n players is given, together with a collection $\{B\}$ of nonempty subsets B of N, called *permissible coalitions*. For each B, $B \in \{B\}$, a number $v(B)$ is given and it is called the *value of the coalition B*.

For the sake of simplicity, we shall assume throughout this paper that all 1-person coalitions are in $\{B\}$ and have a zero value, i.e.,

$$i \in \{B\}, \qquad v(i) = 0. \tag{2.1.a}$$

In addition, we shall also assume that

$$v(B) \geq 0, \qquad B \in \{B\}. \tag{2.1.b}$$

It will turn out later than no essential change will occur if we add to $\{B\}$ all other, nonpermissible coalitions, assigning them the value zero.

A *payoff configuration* (p.c.) will now be defined as an expression of the form

$$(\mathscr{x}; \mathscr{B}) = (x_1, x_2, \ldots, x_n; B_1, B_2, \ldots, B_m), \tag{2.2}$$

where $B_1, B_2, \ldots, B_m$ are mutually disjoint sets of $\{B\}$ whose union is N; i.e.,

$$B_j \cap B_k = \varnothing, j \neq k; \quad \bigcup_{j=1}^{m} B_j = N; \tag{2.3}$$

and the x_i are real numbers which satisfy

$$\sum_{i \in B_j} x_i = v(B_j); \qquad j = 1, 2, \ldots, m. \qquad (2.4)$$

A p.c. is therefore a representation of a possible outcome of the game, in which the players divide themselves into the coalitions $B_1, B_2, \ldots, B_m$, each coalition shares its value among its members, and each player receives the amount x_i, $i = 1, 2, \ldots, n$.

When people are faced with such a game, each one trying to get as high an amount as he thinks he can get, it is reasonable to expect that some of the p.c.'s will never form, e.g., one does not expect that a p.c. will occur with $x_{i_0} < 0$, since the player i_0 alone can secure more by playing as a 1-person coalition. We are willing to make a strong assumption, namely, that the outcome (2.2) will be a *coalitionally rational p.c.* (c.r.p.c.),[2] i.e., for each B, $B \in \{B\}$, $B \subset B_j$, $j = 1, 2, \ldots, m$,

$$\sum_{i \in B} x_i \geq v(B). \qquad (2.5)$$

Thus, we assume that a coalition will not form if some of its members can obtain more by themselves forming a permissible coalition.

The assumption of coalition rationality differs from the assumption of belonging to the core by the restricting condition $B \subset B_j$. This restriction avoids some of the difficulties which arises when dealing with games whose core is empty. (See Luce and Raiffa [3].)

In itself, the coalition rationality assumption is a very strong one, as it forces the game to be essentially superadditive within those coalitions which are actually formed. Indeed, a coalition whose value is less than the sum of the values of disjoint subcoalitions cannot occur in any c.r.p.c., and can as well be declared nonpermissible or have its value be replaced by zero. Moreover, this assumption is open to the same theoretical objections which are discussed at length in Luce and Raiffa [3]. As a matter of fact, our theory can be developed without the coalition rationality assumption, as indicated in Section 10. Nevertheless, as we are only interested in "stable" outcomes, we feel it instructive to make this assumption.

---

[2] In Aumann and Maschler [1], a c.r.p.c. is called a p.c.

Several phenomena can be observed when watching people who are confronted with a game such as described above. Usually, negotiations start, each one tries to get at least as much as he expects, and at the same time there is an attempt to enter into a "safe" coalition. This latter factor applies, in particular, to those coalitions which are planned to operate for a long period. The search for "safety" gives rise to feelings of sympathy and antipathy which play an important role in the final decisions. Guarantees of all kinds are demanded, contracts are signed, etc. *If people do not feel safe enough, they often do not enter a coalition even if they can win more in it.*

The demand for safety is usually considered legitimate and a sound way to convince the partners to get a smaller amount of profit *in order that no one in the coalition will feel deprived.* There is a desire for "fair play," which can be achieved in various ways. Often it is accepted that "if all things are equal" it is "fair" to divide the profits equally. Sometimes people share the profits according to some fixed ratio established by other precedents, etc.

If "all things" are *not* equal, people will still be happy with their coalition if they agree that the "stronger" partners will get more. Thus, during the negotiations, prior to the coalition formation, each player will try to convince his partners that in some sense he is strong. This he can try in various ways, among which an important factor is his ability to show that he has other, perhaps better, alternatives. His partners, besides pointing out their own alternatives, may argue in return that even without his help they can perhaps keep their proposed shares. Thus, a negotiation quite often takes the form of a sequence of "threats" and "counterthreats," or "objections" against "counterobjections." It is this principle that we shall try to formulate mathematically. It seems that a certain kind of stability is reached is reached if all objections can be answered by counterobjections.

Perhaps it is not enough that any objection by one person could be met. It is possible that a subset of the players of a coalition unite, during the negotiation period, and threaten another subset. If we insist on a strong stability, we have to take care also of such threats. This, in fact, will be done.

To be sure, there are other means used during the bargaining period, such as threats based on the so-called "interpersonal comparison of utilities," sanctions in other games, propaganda, etc. These will be ignored in this paper.

The following example will illustrate our purpose. Let $n = 3$, $v(1) = v(2) = v(3) = v(123) = 0$, $v(12) = 100$, $v(13) = 100$, $v(23) = 50$. Consider the p.c.

$$(80, 20, 0; 12, 3). \tag{2.6}$$

Now, player 2 can object by pointing out that in the p.c.

$$(0, 21, 29; 1, 23) \tag{2.7}$$

he and player 3 get more. Player 1 has no counterobjection because he cannot keep his 80 while offering player 3 at least 29. Thus, (2.6) is unstable. On the other hand,

$$(75, 25, 0; 12, 3) \tag{2.8}$$

is stable. An objection of player 2, e.g.,

$$(0, 26, 24; 1, 23), \tag{2.9}$$

can be met by a counterobjection

$$(75, 0, 25; 13, 2); \tag{2.10}$$

or an objection of player 1, e.g.,

$$(76, 0, 24; 13, 2) \tag{2.11}$$

can be met by the counterobjection

$$(0, 25, 25; 1, 23). \tag{2.12}$$

In these counterobjections, the threatened player can keep his share and offer his partners at least what the player who objects offered. It will turn out that the only stable p.c.'s in this game are

$$(0, 0, 0; 1, 2, 3); \ (75, 25, 0; 12, 3); \ (75, 0, 25; 13, 2); \ (0, 25, 25; 1, 23).$$
$$\tag{2.13}$$

Let it be said to once that our paper was largely motivated by the fact that most of our friends, to whom this game was presented, started their considerations from the p.c.'s (2.8) and (2.10). We tried to find what characterizes these p.c.'s and how they can be generalized to more complicated cases.

Let Γ be a game, as described above. Let K be a nonempty subset of the set of players N. A player i will be called a *partner of K in a p.c.* $(x, \mathscr{B})$, if he is a member of a coalition in $\mathscr{B}$ which intersects K. The set $P[K; (x; \mathscr{B})]$ of all the partners of K in $(x; \mathscr{B})$ is, therefore,

$$P[K; (x; \mathscr{B})] \equiv \{i \mid i \in B_j, \, B_j \cap K \neq \varnothing\}. \tag{2.14}$$

Note that $K \subset P[K; (x, \mathscr{B})]$; i.e., each member of K is also a partner of K, contrary to everyday usage. K needs only the consent of its partners in order to get its part of x.

DEFINITION 2.1. Let $(x; \mathscr{B})$ be a coalitionally rational payoff configuration (2.2), (2.5) for a game Γ. Let K and L be nonempty disjoint subsets of a coalition B_j which appears in $(x; \mathscr{B})$. An *objection* of K against L in $(x; \mathscr{B})$ will be a c.r.p.c.

$$(\mathscr{y}; \mathscr{C}) \equiv (y_1, y_2, \ldots, y_n; C_1, C_2, \ldots, C_l) \tag{2.15}$$

for which

$$P[K; (\mathscr{y}; \mathscr{C})] \cap L = \varnothing, \tag{2.16}$$

$$y_i > x_i \quad \text{for all } i, i \in K, \tag{2.17}$$

$$y_i \geq x_i \quad \text{for all } i, i \in P[K; (\mathscr{y}; \mathscr{C})]. \tag{2.18}$$

Verbally, in their objection, players K claim that, without the aid of players L ((2.16)), they can get more in another c.r.p.c. ((2.17)), and the new situation is reasonable because their new partners do not get less than what they got in the previous p.c. ((2.18)).

DEFINITION 2.2. Let $(x; \mathscr{B})$ be a coalitionally rational payoff configuration (2.2), (2.5) in a game Γ, and let $(\mathscr{y}; \mathscr{C})$ be an objection of a set K against a set L in $(x; \mathscr{B})$, $K, L \subset B_j$. A *counterobjection* of L against K is a c.r.p.c.

$$(\mathscr{z}; \mathscr{D}) \equiv (z_1, z_2, \ldots, z_n; D_1, D_2, \ldots, D_k) \tag{2.19}$$

for which

$$P[L; (\mathscr{x}; \mathscr{D})] \not\supset K, \tag{2.20}$$

$$z_i \geq x_i \quad \text{for all } i, i \in P[L; (\mathscr{x}; \mathscr{D})], \tag{2.21}$$

$$z_i \geq y_i \quad \text{for all } i, i \in P[L; (\mathscr{x}; \mathscr{D})] \cap P[K; (\mathscr{y}; \mathscr{C})]. \tag{2.22}$$

Verbally, in their counterobjection, players L claim that they can hold their original properties ((2.21)), promise their partners at least their original share ((2.21)), and if they need partners of K in his objection, they can give them not less than what they were offered in the objection ((2.22)). Sometimes, the members of L have to use the tactics of "divide and rule" by using members of K as partners, but they may not use *all* members of K ((2.20)).

DEFINITION 2.3. A c.r.p.c. $(\mathscr{x}; \mathscr{B})$ is called stable if for each objection of a K against an L in $(\mathscr{x}; \mathscr{B})$ there is a counterobjection of L against K. The *bargaining set* $\mathscr{M}$ of a game Γ is the set of all stable c.r.p.c.'s.

The feeling of "safety" suggested by this definition lies in the assurance that all threats *within a coalition* can be met. It may be felt, perhaps, that there is a lack of symmetry when comparing (2.16) and (2.20), but the situation is not symmetric in the first place. An objection (2.15) can serve, in general, as an objection of K (or another "K") against various groups "L," each one of which has to have a counterobjection.

To be sure, even if there is a desire for stability as demanded in the p.c.'s of the bargaining set, this does not mean that the outcome will belong to $\mathscr{M}$. A player, e.g., may agree to sacrifice some of his profits in order to make sure that he enters a coalition. Other factors, mentioned above, may also cause deviations from $\mathscr{M}$. However, if the demand for stability is strong enough, we hope that the outcome will not be too far from $\mathscr{M}$; in this sense the theory has a normative aspect. Moreover, as the number of the players increases, there arise many possible threats, and, using the concepts involved in the definition, one may compute and show the players where they are "safe" and what threats they do possess. This is another normative aspect.[3]

The bargaining set is never empty. Indeed, $(0, 0, \ldots, 0; 1, 2, \ldots, n)$ always belongs to $\mathscr{M}$.

---

[3] Results close to the bargaining set have been observed in an experimental study [4].

In a coalition of zero value, any objection (if there is one) can be countered by the other players playing as 1-person coalitions.

A dummy always gets zero in a c.r.p.c., therefore he cannot belong to any objecting K. On the other hand, he can always keep his zero by playing alone. He can be of no use for any objection or counterobjection, since the same can be effected without his help. Thus, a dummy has no essential effect on $\mathcal{M}$.

The definition of $\mathcal{M}$ does not use "interpersonal comparisons of utilities" and it is independent of the names of the players.

THEOREM 2.1. *The bargaining set $\mathcal{M}$ of a game Γ can be represented as the set of solutions of a conjunctive-disjunctive[4] system of linear inequalities involving the x_i as unknowns. It is, therefore, a union of a finite number of polyhedral convex sets in the n-space with coordinates $(x_1, x_2, \ldots, x_n)$.*

PROOF.[5] In any finite expression with coordinates which has the form of quantifiers followed by linear inequalities connected by the words "or" and "and," the free variables—if such exist—which satisfy the expression are those and only those which satisfy a certain disjunctive-conjunctive system of linear inequalities. This is a known theorem in logic, but for the sake of completeness we sketch the proof. It is sufficient to prove the theorem when there is only one quantifier. Moreover, we may assume that this quantifier is $\exists$, because $\forall = \sim \exists \sim$. The theorem now follows from the fact that the projection of a polyhedron is a polyhedron.

§3. THE TWO-PERSON GAME

The bargaining set $\mathcal{M}$ for the game:

$$v(1) = v(2) = 0 \qquad v(12) = a \geq 0, \tag{3.1}$$

consists of all possible c.r.p.c.'s; i.e.,

$$\begin{aligned} &(0, 0; 1, 2) \\ &(x_1, x_2; 12) \qquad x_1 + x_2 = a, \quad x_1 \geq 0, \quad x_2 \geq 0. \end{aligned} \tag{3.2}$$

Indeed, there are no possible objections.

---

[4] I.e., a system of linear inequalities connected by the words "or" and "and."
[5] We are indebted to Professor M. Rabin and Professor A. Robinson for pointing this out.

§4. The 3-Person Game. Permissible Coalitions
of Less than Three Players

In this section we shall study the game:

$$v(1) = v(2) = v(3) = 0; \qquad v(12) = a; \quad v(23) = b;$$
$$v(13) = c; \qquad a, b, c \geq 0. \tag{4.1}$$

THEOREM 4.1. *In the game* (4.1), *essentially two cases arise.*
Case A. If a, b, c *satisfy the "triangle inequality"*

$$a \leq b + c, \qquad b \leq a + c, \qquad c \leq a + b, \tag{4.2}$$

then the bargaining set $\mathcal{M}$ *is:*

$$
\begin{pmatrix}
(& 0 & , & 0 & , & 0 & ; 1, 2, 3) \\
\left(\dfrac{a + c - b}{2}, \right. & \dfrac{a + b - c}{2}, & & 0 & ; 12, 3 & \left. \right) \\
\left(\dfrac{a + c - b}{2}, \right. & 0 & , & \dfrac{c + b - a}{2} & ; 13, 2 & \left. \right) \\
\left(0 \right. & , & \dfrac{a + b - c}{2}, & \dfrac{c + b - a}{2} & ; 1, 23 & \left. \right)
\end{pmatrix}
\tag{4.3}
$$

Case B. If, e.g., $a > b + c$, *then the bargaining set* $\mathcal{M}$ *is:*

$$
\begin{pmatrix}
(& 0 & , & 0 & , 0; 1, 2, 3) \\
(c \leq x_1 \leq a - b, & a - x_1, 0; 12, 3 &) \\
(& c & , & 0 & , 0; 13, 2 &) \\
(& 0 & , & b & , 0; 1, 23 &)
\end{pmatrix}
\tag{4.4}
$$

Before proving this theorem, we shall give some illustrations which will throw some light on the nature of the bargaining sets.

EXAMPLE 1. Let $a = 100$, $b = 100$, $c = 50$. The triangle inequality is satisfied, and therefore $\mathcal{M}$ is discrete: $\{(0, 0, 0; 1, 2, 3) \ (25, 75, 0; 12, 3) \ (25, 0, 25; 13, 2) \ (0, 75, 25; 1, 23)\}$.

One can approach this solution also by the following intuitive argument: Suppose that player 1 receives α, then player 2 gets $100 - \alpha$ and he is thus willing to pay player 3 at most $100 - (100 - \alpha) = \alpha$. Thus, player 3 will be willing to pay player 1 at most $50 - \alpha$. If $50 - \alpha > \alpha$, then player 1 will prefer to join player 3. This will cause player 2 to agree to get less. If $50 - \alpha < \alpha$, player 2 will demand more as he will get more from player 3 if player 1 insists on getting α. Thus an equilibrium will be reached only if $\alpha = 50 - \alpha$, in which case $\alpha = 25$.

EXAMPLE 2. The above argument fails in the case $a = 20$, $b = 30$, $c = 100$. Here one obtains $\alpha = 45$ in which case player 2 will lose money. This he can avoid by playing alone. Our bargaining set is no longer discrete: $\{(0,0,0; 1,2,3) (20 \leq x_1 \leq 70, 0, 100 - x_1; 13, 2) (20, 0, 0; 12, 3) (0, 0, 30; 1, 23)\}$. One can reason as follows: Player 1, being in the coalition 13, will not be satisfied in getting less than 20, since otherwise he will do better by joining player 2. Similarly, player 3 will demand at least 30. Fortunately, both demands can be satisfied, and player 2 cannot cause any harm since he is a weak player.

EXAMPLE 3. Let $a = 100$, $b = 100$, $c = 0$. We observe that the bargaining set is again discrete: $\{(0,0,0; 1,2,3) (0, 100, 0; 12, 3) (0, 0, 0; 13, 2) (0, 100, 0; 1, 23)\}$. This solution reflects the character of an "unrestricted competition" in our bargaining set. Indeed, player 2 can practically receive the amount 100 because whatever the positive demand of player 1 will be, player 3 will be "satisfied" in getting less, and vice versa. One observes that our theory does not take into account the psychological threat that player 2 may also "lose" his profit 100, and probably will therefore be willing to pay some amount in order to be in a coalition with player 1 or with player 3. In practical situations several side conditions may come into consideration such as: (1) It may be *customary* not enter a coalition unless a certain minimum amount or percentage of profit is guaranteed in advance. (2) A "Cartel" agreement is decided between player 1 and player 3, in which both of them declare not to enter a coalition with player 2 without getting at least a certain amount of profit. (3) There is enforced a "Cartel" or an "Anti-Cartel" law in the country. (4) It is known that in order to ensure a certain profit, one is willing to give up a certain amount or percentage in order to "push" an equilibrium situation to his side.

PROOF OF THEOREM 4.1. Certainly, $(0, 0, 0; 1, 2, 3) \in \mathscr{M}$.

Next, let us examine under what circumstances a payoff configuration $(x_1, x_2, 0; 12, 3)$ can belong to the bargaining set. It should be coalitionally rational, and therefore it must satisfy

$$x_1 \geq 0, \qquad x_2 \geq 0; \qquad x_1 + x_2 = v(12). \qquad (4.5)$$

LEMMA 1. *A necessary and sufficient condition that player* 1 *has no objection is*:

$$x_1 \geq v(13). \tag{4.6}$$

PROOF. Indeed, if $x_1 \geq v(13)$, then player 1 has no objection either by playing alone (see (4.5)) or by participating in the coalition 13. If $x_1 < v(13)$, then player 1 can suggest the objection

$$\left(\frac{v(13) + x_1}{2}, 0, \frac{v(13) - x_1}{2}; 13, 2 \right). \tag{4.7}$$

This is a coalitionally ration payoff configuration.

LEMMA 2. *A necessary and sufficient condition that player* 1 *has an objection and to each such objection player* 2 *has a counterobjection, is*:

$$x_1 < v(13), \tag{4.8}$$

$$x_1 - x_2 \geq v(13) - v(23) \quad \text{or} \quad x_2 = 0. \tag{4.9}$$

PROOF. Indeed, if (4.8) and (4.9) hold, then, by Lemma 1, player 1 has an objection. This can only be (see (4.5)) of the form

$$(x_1 + \epsilon, 0, v(13) - x_1 - \epsilon; 13, 2), \tag{4.10}$$

where ϵ is a sufficiently small positive number. If $x_2 = 0$, then (4.10) is itself also a counterobjection; otherwise,

$$(0, v(23) - v(13) + x_1 + \epsilon, v(13) - x_1 - \epsilon; 1, 23) \tag{4.11}$$

is a possible counterobjection. By (4.8), player 2 will now receive even more than x_2. If (4.8) does not hold, then there is no objection for player 1, by Lemma 1. If (4.8) holds, but

$$x_2 > 0 \quad \text{and} \quad x_1 - x_2 < v(13) - v(23), \tag{4.12}$$

then player 1 can object by (4.10), choosing ϵ so small that $v(23) - v(13) + x_1 + \epsilon < x_2$. Now, player 2 does not have any counterobjection, either by playing alone or by forming a coalition with player 3.

Summing up, and making the necessary permutations, we obtain:

LEMMA 3. *A necessary and sufficient condition that a payoff configuration* $(x_1, x_2, 0; 12, 3)$ *will belong to the bargaining set* $\mathcal{M}$, *is that* x_1 *and* x_2 *will satisfy* (4.5) *as well as at least one of the following columns*:

$$
\begin{array}{c|c|c}
x_1 \geq v(13) & x_1 < v(13) & x_1 < v(13) \\
& x_2 = 0 & x_1 - x_2 \geq v(13) - v(23)
\end{array}
\tag{4.13}
$$

and also at least one of the following columns:

$$
\begin{array}{c|c|c}
x_2 \geq v(23) & x_2 < v(23) & x_2 < v(23) \\
& x_1 = 0 & x_2 - x_1 \geq v(23) - v(13).
\end{array}
\tag{4.14}
$$

Taking into account that $x_1 + x_2 = a$, these inequalities reduce to

$$
\begin{array}{c|c|c}
0 \leq x_1 \leq a & x_1 < c & 0 \leq x_1 \leq a \\
& & \dfrac{a+c-b}{2} \leq x_1 < c \\
c \leq x_1 & x_1 = a &
\end{array}
\tag{4.15}
$$

$$
\begin{array}{c|c|c}
x_1 \leq a - b & a - b < x_1 & a - b < x_1 \leq \dfrac{a+c-b}{2} \\
& x_1 = 0 &
\end{array}
\tag{4.16}
$$

We now use the assumption $a, b, c \geq 0$, and the inequalities (4.15), (4.16). A detailed calculation yields the following results:

Case A. If a, b, c satisfy the "triangle inequalities" (4.2), then

$$
x_1 = \frac{a+c-b}{2}
\tag{4.17}
$$

is the only solution.

Case B. If $a > b + c$, then each x_1 satisfying

$$
c \leq x_1 \leq a - b,
\tag{4.18}
$$

is a solution, and there are no other solution.

Case C. If $b > a + c$, then $x_1 = 0$ is the only solution.

Case D. If $c > a + b$, then $x_1 = a$ is the only solution.

These are the only possible cases, they exclude each other, and therefore the proof of Theorem 4.1 has been completed.

§5. The General 3-Person Game

Let us add the coalition 123 with its value $v(123) = d \geq 0$ to the game treated in Section 4. This coalition will have no effect on the previous p.c.'s of the bargaining set. Indeed, this coalition cannot be used for objections and counterobjections, because it contains all the players L and K. Thus, it only remains to find out under what condition does a p.c. $(x_1, x_2, x_3; 123)$ belong to the new bargaining set.

As it should be coalitionally rational, it is necessary that x_1, x_2, x_3 satisfy

$$x_1, x_2, x_3 \geq 0; \qquad x_1 + x_2 \geq a, \qquad x_2 + x_3 \geq b,$$

$$x_1 + x_3 \geq c; \qquad x_1 + x_2 + x_3 = d. \tag{5.1}$$

On the other hand, if (5.1) is satisfied, there can be no objection and hence this pair belongs to $\mathcal{M}$.

In order that the inequalities (5.1) have at least one solution, it is necessary and sufficient that

$$d \geq a, b, c, \qquad d \geq \frac{a + b + c}{2}. \tag{5.2}$$

We have thus proved:

THEOREM 5.1. *In the 3-person game for which*

$$v(1) = v(2) = v(3) = 0, \qquad v(12) = a, \qquad v(23) = b,$$

$$v(13) = c, \qquad v(123) = d, \qquad a, b, c, d \geq 0,$$

the bargaining set $\mathcal{M}$ consists of the p.c.'s given by Theorem 4.1, and also of the p.c.'s $(x_1, x_2, x_3; 123)$ which satisfy (5.1). The latter p.c.'s exist if and only if (5.2) is satisfied.

§6. The 4-Person Game. Coalitions of 1 Person and 3 Persons

Consider the 4-person game, in which the permissible coalitions are all the single-person and the three-person coalitions. Let their values be

$$v(1) = v(2) = v(3) = v(4) = 0, \qquad v(123) = a, \qquad v(124) = b,$$

$$v(134) = c, \qquad v(234) = d, \qquad a, b, c, d \geq 0. \tag{6.1}$$

Evidently $(0,0,0,0; 1,2,3,4)$ belongs to the bargaining set $\mathcal{M}$. Similar considerations to those which were used in Section 4 lead to the inequalities which are listed in Appendix 1. These inequalities express a necessary and sufficient condition in order that the payment configuration $(x_1, x_2, x_3; 123, 4)$ belongs to the bargaining set.

We omit the calculations, which are somewhat lengthy but easy, and state the results. There are essentially four different cases:

Case A. If

$$
\begin{aligned}
2a &\le b + c + d, \\
2b &\le a + c + d, \\
2c &\le a + b + d, \\
2d &\le a + b + c,
\end{aligned}
\tag{6.2}
$$

then the bargaining set is

$$
\begin{pmatrix}
0 & 0 & 0 & 0 & ; 1,2,3,4 \\[4pt]
\dfrac{a+b+c-2d}{3} & \dfrac{a+b+d-2c}{3} & \dfrac{a+c+d-2b}{3} & 0 & ; 123,\ 4 \\[8pt]
\dfrac{a+b+c-2d}{3} & \dfrac{a+b+d-2c}{3} & 0 & \dfrac{b+c+d-2a}{3} & ; 124,\ 3 \\[8pt]
\dfrac{a+b+c-2d}{3} & 0 & \dfrac{a+c+d-2b}{3} & \dfrac{b+c+d-2a}{3} & ; 134,\ 2 \\[8pt]
0 & \dfrac{a+b+d-2c}{3} & \dfrac{a+c+d-2b}{3} & \dfrac{b+c+d-2a}{3} & ; 234,\ 1
\end{pmatrix}
\tag{6.3}
$$

Case B. If

$$
\begin{aligned}
2a &> b + c + d, \\
2b &\le a + c + d, & b &\le c + d, \\
2c &\le a + b + d, & c &\le b + d, \\
2d &\le a + b + c, & d &\le b + c,
\end{aligned}
\tag{6.4}
$$

then the bargaining set is

$$
\begin{pmatrix}
0 & 0 & 0 & 0;1,2,3,4) \\
x_1 & a - x_1 - x_3 & x_3 & 0;123,\ 4\ ) \\
\dfrac{b+c-d}{2} & \dfrac{b+d-c}{2} & 0 & 0;124,\ 3 \\
\dfrac{b+c-d}{2} & 0 & \dfrac{c+d-b}{2} & 0;134,\ 2 \\
0 & \dfrac{b+d-c}{2} & \dfrac{c+d-b}{2} & 0;234,\ 1
\end{pmatrix}. \tag{6.5}
$$

Here, x_1 and x_3 satisfy the inequalities

$$
0 \le x_1 \le a - d, \qquad 0 \le x_3 \le a - b, \qquad c \le x_1 + x_3 \le a. \tag{6.6}
$$

Case C. If

$$
\begin{aligned}
2a &> b + c + d, \\
2b &\le a + c + d, \qquad b > c + d, \\
2c &\le a + b + d, \\
2d &\le a + b + c,
\end{aligned} \tag{6.7}
$$

then the bargaining set is

$$
\begin{pmatrix}
0 & 0 & 0 & 0;1,2,3,4) \\
x_1 & a - x_1 - x_3 & x_3 & 0;123,\ 4\ ) \\
(c \le \xi_1 \le b - d & b - \xi_1 & 0 & 0;124,\ 3\ ) \\
c & 0 & 0 & 0;134,\ 2\ ) \\
0 & d & 0 & 0;234,\ 1\ ).
\end{pmatrix} \tag{6.8}
$$

Here, x_1 and x_3 satisfy the inequalities (6.6).

Case D. If

$$
\begin{aligned}
2a &> b + c + d, \\
2b &> a + c + d, \\
a &\ge b,
\end{aligned} \tag{6.9}
$$

the bargaining set is the same as in Case C.

Only Case A is completely discrete; all other cases contain the continuum (6.6). Equations (6.2) can be considered as a generalization of the triangle inequalities. In fact, it follows from (6.2) that any three of the numbers a, b, c, d satisfy the triangle inequalities. Moreover, an equality $a = b + c$, for example, can occur only if $a = d$. The converse does not hold. (E.g., $a = 8$, $b = c = d = 5$.)

It is possible to approach the bargaining set in Case A as follows: If players 1 and 2 get α and β, respectively, then player 3 gets $a - \alpha - \beta$ in the coalition 123. With these values, player 4 will get $b - \alpha - \beta$ in the coalition 124, $c - a + \beta$ in the coalition 134, and $d - a + \alpha$ in the coalition 234. In order that no coalition can exert threats on others, it is necessary and sufficient that

$$b - \alpha - \beta = c - a + \beta = d - a + \alpha. \qquad (6.10)$$

Hence

$$\alpha = \frac{a + b + c - 2d}{3}, \qquad \beta = \frac{a + b + d - 2c}{3}. \qquad (6.11)$$

If

$$2a \geq b + c + d, \qquad a \geq b, c, d, \qquad (6.12)$$

then the coalition 123 is strong and player 4 cannot get more than zero. If we decide to omit him from the game, and look at each of the remaining two persons in a coalition which contained him as a new 2-person coalition which has the same value as before, we get a 3-person game in which $v(123) = a$, $v(12) = b$, $v(23) = d$, $v(13) = c$, $v(1) = v(2) = v(3) = 0$.

A comparison with the previous two sections shows that the bargaining set of the new game is essentially the same as the game treated in Cases B, C, D, and Case A, if equality holds in the first relation of (6.12).

(Note that each of the systems (6.4) and (6.7) implies $a > b, c, d$, (6.9) implies $a \geq b$, $a > c, d$, and (6.7) as well as (6.9) implies $b > c + d$.)

We can therefore conclude:

THEOREM 6.1. *A 4-person game, in which the permissible coalitions are all the 1-person and 3-person coalitions, always has a bargaining set in which all possible partitions into coalitions appear. The set is discrete if and only if (6.2) is satisfied. If (6.12) holds, the bargaining set is essentially the same as the one of the full 3-person game obtained from the original one by deleting a player who belongs to the complement of a maximal valued coalition. This player always get 0.*

REMARK 1. The same situation occurs in a 3-person game in which the only nonpermissible coalition is the 3-person coalition. If the triangle inequality does not hold, then one coalition is strong enough to reduce the game to 2-person game with essentially the same bargaining set. The weak player gets 0.

REMARK 2. The conditions (6.12) are necessary and sufficient for the existence of a c.r.p.c. $(x_1, x_2, x_3, 0; 123, 4)$ such that

$$x_1, x_2, x_3 \geq 0, \qquad x_1 + x_2 \geq b, \qquad x_1 + x_3 \geq c, \qquad x_2 + x_3 \geq d.$$

$$(6.13)$$

Obviously, such p.c.'s cannot be objected against. However, in any c.r.p.c., in which player 4 is in a 3-person coalition and receives more than 0, there exists an objection against player 4 which cannot be countered. Thus, the coalition 123 "dictates" everything; this is why player 4 cannot claim more than 0.

REMARK 3. The following *ad hoc* rule serves for the discrete case: The value of each coalition is equally divided among its members. If a person enters a coalition he gets the sum of "his shares" minus the sum of the "shares" which his partners get from coalitions which do not include him. For example: The first player's shares are $a/3, b/3, c/3$. If he is entering the coalition 123, his partners have their shares $d/3, d/3$ from the coalition 234, which is that coalition that does not contain player 1; therefore, this player gets

$$\frac{a}{3} + \frac{b}{3} + \frac{c}{3} - \frac{d}{3} - \frac{d}{3} \qquad (6.14)$$

if he enters the coalition 123.

The same rule applies also to the 3-person game, with 1- and 2-person coalitions, in the discrete case.

REMARK 4. The discrete case exhibits a game with a nonnegative "3-quota." Each player always receives his "quota" in the bargaining set if he succeeds in becoming a member of a 3-person coalition. Quota games are treated in [5], and the results of this section are further generalized in [6].

§7. EXISTENCE THEOREMS. COUNTEREXAMPLES

DEFINITION 7.1. A permissible coalition in a game Γ will be called *effective*, if it is possible to divide its value among its members in such a way that no permissible subcoalition can alone make more.

Condition (5.2), for example, is a necessary and sufficient condition that the coalition 123 will be effective in the game treated in Section 5.

Clearly, we can assume that all subsets of N are permissible coalitions and that those having a positive value are effective, since we are dealing only with c.r.p.c.'s. The zero-valued coalitions will be called *trivial coalitions*.

The first question which may arise is whether each partition of the players, in which the only trivial coalitions are 1-person coalitions, is represented in $\mathcal{M}$. The answer is *no*.

EXAMPLE 7.1. $n = 5$, the nontrivial coalitions are 12, 35, 134, 2345, with values:

$$v(12) = 10, \qquad v(35) = 85, \qquad v(134) = 148, \qquad v(2345) = 160.$$
$$(7.1)$$

Consider the coalitionally rational payoff configuration

$$(\alpha, \beta, 0, 0, 0; 12, 3, 4, 5), \tag{7.2}$$

where, of course, $0 \leq \alpha \leq 10$, $\alpha + \beta = 10$. Now, player 1 can object by

$$(11, 0, 29, 108, 0; 134, 2, 5). \tag{7.3}$$

This objection is *justified*—i.e., player 2 has no counterobjection—if $\alpha < 10$. Indeed, any attempt of player 2 to keep his positive share β will end with a coalitionally nonrational p.c. Thus, (7.2) can belong to $\mathcal{M}$ only if $\alpha = 10$, $\beta = 0$. But this case is also ruled out, since now player 2 has a justified objection: $(0, 1, 100, 44, 15; 1, 2345)$.

Let Γ be a game, some of the values of the coalitions of which are positive. Is it possible that no p.c. belongs to $\mathcal{M}$ unless all the players get zero? In other words—is it possible that, in spite of some coalitions having a positive value, it would be worthless to enter into such coalitions, if one insists on the stability demanded by the definition of $\mathcal{M}$? This, in fact, may happen as the following example shows.

EXAMPLE 7.2.[6]

$$v(12b) = 1, \quad b = 3, 4, 5, 6.$$
$$v(1ab) = 1, \quad a = 3, 4; \quad b = 5, 6.$$
$$v(2pq) = 1, \quad p = 3, q = 4 \quad \text{or} \quad p = 5, q = 6.$$
$$v(3456) = 1,$$
$$v(B) = 1, \quad \text{B contains at least one of the}$$
$$\qquad\qquad\qquad \text{above-mentioned coalitions}$$
$$v(B) = 0, \quad \text{otherwise.}$$

It is a long but easy computation to verify that for this game $(\boldsymbol{x};$ $\mathcal{B}) \in \mathcal{M}$ implies $x_i = 0$, $i = 1, 2, \ldots, 6$.

The following theorem might be helpful in gaining some more insight into the nature of the bargaining set $\mathcal{M}$.

THEOREM 7.1. *Let Γ be an n-person game, in which 12 is a permissible coalition. Let $\mathcal{B}^0 \equiv 12$, $B_2, \ldots, B_m$ be a fixed partition. Let $(\boldsymbol{x}, \mathcal{B}^0) \equiv (x_1, x_2, \ldots, x_n; \mathcal{B}^0)$ be a c.r.p.c. and let J be the set of all the numbers σ_1, $0 \le \sigma_1 \le v(12)$, such that player 1 has a justified objection[7] against player 2, in $(\sigma_1, v(12) - \sigma_1, x_3, x_4, \ldots, x_n; \mathcal{B}^0)$; then J is an open set relative to the closed interval $[0, v(12)]$.*

PROOF. If

$$(x_1, x_2, \ldots, x_n; 12, B_2, \ldots, B_m) \tag{7.5}$$

is a coalitionally rational payoff configuration, then so is also

$$(x_1 + \epsilon, x_2 - \epsilon, x_3, \ldots, x_n; 12, B_2, \ldots, B_m), \tag{7.6}$$

provided that $-x_1 \le \epsilon \le v(12) - x_1$.

If $x_1 \in J$, then $\delta \equiv v(12) - x_1 > 0$, since otherwise player 2 can counterobject by playing alone.

---

[6] This game was given by von Neumann and Morgenstern [9], pp. 467–469, as an example of a simple game which is not a weighted majority game and for which no main simple solution exists.

[7] By the term "a justified objection" we mean an objection which has no counterobjection.

Let $(\mathscr{y}; \mathscr{C})$ be an objection of player 1 against player 2; then $y_1 > x_1$. Let z_2 be the maximum that player 2 can get by joining a coalition such that his partners (if such exist) get what they are supposed to get in a counterobjection. Obviously such a maximum exists, and $z_2 < x_2$, because $x_1 \in J$. Choose ϵ such that

$$-x_1 \leq \epsilon < \min(\delta, y_1 - x_1, x_2 - z_2); \qquad (7.7)$$

then $x_1 + \epsilon$ will also belong to J. Indeed, (7.6) will be coalitionally rational; $(\mathscr{y}; \mathscr{C})$ will remain an objection which is justified.

Thus, if $x_1 \in J$, then so are all the points on the interval $[0, x_1 + \epsilon]$.

THEOREM 7.2. *Let Γ be an n-person game, in which* 12 *is a permissible coalition and all the permissible coalitions are* 1-, 2-, *and* 3-*person coalitions; then, if $(x; \mathscr{B}^0)$ is a c.r.p.c., there exists a c.r.p.c.*

$$(\xi_1, \xi_2, x_3, x_4, \ldots, x_n; 12, B_2, \ldots, B_m) \qquad (7.8)$$

such that neither player 1 *nor player* 2 *has any justified objection. Here* $\mathscr{B}^0 \equiv 12, B_2, \ldots, B_m$.

PROOF. We proved in Theorem 7.1 that the numbers x_1, for which player 1 has a justified objection, form an open set T_1 with respect to $[0, v(12)]$. Similarly, the numbers x_1 for which player 2 has a justified objection form an open set T_2 with respect to the same interval ($x_3, \ldots, x_n$ remain fixed). We shall show that T_1 and T_2 are disjoint, from which it will follow that there is a point ξ_1 in $[0, v(12)]$, which is neither in T_1 nor in T_2, and therefore (7.8) will satisfy the requirements. (None of the sets is the closed interval because $v(12) \notin T_1$, $0 \notin T_2$.)

Indeed, suppose that

$$(\sigma_1, \sigma_2, x_3, \ldots, x_n; 12, B_2, \ldots, B_m) \qquad (7.9)$$

is a c.r.p.c. in which both players have justified objections. Player 1, in his objection, must join a coalition C which contains more than one person and does not contain player 2. Similarly, player 2 must join, in his objection, a coalition D which consists of more than one person and does not contain player 1. If $C \cap D = \emptyset$, then player 2's objection can serve as a counterobjection for player 1's objection, the latter being therefore not justified. If $C \cap D \equiv E \neq \emptyset$, then E contains one or two members. Without loss of generality, we can assume that the total amount that the players in E got in player 2's objection was not less than what they got from player 1's objection. If E contains one member, player 2's objection is a counterobjection to player 1's objection. If E contains two members, this is not always true, because in order to counterobject, player 2 has to modify, perhaps, his payments to the members of E. By doing so, a payoff configuration may result, which is not coalitionally rational; i.e., one player in E and player 2 can now make more by together forming a coalition. But if this is the case, then this coalition can serve in a counterobjection; e.g., by player 2 taking σ_2 for himself and letting the other player get the rest.

REMARK. The theorem fails to hold if we remove the restriction on the the the number of the players in the permissible coalitions. A counterexample is provided in Example 7.1.

Application. Suppose that each one of two men got a license to build a gasoline station. Each one considers the possibility of taking at most two partners. They expect various profits from the corresponding possible coalitions. The other partners do not have licenses. Of course, the two men consider also their joint coalition. Under these assumptions, Theorem 7.2 says that the coalition of the two licensees is represented in the bargaining set.

§8. THE 4-PERSON GAME IN WHICH ONLY 1- AND 2-PERSON COALITIONS ARE PERMISSIBLE

The inequalities which determine under what condition $(x_1, x_2, x_3, x_4; 12, 34)$ in $\mathcal{M}$, for the game:

$$v(1) = v(2) = v(3) = v(4) = 0, \qquad v(12) = a, \qquad v(23) = b,$$

$$v(34) = c, \qquad v(13) = d, \qquad v(24) = e, \qquad v(14) = f, \quad (8.1)$$

$$a, b, c, d, e, f \geq 0,$$

are given in Appendix 2.

Theorem 7.2 ensures that any partition which contains only one 2-person coalition is represented in $\mathcal{M}$. We shall now study the case of partitions into two couples. Our object is to prove that some such partitions appear in $\mathcal{M}$. It turns out that this can be proved even if we limit ourselves to *maximal* partitions, i.e., to those partitions in which the sum of the values of the coalitions is maximal. This restriction helps us by reducing the number of inequalities which need to be examined.

THEOREM 8.1. *Let* Γ *be the game* (8.1), *where*

$$a + c \geq d + e, \qquad a + c \geq b + f; \tag{8.2}$$

then there always exists a p.c. $(x_1, x_2, x_3, x_4; 12, 34)$ *in the bargaining set* $\mathcal{M}$.

PROOF. We omit the calculations, but state the various cases.
Case A. If

$$a \leq b + d, \qquad b \leq a + d, \qquad d \leq a + b, \qquad 2c \geq b + d - a, \tag{8.3}$$

then

$$\left(\frac{a + d - b}{2}, \frac{a + b - d}{2}, \frac{d + b - a}{2}, \frac{2c + a - b - d}{2}; 12, 34 \right) \in \mathcal{M}. \tag{8.4}$$

If

$$a \leq b + d, \qquad b \leq a + d, \qquad d \leq a + b, \qquad 2c < b + d - a, \tag{8.5}$$

then

$$\left(\frac{a + d - b}{2}, \frac{a + b - d}{2}, c, 0; 12, 34 \right) \in \mathcal{M}. \tag{8.6}$$

Case B. If

$$a > b + d, \qquad c > d + f, \tag{8.7}$$

then

$$(d, a - d, 0, c; 12, 34) \in \mathcal{M}. \tag{8.8}$$

Case C. If

$$a > b + d, \qquad f > c + d, \qquad b + c \geq e, \tag{8.9}$$

then

$$(f - c, a + c - f, 0, c; 12, 34) \in \mathcal{M}. \tag{8.10}$$

(Indeed, (8.9) and (8.2) imply $c \geq e - b \geq e - (a + c - f)$, hence $a +$
$2c \geq e + f$. The rest follows directly.) If

$$a > b + d, \qquad f > c + d, \qquad e > b + c, \qquad (8.11)$$

then

$$\left(\frac{a + f - e}{2}, \frac{a + e - f}{2}, 0, c; 12, 34 \right) \in \mathcal{M}. \qquad (8.12)$$

(Indeed, (8.2) and (8.11) imply $2d + e \leq d + a + c < a + f$. Also $2b +$
$f \leq b + a + c < a + e$.)

   Case D. If

$$a > b + d, \qquad d > f + c, \qquad b + c \geq e, \qquad (8.13)$$

then

$$(a - b, b, 0, c; 12, 34) \in \mathcal{M}. \qquad (8.14)$$

If

$$a > b + d, \qquad d > f + c, \qquad e > b + c, \qquad (8.15)$$

then

$$(a + c - e, e - c, 0, c; 12, 34) \in \mathcal{M}. \qquad (8.16)$$

   Case E. If

$$d > a + b, \qquad d > c + f, \qquad (8.17)$$

then

$$(a, 0, d - a, c + a - d; 12, 34) \in \mathcal{M}. \qquad (8.18)$$

All other cases are either not maximal partitions, or they can be
reduced to these cases by permuting the players:

$$1 \leftrightarrow 3, \qquad 2 \leftrightarrow 4. \qquad (8.19)$$

§9. The Restricted Bargaining Set

In a given game there are in general many stable p.c.'s. Though we do
not possess a criterion for choosing between them, there are cases in
which it is clear that some p.c.'s in $\mathcal{M}$ are "better" than others. We

therefore suggest that the latter should be deleted from $\mathcal{M}$, thus giving rise to the restricted bargaining set $\mathcal{M}^*$.

A p.c. $(x, \mathcal{B})$ in $\mathcal{M}$ should be deleted if one of the following cases occurs:

(i) There exists in $\mathcal{M}$ a p.c. $(x^*; \mathcal{B}^*)$ with

$$x_i^* > x_i; \qquad i = 1, 2, \ldots, n. \tag{9.1}$$

(ii) There exists in $\mathcal{M}$ a p.c. $(x^{**}; \mathcal{B}^{**})$, where the coalitions by $\mathcal{B}^{**}$ are unions of coalitions in $\mathcal{B}$, such that

$$x_i^{**} > x_i \tag{9.2}$$

for all the players i which belong to a union of more than one coalition of $\mathcal{B}$ and

$$x_i^{**} \geq x_i \tag{9.3}$$

for all the other players.

One sees that in the examples given in the previous sections, only those coalitions which have relatively big values (if such exist) will appear in $\mathcal{M}^*$.

§10. POSSIBLE MODIFICATIONS

Inasmuch as our theory tries to cope with "reality," it is flexible enough to allow for some modifications.

For instance, if players are faced with the game treated in Example 7.2, they may claim that the demand for stability is too strong. They would rather relax this demand and still gain something from the game.

One can then offer them the following definition of a bargaining set $\mathcal{M}_1$:

DEFINITION 10.1. A c.r.p.c. $(x; \mathcal{B})$ belongs to the bargaining set $\mathcal{M}_1$, if for any objection K against L, there is somebody in L who can counterobject.

According to this definition, each player in a coalition B_j which contains K, who does not belong to the partners of K, is required to be able to counterobject; but several such players may perhaps be unable to protect their shares simultaneously. Clearly, the resulting bargaining set $\mathcal{M}_1$ *includes* $\mathcal{M}$, since the number of sets which is required to counterobject is reduced. In this case, e.g., the players of the game treated in Example 7.2 may agree to

$$(\tfrac{1}{3}, \tfrac{1}{3}, \tfrac{1}{3}, 0, 0, 0; 123, 4, 5, 6) \tag{10.1}$$

which belongs to $\mathcal{M}_1$.

In some other real-life cases, one can estimate and tell in advance which coalitions may object and which coalitions may counterobject. This leads to various bargaining sets and brings us to the circle of ideas surrounding ψ-stability. (See Luce and Raiffa [3], pp. 163–168, 174–176, 220–236.)

One may limit K to be always one-person and L to be the remaining members of the coalition, except for K's partners. This type of stability of one against the rest, which generates a bargaining set $\mathcal{M}_2$, is still different from the stability demanded in $\mathcal{M}$ as the following example shows:

EXAMPLE 10.1. Consider the game:

$$n = 5, \quad v(i) = 0, \quad v(123) = 30, \quad v(24) = v(35) = 50,$$
$$v(1245) = v(1345) = 60. \tag{10.2}$$

Let

$$(x; \mathcal{B}) = (10, 10, 10, 0, 0; 123, 4, 5). \tag{10.3}$$

If $K = 1$, 2 or 3, then the remaining players which belong to the same coalition and are not among his partners can always counterobject; but the objection for $K = 23$,

$$(0, 11, 11, 39, 39; 1, 24, 35), \tag{10.4}$$

has no counterobjection because player 1 cannot keep his profits. Thus $(x; \mathcal{B}) \in \mathcal{M}_2$ but $\notin \mathcal{M}$.

It is easy to show that $\mathcal{M} \subset \mathcal{M}_2$.

Sometimes people would like to feel safe not only within their coalitions but also from "outside" threats. It may happen, e.g., that several players from *various coalitions* will threaten together other people from these coalitions. A reasonable way to cope with this strong demand for stability would be to allow K and L to belong to several coalitions, provided that K and L are required to intersect the same coalitions. This will bring us to a bargaining set $\mathcal{M}_0$ which is *included* in $\mathcal{M}$. Let us remark that $\mathcal{M}_0 = \mathcal{M}$ for the 2- and 3-person games, as well as for the 4-person game with only 1-, 3-, and 4-person coalitions permissible. If $n = 4$, where 1- and 2-person coalitions are permissible, one has to replace the inequalities of Appendix 2 by those listed in Appendix 3. Fortunately, these inequalities are satisfied in all the examples given in Section 8, and therefore Theorem 8.1 is valid if one replaces $\mathcal{M}$ by $\mathcal{M}_0$.

Finally, we would like to question the assumption of coalitional rationality. If we drop this condition, we may arrive a negative values in the bargaining set, but this does not have to bother us, since $(0, 0, \ldots, 0; 1, 2, \ldots, n)$ will certainly remain in the bargaining set, and therefore we can *demand* that the *restricted* bargaining set will contain only individually rational p.c.'s. However, we shall show in Example 10.2 that the resulting restricted bargaining set may still contain non-coalitionally rational p.c.'s.

EXAMPLE 10.2. Let Γ be the game

$$v(i) = 0, \quad v(12) = v(45) = v(46) = v(56) = v(123) = 30,$$

$$\tag{10.5}$$

$$v(34) = 10.$$

In this game, the non-coalitionally rational p.c.

$$(10, 10, 10, 0, 15, 15; 123, 4, 56) \tag{10.6}$$

is stable if one drops the condition of coalitional rationality. In fact, it then belongs to the restricted bargaining set, since otherwise there exists a p.c. $(x; \mathcal{B})$ in the bargaining set with

$$\sum_{i=1}^{6} x_i > 60. \tag{10.7}$$

This can only occur if the coalition 34 is formed. Since, in addition, $x_3 \geq 10$, $x_4 \geq 0$, player 3 gets 10. This is impossible because in this case, player 4 has a justified objection, due to the fact that $x_1 + x_2 = 30$.

We have thus shown that the restricted bargaining set may contain a non-coalitionally rational p.c., if this condition is dropped from the definition of the bargaining set.

§11. Concluding Remarks

Perhaps, the nearest to our theory is Vickrey's concept of self-policing patterns [8]. His objections—called "heretical imputations"—are similar to ours; however, his counterobjections—named "penalizing policy imputations"—are quite different.

Both the heretical and the penalizing policy imputations are in Vickrey's case *imputations*, whereas this is not the case in our theory. His penalizing policing imputation insists that at least one member of the "heretical coalition" is punished, whereas we only demand that the set L will be able to hold on to its property. However, the main difference lies, perhaps, in the fact that Vickrey is looking for a *set* of imputations—"self-policing patterns"—which are stable as a whole,[8] while our bargaining set consists of payoff configurations, each on of which is stable in itself.

It has been pointed out in Section 10 that if a lack of communication is known to exist, one can incorporate ideas from ψ-stability theory into ours. Both theories stress the dependence of an outcome on the coalition structure which actually forms. However, ψ-stability theory requires the payoffs to the imputations,[9] whereas we require that the outcome satisfies (2.4). (See, e.g., Luce and Raiffa [3], p. 222.). The coalition rationality requirement (2.5) is a special case of a ψ-stability requirement, if one requires that all the subsets of the coalitions in the coalition structure τ are values of $\psi(\tau)$. A similar requirement appears also in Milnor's class L of reasonable outcomes ([3], pp. 240–242).

In many practical situations, the characteristic function is not the best way to describe a game. It would rather be better to apply the "Thrall characteristic function" (see Thrall [7]), which associates with each *coalition-structure* a value for each coalition appearing in that structure. One can try to define the concepts of objections and counterobjections, for such cases, and it is possible to do so in various ways.

It is also possible to apply the notions described in this paper to the Aumann-Peleg characteristic function for cooperative games without side payments [2], essentially without change.

Finally, we should like to point out that our theory gives in many cases answers similar to those appearing in classical theories. Thus, e.g., the bargaining set in the discrete case of the 3-person non-zero-sum

---

[8] In particular, he looks for "strong solutions."
[9] Or at least feasible individually rational n-tuples ([3], p. 226).

game[10] consists essentially of the "central" three points of the nondiscriminatory von Neumann-Morgenstern solution (see [9], pp. 550–554), but does not contain the additional "wiggles" that occur in their solution. The bargaining set for the nondiscrete case is essentially the core. This suggests a pattern in which the bargaining set forms the "central" or "intuitive" part of a von Neumann-Morgenstern solution, whereas the "complications" disappear.

APPENDIX 1

Let Γ be a 4-person game, the coalitions of which, and their values, are given by (6.1). In order that the pair $(x_1, x_2, x_3, 0; 123, 4)$ belongs to the bargaining set $\mathcal{M}$, it is necessary and sufficient that

$$0 \leq x_1, \qquad 0 \leq x_3, \qquad x_1 + x_3 \leq a, \qquad x_2 = a - x_1 - x_3, \quad (A1.1)$$

and that at least one inequality (or equality) in each of the following lines should be satisfied.

$x_1 + x_3 \geq c$	$2x_1 + x_3 \geq a + c - d$	$x_1 + x_3 = a$
$x_3 \leq a - b$	$x_3 - x_1 \leq d - b$	$x_3 = 0$
$x_1 \leq a - d$	$2x_1 + x_3 \leq a + c - d$	$x_1 = 0$
$x_3 \leq a - b$	$x_1 + 2x_3 \leq a + c - b$	$x_3 = 0$
$x_1 + x_3 \geq c$	$x_1 + 2x_3 \geq a + c - b$	$x_1 + x_3 = a$
$x_1 \leq a - d$	$x_3 - x_1 \geq d - b$	$x_1 = 0$

APPENDIX 2

Let Γ be a 4-person game, the coalitions of which, and their values, are given in (8.1). In order that the pair $(x_1, x_2, x_3, x_4; 12, 34)$ belong to the bargaining set $\mathcal{M}$, it is necessary and sufficient that

$$0 \leq x_1 \leq a, \qquad 0 \leq x_3 \leq c, \qquad x_1 + x_2 = a, \qquad x_3 + x_4 = c, \quad (A2.1)$$

and that at least one inequality (or equality) in each line be satisfied.

If the partition 12, 34 is maximal (in the sense of (8.2)), the last column can be omitted.

---

[10] Clearly, one has to modify the characteristic function in the obvious way so as to get superadditivity.

$x_1 = a$	$x_1 + x_3 \geq d$	$2x_1 \geq a + d - b$	$x_1 + x_3 \geq a + c - e$
$x_1 = a$	$x_1 - x_3 \geq f - c$	$2x_1 \geq a + f - e$	$x_1 - x_3 \geq a - b$
$x_1 = 0$	$x_1 - x_3 \leq a - b$	$2x_1 \leq a + d - b$	$x_1 - x_3 \leq f - c$
$x_1 = 0$	$x_1 + x_3 \leq a + c - e$	$2x_1 \leq a + f - e$	$x_1 + x_3 \leq d$
$x_3 = c$	$x_1 + x_3 \geq d$	$2x_3 \geq c + d - f$	$x_1 + x_3 \geq a + c - e$
$x_3 = c$	$x_1 - x_3 \leq a - b$	$2x_3 \geq c + b - e$	$x_1 - x_3 \leq f - c$
$x_3 = 0$	$x_1 - x_3 \geq f - c$	$2x_3 \leq c + d - f$	$x_1 - x_3 \geq a - b$
$x_3 = 0$	$x_1 + x_3 \leq a + c - e$	$2x_3 \leq b + c - e$	$x_1 + x_3 \leq d$

APPENDIX 3

The following inequalities replace those given in Appendix 2, if one desires that $(x_1, x_2, x_3, x_4; 12, 34)$ shall belong to $\mathcal{M}_0$. (See Section 10.) Again, at least one inequality in each line should be satisfied, as well as those given in (A2.1).

$x_1 + x_3 \geq d$	$x_1 = a$ $2x_3 \geq c + d - f$	$x_3 = c$ $2x_1 \geq a + d - b$	$x_1 + x_3 \geq a + c - e$
$x_1 - x_3 \geq f - c$	$x_1 = a$ $2x_3 \leq c + d - f$	$x_3 = 0$ $2x_1 \geq a + f - e$	$x_3 - x_1 \leq b - a$
$x_1 - x_3 \leq a - b$	$x_1 = 0$ $2x_3 \geq b + c - e$	$x_3 = c$ $2x_1 \leq a + d - b$	$x_1 - x_3 \leq f - c$
$x_1 + x_3 \leq a + c - e$	$x_1 = 0$ $2x_3 \leq b + c - e$	$x_3 = 0$ $2x_1 \leq a + f - e$	$x_1 + x_3 \leq d$

BIBLIOGRAPHY

[1] Aumann, R. J. and Maschler, M. "An equilibrium theory for n-person cooperative games," *American Math. Soc. Notices*, Vol. 8 (1961), p. 261.

[2] Aumann, R. J. and Peleg, B. "von Neumann–Morgenstern solutions to cooperative games without side payments," *Bull. Amer. Math. Soc.* 66 (1960), pp. 173–179.

[3] Luce, R. D. and Raiffa, H. *Games and Decisions.* New York: John Wiley and Sons, 1957.

[4] Maschler, M. "An experiment on n-person cooperative games," *Recent Advances in Game Theory, Proceedings of a Princeton University Conference, October 1961.* Privately printed for members of the conference (1962), pp. 49–56.

[5] Maschler, M. "Stable payoff configurations for quota games." in this study.

[6] Maschler, M. "n-Person games with only 1, $n - 1$, and n-person permissible coalitions," *J. Math. Analysis and Applications*, Vol. 6 (1963), pp. 230–256.

[7] Thrall, R. M. "Generalized characteristic functions for n-person games," *Recent Advances in Game Theory, Proceedings of a Princeton University Conference, October 1961*. Privately printed for members of the conference (1962), pp. 157–160.

[8] Vickrey, W. "Self-policing properties of certain imputation sets," *Contributions to the Theory of Games*, Vol. IV; *Annals of Mathematics Studies*, No. 40, Princeton: Princeton University Press, 1959, pp. 213–246.

[9] von Neumann, J. and Morgenstern, O. *Theory of Games and Economic Behavior*, 2nd ed. Princeton: Princeton University Press, 1947.

Robert J. Aumann
Michael Maschler
Princeton University
The Hebrew University of Jerusalem

13

EXISTENCE OF COMPETITIVE EQUILIBRIA IN MARKETS WITH A CONTINUUM OF TRADERS[1]

ROBERT J. AUMANN

An appropriate model for a market with many individually insignificant traders is one with a continuum of traders. Here it is proved that competitive equilibria exist in such markets, even though individual preferences may not be convex. Such a result is not true for markets with finitely many traders.

1. INTRODUCTION

The problem of rigorously establishing the existence of a competitive equilibrium in a market was first brought to the attention of economists by Wald [11]. Since the appearance of his pioneering paper, other authors[2] have established the existence of competitive equilibria under various sets of assumptions. In all this work, it was invariably assumed that the traders have convex preferences.[3] Indeed, if this assumption is abandoned it is easy to give examples of markets that do not possess any competitive equilibria.

Attention has recently been called[4] to the possibility of dispensing with the convexity assumption if the market in question has a large number of traders, no individual one of whom can significantly affect the outcome of trading. In a heuristic, imprecise way it was argued that the preferences of a large number of individually insignificant traders

---

[1] Research partially supported by the Office of Naval Research, Logistics and Mathematical Statistics Branch, under Contract No. N62558-3586. Previous research connected with this paper was supported by the Carnegie Corporation of New York through the Econometric Research Program of Princeton University and by U.S. Air Force Project RAND.

[2] Such as Arrow-Debreu [1], Gale [7], and McKenzie [9].

[3] I.e., that the set of commodity bundles preferred or indifferent to a given bundle is convex.

[4] See the articles by Bator, Farell, Koopmans, and Rothenberg in the *Journal of Political Economy*: (Vol. 67, 1959, pp. 377–391; Vol. 68, 1960, pp. 435–468; Vol. 69, 1961, pp. 478–493).

would have a convex effect in the aggregate, even if none of the individual preferences were convex. A rigorous treatment of this theme was given very recently by Shapley and Shubik [10], though not directly in connection with the competitive equilibrium. Their work will be discussed in Section 8.

In a previous paper [2], we suggested that the most appropriate model for a market with many individually insignificant traders is one with a continuum of traders. Analogous models are used in physics, for example, when the large number of particles in a fluid are replaced for mathematical convenience by a continuum of particles. This raises the question of whether it would be possible to establish the existence of competitive equilibria in markets with a continuum of traders, even when the preferences need not be convex. The purpose of this paper is to give an affirmative answer to that question, and thus to underscore the power and scope of the continuum-of-traders approach to market theory.

We remark that the concept of competitive equilibrium is generally agreed to be significant only in a market with "perfect competition," i.e., one with a large number of individually insignificant traders. The concept makes no sense for a small number of traders. Thus, we show here that when competitive equilibria are at all relevant, convex preferences are not needed to establish their existence.

The proof is based on McKenzie's beautiful existence proof [9] for competitive equilibria in finite markets. Major modifications are required, however, because of the presence of a continuum of traders (which necessitates the use of Banach-space methods) and the nonavailability of convex preferences.

In Section 2 we give a precise statement of the model and the main theorem. Section 3 is devoted to the statement of an auxiliary theorem. In Section 4 the proof of the auxiliary theorem is outlined, and in Section 5 it is completed. In Section 6 the main theorem is deduced from the auxiliary theorem.

Section 7 is devoted to a detailed comparison of our proof with McKenzie's, and Section 8 to a discussion of the relation of our current result to that of our previous paper [2] and to the Shapley-Shubik results [10].

Our result concerns true markets only, i.e., pure exchange economies. Presumably it can be extended to economies with production (at least if one assumes constant returns to scale), but we have not done this.

2. MATHEMATICAL MODEL AND STATEMENT OF MAIN THEOREM

We shall be working in a Euclidean space E^n; the dimensionality n of the space represents the number of different commodities being traded in the market. Superscripts will be used exclusively to denote coordinates. Following standard practice, for x and y in E^n we take $x > y$ to mean $x^i > y^i$ for all i; $x \geq y$ to mean $x^i \geq y^i$ for all i; and $x \geqq y$ to mean $x \geq y$ but not $x = y$. The integral of a vector function is to be taken as the vector of integrals of the components. Superscripts will be used exclusively to denote coordinates. The scalar product $\sum_{i=1}^{n} x^i y^i$ of two members x and y of E^n is denoted $x \cdot y$. The symbol 0 denotes the origin in E^n as well as the real number zero; no confusion will result. The symbol $\setminus$ will be used for set-theoretic subtraction, whereas $-$ will be reserved for ordinary algebraic subtraction.

A *commodity bundle x* is a point in the nonnegative orthant Ω of E^n. The set of *traders* is the closed unit interval $[0, 1]$; it will be denoted T. The words "measure," "measurable," "integral," and "integrable" are to be understood in the sense of Lebesgue. All integrals are with respect to the variable t (which stands for trader), and in most cases the range of integration is all of T. In an integral we will therefore omit the symbol dt and the indication of dependence of the integrand on t, and will specifically indicate the range of integration only when it differs from all of T. Thus $\int x$ means $\int_T x(t) \, dt$. A *null set* is a set of measure 0. *Null sets of traders are systematically ignored throughout the paper.* Thus a statement asserted for "all" traders, or "each" trader, or "each" trader in a certain set, is to be understood to hold for all such traders except possibly for a null set of traders.

An *assignment* (of commodity bundles to traders) is an integrable function on T to Ω. There is a fixed *initial assignment i; intuitively, $i(t)$ is the bundle with which trader t comes to market. We assume

$$\int i > 0. \tag{2.1}$$

Intuitively, this asserts that no commodity is totally absent from the market.

For each trader t there is defined on Ω a relation $\succsim_t$ called *preference-or-indifference*. This relation is assumed to be a *quasi-order*, i.e., transitive, reflexive, and complete.[5] From $\succsim_t$ we define relations $\succ_t$ and $\sim_t$ called *preference* and *indifference*, respectively, as follows:

$$x \succ_t y \quad \text{if} \quad x \succsim_t y \quad \text{but not} \quad y \succsim_t x;$$

$$x \sim_t y \quad \text{if} \quad x \succsim_t y \quad \text{and} \quad y \succsim_t x.$$

The following assumptions are made:

Desirability (*of the commodities*): $x \geq y$ implies $x \succ_t y$.   (2.2)

Continuity (*in the commodities*): For each $y \in \Omega$, the sets $\{x : x \succ_t y\}$ and $\{x : y \succ_t x\}$ are open (relative to Ω).   (2.3)

Measurability: If x and y are assignments, then the set $\{t : x(t) \succ_t y(t)\}$ is measurable.   (2.4)

The intuitive content of these assumptions should be fairly clear from their names. Note that together with the assumption that $\succsim_t$ is a quasi-order, the continuity assumption (2.3) yields the existence of a continuous utility function $\nu_t(x)$ on Ω for each fixed trader t [4]. Then the measurability assumption[6] (2.4) says that the ν_t can be chosen so that $\nu_t(x)$ is simultaneously measurable in t and x.

An *allocation* is an assignment x such that $\int x = \int i$. A *price vector* is a member p of R^n such that $p \geq 0$; though it is in Ω, it should not be thought of as a commodity bundle. A *competitive equilibrium* is a pair consisting of a price vector p and an allocation x, such that for all traders t, $x(t)$ is a maximal with respect to $\succ_t$ in the "budget set" $B_p(t) = \{x \in \Omega : p \leq p \cdot i(t)\}$.

MAIN THEOREM: *Under the conditions of this section, there is a competitive equilibrium.*

3. STATEMENT OF AUXILIARY THEOREM

To prove the main theorem, we first establish an auxiliary theorem, which has some interest in its own right. Let us define a *market $\mathscr{M}$* to consist of a positive integer n (the number of commodities), an initial

---

[5] A relation $\mathscr{R}$ is called *transitive* if $x\mathscr{R}y$ and $y\mathscr{R}z$ imply $x\mathscr{R}z$; *reflexive* if $x\mathscr{R}x$ for all x; and *complete* if for all x and y, either $x\mathscr{R}y$ or $y\mathscr{R}x$.

[6] In this context (but not in [2]), the measurability assumption is equivalent to the assumption that $\{t : x \succ_t y\}$ is measurable for all x and y in Ω.

assignment i, and preference-or-indifference relation $\succsim_t$ on Ω for each of the traders t. The markets that we consider here differ from those described in the previous section in a number of ways. First, condition (2.1) on the initial assignments is strengthened to read

$$i(t) > 0 \quad \text{for all } t. \tag{3.1}$$

This means that a positive amount of each commodity is initially held by each trader.

Second, a bundle x is said to *saturate*, or more explicitly, to *saturate trader t's desire*, if $x \succsim_t y$ for all $y \in \Omega$. Assumption 2.2 is weakened to read as follows:

Weak Desirability: Unless y saturates, $x > y$ implies $x \succ_t y$. (3.2)

Notice that this is a double weakening of (2.2); the hypothesis $x \geq y$ is replaced by $x > y$, and allowance is made for saturation (saturation is impossible under (2.2)).

Third, under the auxiliary theorem we do not only permit saturation, we specifically require it. Let v be an assignment. We say that trader t's desire is *commodity-wise saturated* at $v(t)$ if for all bundles x and commodities i such that $x^i \geq v^i(t)$, we have

$$x \sim_t (x^1, \ldots, x^{i-1}, v^i(t), x^{i+1}, \ldots, x^n).$$

In other words, changing the value of the ith coordinate above $v^i(t)$ does not change the indifference level. Intuitively, this means that desire for the ith commodity is saturated when the quantity of that commodity is $v^i(t)$, although trader t may still want more of other commodities j, of which he holds less than $v^j(t)$. To rephrase the condition, let $V(t) = \{x \in \Omega : x \leq v(t)\}$ be the "hyper-rectangle" of bundles that are $\leq v(t)$, and define a mapping v_t from Ω into $V(t)$ as follows: $v_t(x)$ is the bundle formed from x by replacing by $v^i(t)$ all coordinates x^i of x that exceed $v^i(t)$. Then commodity-wise saturation at $v(t)$ asserts that $v_t(x) \sim_t x$. It follows that the entire preference order is determined by its behavior in the hypercube $V(t)$ since $x \succsim_t y$ if and only if $v_t(x) \succsim_t v_t(y)$. A preference order with commodity-wise saturation is illustrated in Figure 1.

The existence of a $v(t)$ that commodity-wise saturates desire is intuitively very acceptable; it simply means that there is an upper bound

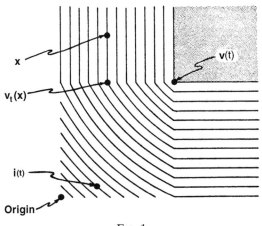

x

$v_t(x)$

i(t)

Origin

v(t)

FIG. 1

on the amount of a commodity that can be profitably used by an individual, no matter what other commodities are or are not available. The demand that v be an assignment, i.e., integrable, means that "the market as a whole can be commodity-wise saturated"; more precisely, it means that there is a bundle (namely $\int v$) that can be distributed among the traders in such a way as to commodity-wise saturate each trader's desire. We now assume

> There is an assignment v such that each trader t's desire is commodity-wise saturated at $v(t)$. (3.3)

Finally, we need the following assumption:

> *Saturation restriction*: x cannot saturate unless $x > i(t)$.   (3.4)

AUXILIARY THEOREM: *Let $\mathcal{M}$ be a market satisfying the assumptions of this section as well as (2.3) and (2.4). Then $\mathcal{M}$ has a competitive equilibrium.*

4.   OUTLINE OF THE PROOF OF THE AUXILIARY THEOREM

The starting point of the proof is the *preferred set* $C_p(t)$, defined for each trader t and each price vector p to be the set of commodity bundles preferred or indifferent to all elements of the budget set $B_p(t)$; for-

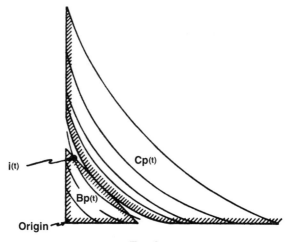

FIG. 2

mally,

$$C_p(t)\{x \in \Omega: \quad \text{for all} \quad y \in \mathbf{B}_p(t), x \succsim_t y\}$$

(see Figure 2). Next, define

$$\int C_p = \left\{\int x: x \text{ is an assignment such that } x(t) \in C_p(t) \quad \text{for all } t\right\};$$

this is called the *aggregate preferred set*. $\int C_p$ is the set of all aggregate
bundles that can be distributed among the traders in such a way that
each trader is at least as satisfied as he is when he sells his initial bundle
and buys the best (by his standards) that he can with the proceeds, at
prices p.

Since we have made no convexity assumption on the preferences, the
individual preferred sets $C_p(t)$ need not be convex. The *aggregate*
preferred set $\int C_p$, on the other hand, *is* convex; as we shall see, that
fact holds only because there is a continuum of traders, and it consti-
tutes the nub of the proof. By using the convexity of the aggregate
preferred set $\int C_p$, we shall be able to show that there is a unique point
$c(p)$ in $\int C_p$ that is nearest to $\int i$; set $h(p) = c(p) - \int i$.

Let P be the simplex of price vectors normalized so that their sum is
1, i.e., $P = \{p \in \Omega: \Sigma_{i=1}^n p^i = 1\}$. The central idea of the proof is to use
h to construct a continuous function f from P to itself, and then to

apply Brouwer's fixed point theorem;[7] the resulting fixed point—denoted q—turns out to be an equilibrium price vector. The function f is defined by

$$f(p) = \frac{p + h(p)}{1 + \sum\limits_{i=1}^{n} h^i(p)}.$$

We shall show later that $h(p) \geq 0$. Therefore, the denominator in the definition of f does not vanish, and so $f(p) \in P$ for all $p \in P$. Suppose q is a fixed point of f. Then

$$q\left(1 + \sum\limits_{i=1}^{n} h^i(q)\right) = q + h(q),$$

i.e.,

$$h(q) = \alpha q, \tag{4.1}$$

where, because $h(p) \geq 0$,

$$\alpha = \sum\limits_{i=1}^{n} h^i(q) \geq 0.$$

We wish to show that

$$h(q) = 0. \tag{4.2}$$

Indeed, suppose (4.2) is false. From the definition of h and the convexity of $\int C_p$ it follows that for all p, the hyperplane through $h(p) + \int i$ perpendicular to $h(p)$ supports[8] $\int C_p$. Applying this for $p = q$, we obtain

$$\left(y - \int i\right) \cdot h(q) \geq h(q) \cdot h(q)$$

[7] Brouwer's theorem asserts that every continuous single-valued function f from P to itself has a fixed point, i.e., a point p such that $f(p) = p$. For a proof, see Dunford and Schwartz [5, Sec. V. 12, p. 468].

[8] This is a standard method of constructing a supporting hyperplane. An explicit proof is given by McKenzie [9, Lemma 7 (1), p. 61].

for all $y \in \int C_q$. Because (4.2) is false, $\alpha > 0$; so by (4.1), we obtain

$$\left(y - \int i\right) \cdot \alpha q \geqq \alpha^2 q \cdot q,$$

and hence

$$\left(y - \int i\right) \cdot q \geqq \alpha(q \cdot q) > 0 \quad \text{for all } y \in \int C_q. \tag{4.3}$$

Now if for each t we let $x(t)$ be a point in the budget set $B_q(t)$ that is maximal with respect to t's preference order, then on the one hand we have $(x(t) - i(t)) \cdot q \leqq 0$, and on the other hand $x(t) \in C_q(t)$. Hence, by integrating we obtain $(\int x - \int i) \cdot q \leqq 0$, and $\int x \in \int C_q$; this contradicts (4.3), and establishes (4.2).

Equation (4.2) says that $\int i \in \int C_q$, i.e., there is an assignment x such that $\int x = \int i$ and $x(t) \in C_q(t)$ for all t. Thus, x is an allocation, and $x(t)$ is preferred or indifferent to all elements of $B_q(t)$. To complete the proof that (q, x) is a competitive equilibrium, it is only necessary to show that $x(t)$ is in $B_q(t)$ for all t. Suppose now that $q \cdot x(t) < q \cdot i(t)$ for some t. Then x does not saturate (because of the saturation restriction (3.4)), and so from the desirability assumption (3.2), it follows that $x(t) + (\delta, \ldots, \delta) \succ_t x(t)$ for $\delta > 0$. But for δ sufficiently small, we shall still have

$$q \cdot (x(t) + (\delta, \ldots, \delta)) = q \cdot x(t) + \delta < q \cdot i(t),$$

so $x(t) + (\delta, \ldots, \delta) \in B_q(t)$; this contradicts $x(t) \in C_q(t)$. So $q \cdot x(t) < q \cdot i(t)$ is impossible, and we conclude that $q \cdot x(t) \geqq q \cdot i(t)$ for all t. If the $>$ sign would hold for some t, we could deduce $\int q \cdot x > \int q \cdot i$, contradicting $\int x = \int i$. So $q \cdot x(t) = q \cdot i(t)$ for all t, and it follows that $x(t) \in B_q(t)$ for all t. So (q, x) is a competitive equilibrium.

The foregoing proof, which follows McKenzie's ideas [9] rather closely, is incomplete in two respects: The required properties of $h(p)$—existence, uniqueness, continuity, and nonnegativity—have not been established; and it has not been shown that the x whose integral $\int x$ contradicts (4.3) may be chosen to be measurable. These items will be taken up in the next section.

5. COMPLETION OF THE PROOF OF THE AUXILIARY THEOREM

In this section we make considerable use of the theory of integrals of set-valued functions, as developed in [3]. Before starting the results from [3] that are used in the sequel, we recall the necessary definitions.

Let F be a function defined on T whose values are subsets of Ω. Define

$$\int F = \left\{ \int f : f \text{ is integrable and } f(t) \in F(t) \text{ for all } t \right\}.$$

F is called *Borel-measurable* if its *graph* $\{(x, t): x \in E^n, x \in F(t)\}$ is a Borel subset of $\Omega \times T$. F is called *integrably bounded* if there is an integrable point-valued function b such that for all t, $x \in F(t)$ implies $x \leq b(t)$. For each t, $F^*(t)$ denotes the convex hull of $F(t)$.

For each p in P, let F_p be a subset of Ω. F is said to be *upper-semicontinuous in p* if for each convergent sequence $p_1, p_2, \ldots$ in P and each convergent sequence $x_1, x_2, \ldots$ in Ω such that $x_1 \in F_{p_1}$, $x_2 \in F_{p_2}, \ldots$, we have $\lim x_k \in F_{\lim p_k}$. It is *lower-semicontinuous in p* if for each convergent sequence $p_1, p_2, \ldots$ in P, every point in $F_{\lim p_k}$ is the limit of a sequence $x_1, x_2, \ldots$ in Ω such that $x_1 \in F_{p_1}, x_2 \in F_{p_2}, \ldots$. It is *continuous* if it is both upper- and lower-semicontinuous.

If $F_1, F_2, \ldots$ are subsets of E^n, then $\limsup F_k$ is defined to be the set of all x in E^n such that every neighborhood of x intersects infinitely many F_k.

The following lemmas are proved in [3]:

LEMMA 5.1: $\int F$ *is convex.*

LEMMA 5.2: *If F is Borel-measurable, and $F(t)$ is nonempty for each t, then there is a measurable function f such that $f(t) \in F(t)$ for all t.*

LEMMA 5.3: *If $F_1, F_2, \ldots$ is a sequence of set-valued functions that are all bounded by the same integrable point-valued function, then $\int \limsup F_k \supset \limsup \int F_k$.*

LEMMA 5.4: *If $F_p(t)$ is continuous in p for each fixed t and Borel-measurable in t for each fixed p in P, and if all the F_p are bounded by the same integrable point-valued function, then $\int F_p$ is continuous in p.*

We now wish to establish the existence, uniqueness, continuity, and nonnegativity of the function h. In principle, the first three of these properties will follow from the closedness and nonemptiness, convexity, and continuity (in p) of $\int C_p$ respectively; nonnegativity will follow from weak desirability. In carrying out the proofs, however, the unboundedness of the $C_p(t)$ causes difficulties. To circumvent these difficulties, we shall find a bounded "substitute" for C_p.

Let v be a commodity-wise saturating assignment (i.e., an assignment satisfying (3.3)), and recall the notation $V(t) = \{x: x \leq v(t)\}$. We shall work with sets $V(t) \cap C_p(t)$, which we shall denote $D_p(t)$, passing back to the consideration of C_p itself only at the very end of the section. Note that the $D_p(t)$ are integrably bounded, uniformly in p, by the function v.

LEMMA 5.5: *For each t, $D_p(t)$ is continuous in p.*

PROOF: A similar lemma was proved by McKenzie [9, Lemma 4, pp. 57, 68]; we repeat the proof for the sake of completeness. Let $p_1, p_2, \ldots \in P$ have limit p. Suppose first that $x_k \in D_{p_k}(t)$ are such that $\lim x_k = x$. Certainly $x \in V(t)$; so if $x \notin D_p(t)$, then there is $y \in B_p(t)$ such that $y \succ_t x$. Then $p \cdot y \leq p \cdot i(t)$, and since assumption (3.1) asserts that $i(t) > 0$, it follows that $p \cdot i(t) > 0$. So we can find a z that is sufficiently close to y so that we still have $z \succ_t x$ (by continuity (2.3)), but $p \cdot z < p \cdot i(t)$. Then for k sufficiently large, we shall still have $p_k \cdot z < p_k \cdot i(t)$. Again applying continuity (2.3), we deduce from $z \succ_t x$ that for k sufficiently large $z \succ_t x_k$; but this contradicts $x_k \in D_{p_k}(t)$. Hence $x \in D_p(t)$, and upper-semicontinuity is proved.

Next, let $x \in D_p(t)$. If x saturates, then it is a member of all $D_{p_k}(t)$, so we can set $x_k = x$ in the definition of lower-semicontinuity. Assume therefore that x does not saturate. Let x_k be a point in $D_{p_k}(t)$ closest to x; the existence of x_k follows from the closedness of $D_{p_k}(t)$, which in turn follows from upper-semicontinuity. For arbitrary $\delta > 0$, set $y_\delta = v_t(x + (\delta, \ldots, \delta))$; then $y_\delta \sim_t x + (\delta, \ldots, \delta) \succ_t x$, by (3.2). Either for all δ, $y_\delta \in D_{p_k}(t)$ for all sufficiently large k, or else for some δ, there are infinitely many k such that $y_\delta \notin D_{p_k}(t)$. In the first case we have for all δ, by the definition of x_k, that $\|x_k - x\| \leq \|y_\delta - x\| \leq \delta\sqrt{n}$, where $\| \; \|$ represents the euclidean norm (i.e., the distance from the origin). Since δ can be chosen arbitrarily small, this shows that $x_k \to x$, and establishes lower-semicontinuity. In the second case, we can assume without loss of generality that $y_\delta \notin D_{p_k}(t)$ for all k. Then for each k there is a $z_k \in B_{p_k}(t \cap V(t)$ such that $z_k \succ_t y_\delta$. Since the z_k are all in $V(t)$, they have a limit point z; again without loss of generality, we can let it be the limit. Since $p_k \to p$, and $p_k \cdot z_k \leq p_k \cdot i(t)$, it follows that $p \cdot z \leq p \cdot i(t)$, i.e., $z \in B_p(t)$. On the other hand, from continuity (2.3) it follows that $z \succsim_t y$; since $y_\delta \succ_t x$, it follows that $z \succ_t x$. But this contradicts $x \in D_p(t) \subset C_p(t)$, and completes the proof of the lemma.

The proof of upper-semicontinuity in this lemma is the only place where use is made of $i(t) > 0$ (3.1), rather than the far weaker $\int i > 0$ (2.1).

LEMMA 5.6: C_p *and* D_p *are Borel-measurable for each fixed p.*

PROOF: Since every measurable function differs on at most a null set from a Borel-measurable function, we may assume that v and i are Borel-measurable. That statement "$x \in D_p(t)$" is equivalent to "$x \leqq v(t)$ and $x \in C_p(t)$." The statement "$x \in C_p(t)$" is equivalent to "for all $y \in B_p(t)$, $x \succsim_t y$"; because of continuity (2.3), this is equivalent to "for each rational point[9] $r \in B_p(t)$, $x \succsim_t r$." For fixed r, "$x \succsim_t r$" is equivalent to "not $r \succ_t x$." Because of continuity, "$r \succ_t x$" is equivalent to "there is a rational point s in Ω such that $s \geqq x$ and $r \succ_t s$." Hence $\{(x,t): x \succsim_t r\}$, which equals

$$\Omega \times T \setminus \bigcup_{\text{rational } s \text{ in } \Omega} [\{x: s \geqq x\} \times \{t: r \succ_t s\}],$$

is a Borel set. Hence $\{(x,t): x \in C_p(t)\}$, which equals

$$\bigcap_{\text{rational } r \text{ in } \Omega} [(\Omega \times \{t: p \cdot r > p \cdot i(t)\}) \cup \{(x,t): x \succsim_t r\}],$$

is a Borel set, and this proves that C_p is Borel-measurable. Hence $\{(x,t): x \in D_p(t)\}$, which equals

$$\{(x,t): x \leqq v(t)\} \cap \{(x,t): x \in C_p(t)\},$$

is a Borel set, and the proof is complete.

COROLLARY 5.7: $\int D_p$ *is closed, nonempty, convex, and continuous in p.*

PROOF: $\int v \in \int D_p$, so nonemptiness if proved. Convexity follows from Lemma 5.1. Since D_p is uniformly integrably bounded by v, continuity follows from Lemmas 5.4, 5.5, and 5.6. Since the values of a continuous set-valued function are always closed, the corollary is proved.

For each p in P, let $d(p)$ be the point in $\int D_p$ that is closest to $\int i$. Such a point exists because $\int D_p$ is nonempty and closed; it is unique because $\int D_p$ is convex.

---

[9] I.e., point with rational coordinates.

LEMMA 5.8: *$d(p)$ is a continuous (point-valued) function of p.*

PROOF: A similar lemma was proved by McKenzie [9, Lemma 10, p. 62]; we repeat the proof for the sake of completeness. Let $p_k \to p$, and let x be a limit point of $d(p_k)$. From the upper-semicontinuity of $\int D_p$ it follows that $x \in \int D_p$. Suppose that there is a point $y \in \int D_p$ such that $\|y - \int i\| < \|x - \int i\|$. By the lower-semicontinuity of $\int D_p$, there is a sequence of points $y_k \in \int D_{p_k}$ converging to y. Let $\{d(p_{k_j})\}$ be a subsequence of $\{d(p_k)\}$ converging to x. Since the norm is continuous, it follows that for j sufficiently large,

$$\left\| y_{k_j} - \int i \right\| < \left\| d(p_{k_j}) - \int i \right\|,$$

contradicting the definition of $d(p_{k_j})$. Hence $x = d(p)$. So the only limit point of $\{d(p_k)\}$ is $d(p)$, and the lemma is proved.

LEMMA 5.9: *For each p in P, $d(p) \geq \int i$.*

PROOF: If not, then $d(p)$ has a coordinate—without loss of generality, we can let it be the first—such that $d^1(p) < \int i^1$. Now $d(p) = \int x$, where $x(t) \in D_p(t)$ for all t. Let $y(t) = (v^1(t), x^2(t), \ldots, x^n(t))$. Then $y(t) \geq x(t)$ and $y(t) \leq v(t)$; therefore $y(t) \in D_p(t)$ for all t. Therefore

$$\left(\int v^1, d^2(p), \ldots, d^n(p) \right) = \int y \in \int D_p.$$

Now $d^1(p) < \int i^1$, and by the saturation restriction (3.4), $\int i^1 < \int v^1$; so there is an α with $0 < \alpha < 1$ such that $\alpha \int v^1 + (1 - \alpha) d^1(p) = \int i^1$. Setting $z = \alpha y + (1 - \alpha) x$ and $z = \int z$, we obtain $z \in \int D_p$ (by the convexity of $\int D_p$), and $z = (\int i^1, d^2(p), \ldots, d^n(p))$. Then from $d^1(p) < \int i^1$, we deduce

$$\left\| z - \int i \right\|^2 = \sum_{i=2}^{n} \left(d^i(p) - \int i^i \right)^2$$

$$< \sum_{i=1}^{n} \left(d^i(p) - \int i^i \right)^2 = \left\| d(p) - \int i \right\|^2.$$

Thus z is closer than $d(p)$ to $\int i$, a contradiction. This proves the lemma.

Let $g(p) = d(p) - \int i$. We have established for $g(p)$ all the properties that we set out to establish for $h(p)$: existence, uniqueness, continuity, and nonnegativity (the last by Lemma 5.9). So with the following lemma we achieve our aim:

LEMMA 5.10: $g(p) = h(p)$.

PROOF: Fix p, and write $g = g(p)$, $h = h(p)$, $c = c(p)$, $d = d(p)$. If $g = 0$ there is nothing to prove. Otherwise, by the definition of g, the hyperplane through d perpendicular to g supports $\int D_p$ (see footnote 8). This means that

(i) $$x \cdot g \geq \|g\|^2 \quad \text{for all } x \in \int D_p - \int i.$$

Suppose there is a point in $\int C_p$ there is nearer to $\int i$ than d is. This means that there is a point y in $\int C_p - \int i$ that is nearer to 0 than g is. Then

(ii) $$\|y\|^2 < \|g\|^2.$$

Furthermore $\|y\|^2 - 2y \cdot g + \|g\|^2 = \|y - g\|^2 > 0$. Hence $\|y\|^2 > y \cdot g + [y \cdot g - \|g\|^2]$. If $y \cdot g - \|g\|^2 \geq 0$, then it follows that $\|y\|^2 > y \cdot g \geq \|g\|^2$, contradicting (ii). Hence

(iii) $$y \cdot g < \|g\|^2.$$

Formula (iii) expresses the geometrically obvious fact that any point nearer than d to $\int i$ must be on the near side of the hyperplane through d perpendicular to g.

Now $y = \int x - \int i$, where $x(t) \in C_p(t)$ for all t. Then by commodity-wise saturation, $v_t(x(t)) \in D_p(t)$ for all t. Furthermore $v_t(x(t)) \leq x(t)$, and $v_t(x(t)) \leq v(t)$. Setting $z(t) = v_t(x(t))$, we obtain $\int z \in \int D_p$ and $\int z - \int i \leq y$. Since $g \geq 0$ (Lemma 5.9), it follows that $(\int z - \int i) \cdot g \leq y \cdot g$. Hence by (iii), $(\int z - \int i) \cdot g < \|g\|^2$. But since $\int z - \int i \in \int D_p - \int i$, it follows from (i) that $(\int z - \int i) \cdot g \geq \|g\|^2$, and this is the contradiction that proves our lemma.

It remains to show that a measurable x may be chosen whose integral will contradict (4.3). According to Section 4, it is sufficient to show that there is a measurable x such that for all t, $x(t)$ is maximal in $B_q(t)$ with respect to t's preference order. Let $X(t)$ be the set of all maximal points in $B_q(t)$. As in the proof of Lemma 5.6, we may assume that i is

Borel-measurable. Then

$$\{(x,t):\ x \in B_q(t)\} = \{(x,t):\ q \cdot x \leqq q \cdot i(t)\}$$

$$= \Omega \times T \smallsetminus \bigcup_\theta [\{x:\ q \cdot x > \theta\} \times \{t:\ \theta > q \cdot i(t)\}],$$

where θ runs over the rational numbers. Hence the left side is a Borel set. Applying Lemma 5.6 we deduce that $\{(x,t):\ x \in X(t)\}$, which equals

$$\{(x,t):\ x \in B_q(t)\} \cap \{(x,t):\ x \in C_q(t)\},$$

is a Borel set. Hence X is Borel-measurable.

Next, we show that $X(t)$ is nonempty for each t. From the compactness of $V(t) \cap B_q(t)$ and the continuity condition (2.3) for preferences, it follows that $V(t) \cap B_q(t)$ has a maximal element y. Then because of commodity-wise saturation, y is also maximal in $B_q(t)$. Indeed, suppose $z \in B_q(t)$ is such that $z \succ_t y$. Now $z \in B_q(t)$ means $q \cdot z \leqq q \cdot i(t)$; therefore $q \cdot v_t(z) \leqq q \cdot z \leqq q \cdot i(t)$, and therefore also $v_t(z) \in B_q(t)$. But by definition, $v_t(z) \in V(t) \cap B_q(t)$. Finally, $v_t(z) \sim_t z \succ_t y$. Thus $v_t(z)$ contradicts the maximality of z in $V(t) \cap B_q(t)$, proving the existence of a maximal element in $B_q(t)$.

From Lemma 5.2 we may now deduce the existence of an appropriate x. This completes the proof of the auxiliary theorem.

6. PROOF OF THE MAIN THEOREM

The general idea is to approximate a given market $\mathscr{M}$ satisfying the conditions of the main theorem by a sequence of markets $\mathscr{M}_k$ satisfying the conditions of the auxiliary theorem. Then by the auxiliary theorem, the $\mathscr{M}_k$ have competitive equilibria (q_k, y_k); from these competitive equilibria we shall construct a pair (q, y) that is a competitive equilibrium in the original market $\mathscr{M}$.

To define the markets $\mathscr{M}_k$, we must specify their initial assignments i_k and their preference orders $\preceq_t^k$; the number of commodities is taken to be n in all the $\mathscr{M}_k$. Let δ_k be a monotone sequence of numbers tending to 0, and define

$$i_k(t) = i(t) + (\delta_k, \dots, \delta_k).$$

To define the preference orders, let γ_k be a monotone sequence of numbers tending to ∞ such that $\gamma_1 > \delta_1$, let

$$v_k(t) = i(t) + (\gamma_k, \ldots, \gamma_k),$$

and let "hyper-rectangles" $V_k(t)$ and functions $v_{k,t}$ from Ω onto $V_k(t)$ be defined as in Section 3, with v_k in place of v. Now define the preference orders by

$$x \succeq_t^k y \quad \text{if and only if} \quad v_{k,t}(x) \succeq_t v_{k,t}(y).$$

It may be verified that the $\mathcal{M}_k$ satisfy the conditions of the auxiliary theorem, with v_k as the commodity-wise saturating assignment. Furthermore, note that the preference orders in $\mathcal{M}_k$ coincide with those in $\mathcal{M}$ for all x and y such that x and y are $\leq i(t) + (\gamma_k, \ldots, \gamma_k)$.

Let (q_k, y_k) be a competitive equilibrium of $\mathcal{M}_k$. Because of the compactness of P, the sequence $\{q_k\}$ has a convergent subsequence, and we may suppose without loss of generality that this subsequence is the original sequence. Let $q = \lim_k q_k$. The following is the crucial lemma of this section:

LEMMA 6.1: $q > 0$.

PROOF: Suppose, on the contrary, that some coordinate of q vanishes, say $q^1 = 0$. First we establish

(i)   if $q \cdot i(t) > 0$, then $\{y_k(t)\}$ has no limit point as $k \to \infty$.

Indeed, suppose y were such a limit point; without loss of generality, assume that it is actually the limit. Now because (q_k, y_k) is a competitive equilibrium in $\mathcal{M}_k$, we have $q_k \cdot y_k(t) \leq q_k \cdot i_k(t)$. Using this and the saturation restriction (3.4) in $\mathcal{M}_k$ we deduce that $y_k(t)$ does not saturate. Hence if $q_k \cdot y_k(t) < q_k \cdot i_k(t)$, then by weak desirability (3.2) in $\mathcal{M}_k$, it would be possible to find a member of $B_{q_k}(t)$ preferred to $y_k(t)$, contradicting the definition of competitive equilibrium. Thus $q_k \cdot y_k(t) = q_k \cdot i(t)$, and so from the hypothesis of (i) we obtain

(ii) $q \cdot y = \lim_k q_k \cdot y_k(t) = \lim_k q_k \cdot i_k(t) = q \cdot i(t) > 0.$

Hence there is a coordinate j such that $y^j > 0$ and $q^j > 0$; assume without loss of generality that $j = 2$. Now by desirability (2.2), $y + \{1, 0, \ldots, 0\} \succ_t y$. If for sufficiently small $\delta > 0$ we define $z = y + \{1, -\delta, 0, \ldots, 0\}$, then $z \in \Omega$, and by continuity we deduce $z \succ_t y$. Again using continuity, we obtain $z \succ_t y_k(t)$ for k sufficiently large. Since (q_k, y_k) is a competitive equilibrium in $\mathcal{M}_k$, we obtain $q_k \cdot z > q_k \cdot i_k(t)$. Letting $k \to \infty$ and applying (ii), we get

(iii) $q \cdot z = \lim_k q_k \cdot z \geq \lim q_k \cdot i_k(t) = q \cdot y.$

But since $q^1 = 0$ and $q^2 > 0$, we have

$$q \cdot z = q \cdot y + q^1 - \delta q^2 = q \cdot y - \delta q^2 < q \cdot y,$$

contradicting (iii). This proves (i).

Since $q \in P$ and $\int i > 0$ (2.1), it follows that $\int q \cdot i = q \cdot \int i > 0$. Let $S = \{t: q \cdot i(t) > 0\}$; then S is nonnull, and we denote its measure by $\mu(S)$. Define

$$A = \left\{ x \in \Omega \colon \sum_{i=1}^{n} x^i \leq 2 \int \sum_{j=1}^{n} i^j \Big/ \mu(S) \right\}.$$

For $t \in S$, it follows from (i) and the compactness of A that $y_k(t) \in A$ for at most finitely many k; that is, for each $t \in S$ there is an integer $k(t)$ such that $\sum_i y_k^i(t) > 2 \int \sum_j i^j / \mu(S)$ for $k \geq k(t)$. Hence for $t \in S$,

(iv) $\liminf_k \sum_i y_k^i(t) \geq 2 \int \sum_j i^j / \mu(S).$

Because y_k is an allocation in $\mathcal{M}_k$, we have

(v) $\lim_k \int \sum_i y_k^i = \lim_k \int \sum_i i_k^i = \lim_k \left[\int \sum_i i^i + n \delta_k \right] = \int \sum_i i^i.$

But by Fatou's Lemma[10] and (iv),

$$\lim_k \int \sum_i y_k^i \geq \int \liminf_k \sum_i y_k^i \geq \int_S \liminf_k \sum_i y_k^i$$

$$\geq \int_S \left[2 \int \sum_j i^j / \mu(S) \right] = \left[2 \int \sum_j i^j \right] \int_S 1 / \mu(S)$$

$$= 2 \int \sum_j i^j > \int \sum_j i^j,$$

where the last inequality follows from $\int i > 0$ (2.1). This contradicts (v) and proves Lemma 6.1.

Since $q_k \to q > 0$, there is a $\delta > 0$ such that $q_k^i \geq \delta$ for k sufficiently large and all i. Without loss of generality, we may assume that $q_k^i \geq \delta$ for all i and k, and that $i_k^i(t) \leq i^i(t) + \delta$ for all i, k, and t. Hence for all i, k, and t,

$$\delta y_k^i(t) \leq q_k \cdot y_k(t) \leq q_k \cdot i_k(t) \leq q_k \cdot i(t) + \delta \leq \sum_{j=1}^n i^j(t) + \delta.$$

Thus we obtain

$$y_k^i(t) \leq 1 + \sum_{j=1}^n \frac{i^j(t)}{\delta}. \tag{6.2}$$

For each t, let $Y(t)$ be the set of limit points of $y_k(t)$ as $k \to \infty$. Let $Y_k(t)$ be the set consisting of the single point $y_k(t)$; then $Y(t) = \limsup Y_k(t)$. By (6.2), all the Y_k are bounded by the same integrable function. Hence by Lemma 5.3,

$$\int i = \lim \int i_k = \lim \int y_k \in \limsup \int Y_k \subset \int \limsup Y_k = \int Y.$$

Let y be such that $y(t) \in Y(t)$ for all t, and

$$\int y = \int i. \tag{6.3}$$

We shall show that (q, y) is a competitive equilibrium in $\mathcal{M}$.

[10] Fatou's Lemma states that if φ_k are nonnegative measurable real functions, then $\liminf_k \int \varphi_k \geq \int \liminf_k \varphi_k$. See [5, III. 6.19, p. 152].

To this end we must demonstrate that y is an allocation, that $y(t)$ belongs to $B_q(t)$ for all t, and that $y(t)$ is maximal in $B_q(t)$ for all t, i.e., that no member of $B_q(t)$ is preferred to $y(t)$. We have already shown that y is an allocation (6.3). Next, since $Y(t) \in \boldsymbol{Y}(t)$, it follows that $Y(t)$ is a limit point of $\{y_k(t)\}$, say $y(t) = \lim_{m \to \infty} y_{k_m}(t)$. Since

$$q_{k_m} \cdot y_{k_m}(t) \leqq q_{k_m} \cdot i_{k_m}(t),$$

we deduce by letting $m \to \infty$ that $q \cdot y(t) \leq q \cdot i(t)$, and so for all t,

$$y(t) \in B_q(t). \tag{6.4}$$

Finally, suppose that for t in a set of positive measure, there is a $z \in B_q(t)$ such that $z \succ_t y(t)$. Clearly $z \neq 0$; suppose without loss of generality that $z^1 > 0$. If for $\delta > 0$ sufficiently small we define $z_\delta = z - (\delta, 0, \dots, 0)$, then we still have

$$z_\delta \succ y(t). \tag{6.5}$$

Moreover, since

$$\lim_k q_k \cdot z_\delta = q \cdot z - q^1 \delta < q \cdot z \leq q \cdot i(t) = \lim q_k \cdot i_k(t),$$

it follows that

$$q_k \cdot z_\delta < q_k \cdot i_k(t)$$

for all sufficiently large k, say for $k > k_0$. Now since $y(t)$ is a limit point of $\{y_k(t)\}$, there is a subsequence $\{y_{k_m}(t)\}$ converging to $y(t)$; hence for m sufficiently large,

$$z_\delta \succ_t y_{k_m}(t),$$

by (6.5). If we also pick m so large so that $k_m \geqq k_0$, then z_δ contradicts the maximality of $y_{k_m}(t)$ in the budget set $\{x \colon q_{k_m} \cdot x \leqq q_{k_m} \cdot i_{k_m}(t)\}$. Thus the supposition $z \succ_t y(t)$ has led to a contradiction, and we conclude that $y(t)$ is maximal in $B_q(t)$ for all t. Together with (6.3) and (6.4), this completes the proof that (q, y) is a competitive equilibrium, and with it the proof of the main theorem.

7. Comparison with McKenzie's Proof

The differences between this proof and McKenzie's are caused by the different initial equipment: we have no convexity assumption to work with, and we have a continuum of traders rather than a finite number.

McKenzie needs the convexity assumption in only one place, to show that the aggregate preferred set (in his case the sum of the individual preferred sets) is convex. This is needed to define $h(p)$ uniquely, and follows from the convexity of the individual preferred sets. In a finite model there is no getting around this: no intuitive assumption other than individual convexity would lead to the convexity of the aggregate preferred set.

In a continuous model, however, this is superfluous, because of Lemma 5.1; this says that the integral of any set-valued function over a nonatomic measure space (in our case the unit interval) is convex, even if the individual values of the function are not convex. In particular, the aggregate preferred set, as the integral of the (possibly nonconvex) individual preferred sets, is convex.

Because of the presence of a continuum of traders, the space of assignments is no longer a subset of a finite-dimensional euclidean space, but of an infinite-dimensional function space. This necessitates the use of completely new methods to justify the passage from properties proved for individual traders to the corresponding properties for the aggregate of all traders. Consider, for example, the continuity of the aggregate preferred set as a function of the price vector. In the finite case, this follows trivially from the continuity of the individual preferred sets. Here, on the other hand, it involves Lemma 5.4, which is comparatively deep. In fact, Lemmas 5.1–5.4, which have been separately published, were originally proved for the purpose of this paper, and they embody the chief mathematical difficulties. The proofs of these lemmas involve Lyapunov's theorem on the range of a vector measure [8], and the methods of functional analysis (Banach spaces) and topology.

Another significant difference between this proof and McKenzie's is in the matter of boundedness. In the proof of the auxiliary theorem, the set of bundles under consideration must be in some sense bounded in order to establish the continuity—and indeed the existence—of the individual preferred sets. McKenzie does this by noting that no individual trader can have more goods than the whole market. This is not available here, because no matter how large an individual trader's bundle is, it is still infinitesimal compared with the whole market. We therefore used the notion of commodity-wise saturation, which does the job of bounding for us. In the passage from the auxiliary to the main theorem we do not have commodity-wise saturation, but need boundedness so that the sequence of competitive equilibria of the auxiliary markets $\mathcal{M}_k$ should converge. Here we first deduce from the desirability assumption (2.2) that all prices must be nonvanishing, and this bounds the bundles under consideration to a finite simplex.

8. THE CORE

Intimately connected with the concept of competitive equilibrium is that of *core*. This is the set of all allocations with the property that no "coalition" of traders can assure each of its members of a more desirable bundle by trading within itself only, without recourse to traders not in the coalition. Formally (in our model), an allocation x is in the core if there is no measurable nonnull set S of traders, for whom there is an allocation y such that $y(t) \succ_t x(t)$ for all $t \in S$ and $\int_S y = \int_S i$. In a finite market, the integral should be replaced by a sum.

In a finite market with convex preferences, the core is never empty; but when the preferences are not convex the core may be empty. As with the competitive equilibrium, it might be conjectured that this "pathology" would "tend to disappear" as the number of traders increases. Investigating this possibility, Shapley and Shubik [10] showed that though the core itself may remain empty for any (finite) number of traders, it is possible to define a kind of approximation to the core called an ε-core; and that for any positive ε, if the number n of traders is allowed to increase in a certain way, the ε-core will become nonempty for sufficiently large n. They concluded that, heuristically speaking, the true core lies "just below the surface" for sufficiently large n. The assumptions on which their theorem is based are comparatively strong: They assumed transferable utilities, that all traders have the same utility function, and that there is a fixed finite number of distinct types of traders (where two traders are of the same "type" if they have the same initial bundles).

We shall now describe how the concepts of core and competitive equilibrium are related. Let us define an *equilibrium allocation* to be an allocation that forms a competitive equilibrium when paired with an appropriate price vector. For finite markets, the core always contains the set of equilibrium allocations, but the two sets do not usually coincide. A long-standing conjecture states, however, that as the number of players in a market increases, the core of the market "tends," in some sense, to the set of equilibrium allocations. Recently this conjecture has been formalized and proved in a number of different ways.[11] In

---

[11] See [2] for a brief survey of these developments.

[2] we showed that for a market with a *continuum* of traders, the core actually *equals* the set of equilibrium allocations. This was shown under conditions that are even weaker than those of this paper.[12] A question that was left open was the *existence* of a competitive equilibrium, or equivalently, the nonemptiness of the core; though it had been shown that the two sets coincide, the possibility that both vanish was left open. From the theorem of this paper, it now follows that the core is nonempty as well. This agrees well with the Shapley-Shubik result (which was, however, obtained under considerably stronger assumptions): Since the ε-core is nonempty for large n, it is to be expected that the true core is nonempty for "infinite n."

The Hebrew University of Jerusalem

REFERENCES

[1] Arrow, K. J., and G. Debreu: "Existence of an Equilibrium for a Competitive Economy," *Econometrica*, Vol. 22, 1954, pp. 265–290.
[2] Aumann, R. J.: "Markets with a Continuum of Traders," *Econometrica*, Vol. 32, 1964, pp. 39–50.
[3] ———: "Integrals of Set-Valued Functions," *Journal of Mathematical Analysis and Applications*, Vol. 12, 1965, pp. 1–12.
[4] Debreu, G.: *Theory of Value*, John Wiley and Sons, Inc., New York, 1959.
[5] Dunford, N., and J. T. Schwartz, *Linear Operators, Part I*, Interscience Publishers, Inc., New York, 1958.
[6] Eggleston, H. G.: *Convexity*, Cambridge University Press, 1958.
[7] Gale, D.: "The Law of Supply and Demand," *Mathematics Scandinavica*, Vol. 3, 1955, pp. 155–169.
[8] Lyapunov, A.: "Sur les Fonctions-vecteurs complètements additives," *Bull. Acad. Sci. URSS sér. Math*, Vol. 4, 1940, pp. 465–478.
[9] McKenzie, L. W.: "On the Existence of General Equilibrium for a Competitive Market," *Econometrica*, Vol. 27, 1959, pp. 54–71.
[10] Shapley, L. S., and M. Shubik: *The Core of an Economy with Nonconvex Preferences*. The RAND Corporation, RM-3518-PR, February, 1963.
[11] Wald, A.: "Uber einige Gleichungssysteme der mathematischen Ökonomie," *Zeitschrift für Nationalökonomie*, Vol. 7, 1936, pp. 637–670. Translated as "On Some Systems of Equations of Mathematical Economics," *Econometrica*, Vol. 19, 1951, pp. 368–403.

---

[12] The model of [2] differs from that of this paper in that there we started out directly with preference relation $\succ_t$ rather than deriving them from preference-or-indifference relations $\succsim_t$; furthermore, unlike here, we there made no assumptions of total or even partial order for the preference relations (for example, transitivity was not assumed). Otherwise, the two models are identical.

14

THE CORE OF AN *n*-PERSON GAME

Herbert E. Scarf[1]

Sufficient conditions are given for a general *n*-person game to have a
nonempty core. The conditions are a consequence of convexity of preferences if the game arises from an exchange economy. The proof of sufficiency is based on a finite algorithm, and makes no use of fixed point
theorems.

1. Introduction

The problems of distribution in an economic system may be analyzed
either by means of the behavioral assumptions of a competitive model
or by the more flexible techniques of *n*-person game theory. In the
competitive model, consumers are assumed to respond to a set of prices
by maximizing utility subject to a budget constraint and producers by
maximizing profit. Consistent production decisions and an allocation of
commodities are obtained by the determination of a set of prices at
which all markets are in equilibrium.

The analysis of these problems by means of *n*-person game theory
requires us to specify the production and distribution activities that are
available to an arbitrary coalition of economic agents. It is frequently
sufficient to summarize the detailed strategic possibilities open to a
coalition by the set of possible utility vectors that can be achieved by the
coalition. For example, in a pure exchange economy each coalition will
have associated with it the collection of all utility vectors that can be
obtained by arbitrary redistributions of the resources of that coalition.

The core of an *n*-person game is a generalization of Edgeworth's
contract curve. A vector of utility levels is suggested which is feasible for
all of the players acting collectively, and an arbitrary coalition is
examined to see whether it can provide higher utility levels for *all* of its
members. If this is possible, the utility vector which was originally

[1] The research reported in this paper was carried out under a grant from the National
Science Foundation. I would like to thank Robert Aumann, Tjalling Koopmans, Lloyd
Shapley, and the referees of *Econometrica* for their perceptive advice on a number of
points.

suggested is said to be blocked by the coalition. The core of the *n*-person game consists of those utility vectors which are feasible for the entire group of players and which can be blocked by no coalition.

As we have seen during the last several years, there is an intimate connection between these two methods of analysis. If the conventional assumptions of the competitive model are made, such as convexity of preferences and convexity and constant returns to scale for the production set, then there will be a price system at which all markets are in equilibrium and a resulting assignment of commodity bundles to consumers. The utility vector associated with this competitive equilibrium may be shown to be in the core. Even further, if the number of consumers tends to infinity in a suitable way, the set of possible utility vectors in the core becomes smaller and tends, in the limit, to those utility vectors associated with competitive equilibria [3].

We do not, of course, expect a competitive equilibrium if the classical assumptions of the competitive model are not made. On the other hand the formulation of the problem of distribution by means of *n*-person game theory is sufficiently flexible to accommodate any number of departures from the classical model. The set of possible utility vectors achievable by a coalition can be discussed in the presence of increasing returns to scale in production, public ownership of some commodities, and social rather than exclusively private goods, to name only a few departures from the classical model. This raises the question of determining conditions which are sufficient to guarantee the existence of a utility vector in the core, and which are described directly in terms of the structure of an *n*-person game rather than appealing indirectly to the existence of competitive equilibria.

In order to see the form that such conditions might take, let us being by examining a game with three players. In this case there are seven possible coalitions; the three one-player coalitions, the three two-player coalitions, and the coalition of all three players. Each such coalition will be able to obtain a set of utility vectors depending on the strategies available to its members. It will be useful to denote by V_S the set of those vectors achievable by the coalition S. $V_{(123)}$ will be represented geometrically by a set of vectors in three space, $V_{(12)}$ will lie in the plane determined by the coordinate axes 1 and 2, and in general V_S will lie in that linear subspace of three space whose coordinates correspond to the members of S. The sets V_S will be assumed to have several technical properties such as being closed and containing any point whose coordinates are less than or equal to those of a point in V_S.

For this game to have a core which is not empty, $V_{(123)}$ must be sufficiently large so as to contain a vector which cannot be blocked by any coalition. One meaning of the term "sufficiently large" can be obtained by assuming that it is to the advantage of a disjoint collection of coalitions to combine. For example if $u_1 \in V_{(1)}$ and $(u_2, u_3) \in V_{(23)}$ then I will assume that $(u_1, u_2, u_3) \in V_{(123)}$, and similarly for all other partitions of the set of three players. The assumption that the game is superadditive, in this sense, is quite natural for most economic models. It is, however, not sufficient to guarantee the existence of a vector in the core, and one additional relationship is required.

Let us assume for a moment that the game derives from a market model in which the three players exchange the commodities which they initially own. The preferences of the ith player will be represented by a utility function $u_i(x^i)$, with x^i the commodity bundle received by this player. The commodity bundle initially owned by the ith player will be denoted by ω^i. With this notation the set $V_{(123)}$ is described by

$$V_{(123)} = \left\{ (u_1, u_2, u_3) | u_j \leq u_j(x^j) \text{ for some } (x^1, x^2, x^3) \right.$$
$$\left. \text{with } x^1 + x^2 + x^3 = \omega^1 + \omega^2 + \omega^3 \right\},$$

the set $V_{(12)}$ by

$$V_{(12)} = \left\{ (u_1, u_2) | u_j \leq u_j(x^j) \text{ for some } (x^1, x^2) \right.$$
$$\left. \text{with } x^1 + x^2 = \omega^1 + \omega^2 \right\},$$

with a similar definition for every set V_S.

This game is clearly superadditive in the sense given above, even in the absence of convex preferences. We know, however, that a market game without convex preferences need not have a core [3], and we should therefore look for some way of translating the convexity of preferences into a relationship that can be stated solely in terms of the sets V_S, in order to find the missing condition. Let us proceed in the following way. Assume that we are given a vector (u_1, u_2, u_3) which is arbitrary except that it satisfies the following three conditions:

$$(u_1, u_2) \in V_{(1,2)},$$
$$(u_2, u_3) \in V_{(2,3)},$$
$$(u_1, u_3) \in V_{(1,3)}.$$

In the market economy this means that there are commodity bundles (x^1, x^2), (y^2, y^3), and (z^1, z^3) with

$$x^1 + x^2 \quad = \omega^1 + \omega^2,$$

$$y^2 + y^3 = \quad \omega^2 + \omega^3,$$

$$z^1 \quad + z^3 = \omega^1 \quad + \omega^3,$$

and

$$u_1(x^1) \geq u_1, \quad u_1(z^1) \geq u_1,$$

$$u_2(x^2) \geq u_2, \quad u_2(y^2) \geq u_2,$$

$$u_3(y^3) \geq u_3, \quad u_3(z^3) \geq u_3.$$

But then

$$\frac{x^1 + z^1}{2}, \quad \frac{x^2 + y^2}{2}, \quad \frac{y^3 + z^3}{2}$$

represents a feasible trade for all three players since these vectors total to $\omega^1 + \omega^2 + \omega^3$. If the preferences of the three consumers are convex then the utility levels associated with this trade can be described quite easily, since convexity implies that

$$u_1\left(\frac{x^1 + z^1}{2}\right) \geq \min[u_1(x^1), \quad u_1(z^1)] \geq u_1,$$

$$u_2\left(\frac{x^2 + y^2}{2}\right) \geq \min[u_2(x^2), \quad u_2(y^2)] \geq u_2,$$

$$u_3\left(\frac{y^3 + z^3}{2}\right) \geq \min[u_3(y^3), \quad u_3(z^3)] \geq u_3.$$

In other words the vector (u_1, u_2, u_3) is obtainable by the three-player coalition and is therefore in $V_{(1,2,3)}$.

This rather curious translation of convexity connecting the three two-player coalitions to the coalition of all players is, in conjunction with the assumptions of superadditivity, sufficient for the existence of a vector in the core of a three-person game. In order to discuss the appropriate generalization of these conditions let us turn to a more formal definition of an n-person game following Aumann and Peleg [1].

The set of n players will be denoted by N and an arbitrary coalition by S. For each set S, E^S will mean the Euclidean space of dimension equal to the number of players in S and whose coordinates have as subscripts the players in S. If u is a vector in E^N then u^S will be its projection onto E^S.

We shall associate with each coalition S a set V_S, in E^S, which represents the set of possible utility vectors that can be obtained by that coalition. The members of S may have to engage in a variety of activities, depending on the nature of the n-person game, in order to obtain a particular vector in V_S. For our purposes, however, all that is required is a summary of the utility vectors achievable by each coalition.

It will be useful to make the following assumptions about the sets V_S.

1. For each S, V_S is a closed set.
2. If $u \in V_S$ and $y \in E^S$ with $y \le u$, then $y \in V_S$.
3. The set of vectors in V_S in which each player in S receives no less than the maximum that he can obtain by himself is a nonempty, bounded set.

These conditions are all quite mild and need no particular comment. They are slightly different from the conditions assumed by Aumann and Peleg; in particular the sets V_S need not be convex.

We have already seen how a three-person exchange model gives rise to a game in this form. In an exchange economy with n consumers, where the ith player's utility function is given by $u_i(x^i)$ and his initial holdings by ω^i, a vector $u \in E^S$ will be in V_S if we can find commodity bundles x^i with $\Sigma_{i \in S} x^i = \Sigma_{i \in S} \omega^i$ and $u_i(x^i) \ge u_i$ for all $i \in S$. Production may be introduced by assuming that each coalition has the ability to transform commodities according to some production set, though this is by no means the only way to incorporate production into an n-person game theory model.

As another example consider the classical case of an n-person game with transferable utility described by a number f_S associated with each coalition. What this means is that a vectors $u \in E^S$ may be obtained by the coalition S if $\Sigma_{i \in S} u_i \le f_S$, so that the sets V_S consist of half spaces defined by hyperplanes whose normal vectors have components either zero or one.

Let u be a point in V_N and u^S its projection onto E^S. The vector u is blocked by the set S if we can find a point $y \in V_S$ with $y > u^S$, or in other words if the coalition S can obtain a higher utility level for each of its members than that given by the vector u. A point $u \in V_N$ will be in the core if it cannot be blocked by any set S.

In order to determine the appropriate generalization of our extra condition in the three-person case, we must have recourse to the concept of a balanced collection of coalitions studied by Shapley [8], Peleg [7], and Bondareva [2] in the context of a game with transferable utility.

DEFINITION: Let $T = \{S\}$ be a collection of coalitions in an *n*-person game. T is said to be a *balanced collection* if it is possible to find nonnegative weights δ_S, for each coalition in T, such that

$$\sum_{\substack{S \in T \\ S \supset \{i\}}} \delta_S = 1 \quad \text{for each } i.$$

In other words, the weights δ_S are to have the property that if any individual is selected, the sum of the weights corresponding to those coalitions in T which contain the individual, must be equal to one. Another way to phrase the definition is by saying that the characteristic function of the set of all players is a nonnegative linear combination of the characteristic functions of the coalitions in a balanced collection.

Balanced collections of coalitions do represent a generalization of the collection of all two-players coalitions studied in the three-person game, since $\delta_{(12)} = \delta_{(13)} = \delta_{(23)} = \frac{1}{2}$ will serve as an appropriate system of weights. It is somewhat unfortunate, given the importance of balanced collections in the study of the core, that we have no really intuitive definition for determining when a given collection is balanced.

This concept permits us to extend to an *n*-person game the additional requirement imposed in the three-person case.

DEFINITION: An *n*-person game is *balanced* if for every balanced collection T, a vector u must be in V_N if $u^S \in V_S$ for all $S \in T$.

The main theory of this paper may now be stated.

THEOREM 1: *A balanced n-person game always has a nonempty core.*

It should be noticed that all of the concepts that have been introduced are purely ordinal in character; they are invariant if a continuous monotonic transformation is applied to the utility of any individual. In fact the discussion could be carried out on an abstract level with the outcomes for each individual represented by arbitrary ordered sets.

2. Some Examples of Balanced Games

The condition that a game be balanced is undoubtedly quite obscure and it will be useful to examine some examples. A market game with convex preferences will always be balanced, for let T be an arbitrary balanced collection and u a vector with $u^S \in V_S$ for each S in T. This means that for each such coalition there is a way of redistributing its assets so as to obtain the vector u^S. If the redistribution gives player i (assuming that he is a member of S) the commodity bundle x_S^i, then

$$\sum_{i \in S} x_S^i = \sum_{i \in S} \omega^i \quad \text{and} \quad u_i(x_S^i) \geq u_i.$$

In order to show that the game is balanced we need to construct an allocation $x^1, \ldots, x^n$ with $\sum_1^n x^i = \sum_1^n \omega^i$ and $u_i(x^i) \geq u_i$ for all i. This allocation may be constructed in terms of the weights δ_S used in the definition of a balanced collection.

For each player i, let us define x^i as $\sum_{S \in T, S \supset \{i\}} \delta_S x_S^i$. By the definition of δ_S, each x^i is a *convex* combination of x_S^i with S ranging over those sets in T which contain the ith player. If the preferences are assumed to be convex then $u_i(x^i)$ is greater than or equal to the smallest of the numbers $u_i(x_S^i)$, and is therefore greater than or equal to u_i. We have, by this device, constructed an assignment of commodity bundles which provides a utility level for each player no less than his corresponding component of u. In order to show that $u \in V_N$ we need only verify that $\sum_1^n x^i = \sum_1^n \omega^i$. But

$$\sum_1^n x^i = \sum_1^n \sum_{\substack{S \supset \{i\} \\ S \in T}} \delta_S x_S^i$$

$$= \sum_{S \in T} \delta_S \sum_{i \in S} x_S^i$$

$$= \sum_{S \in T} \delta_S \sum_{i \in S} \omega^i$$

$$= \sum_1^n \omega^i \sum_{\substack{S \in T \\ S \supset \{i\}}} \delta_S = \sum_1^n \omega^i.$$

This argument demonstrates that an exchange economy with convex preferences will always give rise to a balanced n-person game, and assuming the validity of the main result of this paper, such a game will always have a nonempty core. It is interesting that no addition assumptions are required such as strict monotonicity of the preferences, or strict positivity of the initial holdings. (Of course, in an exchange model, some additional assumptions are required in order to pass to the limit and obtain the existence of competitive equilibria.)

In our second example, an n-person game with transferable utility, it is quite easy to verify that a balanced game has a nonempty core without using the more subtle techniques to be developed later. In fact, a game with transferable utility has a core if and only if it is balanced.

If the sets V_S consist of those vectors in E^S with $\sum_{i \in S} u_i \leq f_S$, the vector $(u_1, \ldots, u_n)$ will be in the core if

$$\sum_{i=1}^{n} u_i \leq f_N \quad \text{and}$$

$$\sum_{i \in S} u_i \geq f_S \quad \text{for all subsets } S.$$

The first inequality implies that $u \in V_N$ and the second set that u cannot be blocked by any coalition S. In other words the game will have a core if the linear programming problem

$$\min \sum_{1}^{n} u_i$$

$$\sum_{i \in S} u_i \geq f_S, \quad \text{for all } S,$$

has a solution in which the objective function is equal to f_N. The dual variables may be denoted by δ_S, one for each subset, and the dual linear programming problem is

$$\max \sum_{S} \delta_S f_S$$

$$\sum_{S \supset \{j\}} \delta_S = 1, \quad \text{and} \quad \delta_S \geq 0.$$

(We have equality here, since the variables in the primal problem are unrestricted in sign.) Let $\{\hat{\delta}_S\}$ be a solution of the dual problem and $\{\hat{u}_i\}$ a solution to the primal problem. Then the collection T of those coalitions for which $\hat{\delta}_S > 0$ is, by definition, a balanced collection, and the solution of the primal problem provides us with a vector $\hat{u}$ such that $\Sigma_{i \in S} \hat{u} = f_S$ for all $S \in T$, since the positivity of a dual variable forces the primal constraints to be equalities. But then $\hat{u}^S \in V_S$ for all $S \in T$ and if the game is balance, this implies that $\hat{u} \in V_N$ or $\Sigma_1^n \hat{u}_i \leq f_N$. This shows that a balanced game has a nonempty core in the case of transferable utility. As we shall see, the general n-person game, without the assumption of transferable utility, requires more elaborate techniques than those of linear programming.

3. A COMBINATORIAL PROBLEM WHICH IMPLIES THEOREM 1

The proof of Theorem 1 will be divided into two parts. We begin by selecting from each set V_S (with S a *proper* subset of N), a finite number of vectors $u^{1,S}, u^{2,S}, \ldots, u^{k_S,S}$, that gives no player in S less than the maximum he can achieve by himself. If the game is balanced, the algorithm will then calculate a vector in V_N which cannot be blocked by any proper coalition using a vector from this finite list to block. The passage to the limit, which involves selecting an infinite, dense sequence of vectors from each V_S will be discussed in a later section.

Essentially we are approximating each set V_S by a set with a finite number of "corners," as illustrated by Figure 1. Since this approximation does not modify the property that the game is balanced, we may, for the moment, restrict our attention to games in which the sets V_S do, in fact, have a finite numbers of corners for each proper subset S.

It is useful to summarize the data of a finite game by means of a matrix C with n rows, the number of players in the game, and $\Sigma_S k_S$ columns, one column for each of the vectors involved in defining the

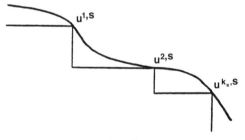

FIG. 1

game. The rows of C will be indexed by i and the columns will require a pair of subscripts (j, S), so that a typical entry in C will be denoted by c_{ijS}. If player i is contained in the coalition S, then c_{ijS} is defined to be the ith component of the vector $u^{j,S}$. It will be useful to define c_{ijS} to be equal to some large number M if player i is not a member of S. The particular choice of M is irrelevant to the actual calculation as long as it is selected to be larger than any of the components of the vectors $u^{j,S}$.

Let us also define a matrix A with n rows and $\Sigma_S k_S$ columns by $a_{ijS} = 1$ if player i is in coalition S, and zero otherwise. A is the incidence matrix of players versus sets, with the column representing S appearing as many times as there are corners in V_S.

Let me begin with an example of a three-person game in order to illustrate the problem. In this example, the set V_S for a typical two-player coalition will be assumed to have two corners. The matrix C is given by

$$
\begin{array}{ccccccccc}
(1) & (2) & (3) & (1,12) & (2,12) & (1,13) & (2,13) & (1,23) & (2,23)
\end{array}
$$

$$
\begin{bmatrix}
0 & M & M & 6 & 2 & 12 & 3 & M & M \\
M & 0 & M & 6 & 8 & M & M & 7 & 2 \\
M & M & 0 & M & M & 2 & 8 & 5 & 9
\end{bmatrix}.
$$

In this example each player, by himself, can obtain a maximum utility of zero. In general, information about V_N need not be included in the matrix C.

It is useful to examine the problem from a geometric point of view. I have drawn, in Figure 2, the set, which may be called V, of those points in the nonnegative orthant which are necessarily contained in $V_{(123)}$ if the game is balanced. V contains those points on the coordinate planes which are achievable by the two-player coalitions, since a two-player coalition and its complementary one-player coalition form a balanced collection. V also contains those vectors (u_1, u_2, u_3) with $(u_1, u_2) \in V_{(12)}$, $(u_1, u_3) \in V_{(13)}$, and $(u_2, u_3) \in V_{(23)}$. From the assumption that the game is balanced we know only that $V_{(123)}$ contains V; it may be considerably larger.

No point in the nonnegative orthant can be blocked by a one-player coalition. Is there a point in V which can be blocked by no two-player coalition? In other words is there a point in V whose three projections are on the boundaries of the sets of utility levels achievable by the two-player coalitions? The reader may verify that there is only one such point, $u = (3, 6, 5)$. Of course this vector need not be in the core if it is

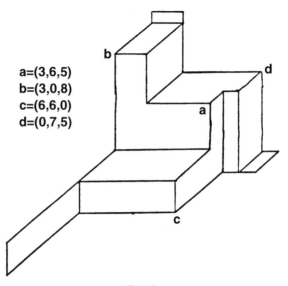

a=(3,6,5)
b=(3,0,8)
c=(6,6,0)
d=(0,7,5)

FIG. 2

not Pareto optimal, but then any Pareto optimal point in $V_{(123)}$ which is greater than or equal to u, is in the core.

The vector $(3, 6, 5)$ is generated by the points $(6, 6, 0)$, $(3, 0, 8)$, and $(0, 7, 5)$ in the sense that if we form the square submatrix of C, corresponding to these points,

$$\begin{bmatrix} 6 & 3 & M \\ 6 & M & 7 \\ M & 8 & 5 \end{bmatrix},$$

and define u_i to be the minimum of the ith row of this square submatrix, then $(u_1, u_2, u_3) = (3, 6, 5)$. The fact that u cannot be blocked is clear from the C matrix, for if u were blocked by S, then there would be a column j, S with $c_{ijS} > u_i$ for all i, and the reader may verify that no such column exists.

Analytically the argument that u is in $V_{(123)}$, if the game is balanced, depends on the observation that

$$(3, 6, 5)^{(12)} = (3, 6) \leq (6, 6) \in V_{(12)},$$

$$(3, 6, 5)^{(13)} = (3, 5) \leq (3, 8) \in V_{(13)},$$

$$(3, 6, 5)^{(23)} = (6, 5) \leq (7, 5) \in V_{(23)},$$

and that the three two-element sets form a balanced collection.

In the general case we also consider a square submatrix of C, and define u_i to be the minimum of the entries in the ith row of this submatrix. For the vector u to lead us to a point in the core two properties are required. First of all we want u to be blocked by no coalition, and this means that for every column in the C matrix at least one entry must be less than or equal to the corresponding entry in the u vector. Of course not every square submatrix of C will product a u vector with this property, and part of the algorithm will be concerned with determining submatrices of this sort.

In order to conclude that the vector u is in V_N a second condition will have to be imposed on the columns defining the submatrix of C.

Let T be the collection of those coalitions S which appear in at least one column of the square submatrix. For each S in T the vector u^S is surely in V_S, since it is less than or equal to one of the corners appearing in V_S. In order to conclude that $u \in V_N$ it is sufficient that the collection T be balanced, or in other words that there exist nonnegative numbers δ_S, zero for those S not in T, and such that

$$\sum_{S \supset \{i\}} \delta_S = 1, \quad \text{for } i = 1, \dots, n.$$

But this is equivalent to saying that the equations $Ax = e$ (with e the vector all of whose components are 1) have a nonnegative solution, with $x_{jS} = 0$, for any column j, S not appearing in the square submatrix of C, since we may take $\delta_S = \Sigma_j x_{jS}$.

In other words we look for a feasible basis in the sense used in linear programming for the equations $Ax = e$. The n columns of this feasible basis give rise to a square submatrix of C, and u_i is defined to be the minimum of the ith row of this submatrix. The feasible basis is to be selected so that for every column in the C matrix at least one entry is less than or equal to the corresponding entry in the u vector.

It is very useful to generalize the problem by considering an arbitrary matrix A, with n rows and m columns, rather than a repeated incidence matrix, a C matrix of the same dimensions as A, and an arbitrary vector b. In this more general case the columns of both the A and C matrix will have the subscript j rather than the more cumbersome subscript (j, S) appropriate to a repeated incidence matrix. We look for a feasible basis for the equations $Ax = b$, and for each such basis we define $u_i = \min\{c_{ij} | \text{for all } j \text{ appearing in the feasible basis}\}$. Will there be a basis, so that for every column k, there is at least one i with $u_i \geq c_{ik}$?

In order to guarantee an affirmative answer to this more general question, the matrices A and C will be assumed to have the properties described in the following definition.

DEFINITION: Let A and C be two n by m matrices of the form:

$$A = \begin{bmatrix} 1 & \cdots & 0 & a_{1,n+1} & \cdots & a_{1,m} \\ \vdots & & & \vdots & & \\ 0 & \cdots & 1 & a_{n,n+1} & \cdots & a_{n,m} \end{bmatrix},$$

$$C = \begin{bmatrix} c_{1,1} & \cdots & c_{1,n} & c_{1,n+1} & \cdots & c_{1,m} \\ \vdots & & & \vdots & & \\ c_{n,1} & \cdots & c_{n,n} & c_{n,n+1} & \cdots & c_{n,m} \end{bmatrix}.$$

We say that A and C are in standard form if

1. for each row i, c_{ii} is the minimum of the elements in its row, and
2. for each nondiagonal element c_{ij} in the square submatrix of C consisting of the first n columns, and for each column $k > n$, we have $c_{ij} \geq c_{ik}$.

The reader may easily verify that the matrices A and C arising from a finite game are in standard form if the vectors $u^{j,S}$ representing the corners, give each player in S no less than the maximum that he can achieve by himself, and if the constant M is selected to be larger than any of the components of these vectors.

THEOREM 2: *Let A and C be two n by m matrices in standard form. Let b be a nonnegative vector such that the set $\{x \mid x \geq 0 \text{ and } Ax = b\}$ is bounded. Then there is a feasible basis for the equations $Ax = b$, so that if we define $u_i = \min c_{ij}$, for all columns j in this basis, then for every column k, there is an index i with $u_i \geq c_{ik}$.*

From the previous discussion it is clear that Theorem 2 implies that a balanced game has a nonempty core if each V_S has a finite number of corners.

4. AN ALGORITHM FOR THE PROBLEM OF THEOREM 2

The problem posed in Theorem 2 is not remotely a linear programming problem even though only linear inequalities are involved. Any attempt to cast this problem in a linear programming form would run into the difficulty that not all of the relevant inequalities are to be satisfied simultaneously. An attempt to use integer programming methods would neither provide an existence theorem nor take advantage of the special structure of the problem. The algorithm of this paper, which is based on the ingenious procedure discovered by Lemke and Howson [4, 5] for the solution of a two-person nonzero-sum game, provides a method for calculating a solution to this problem, and since the algorithm terminates in a finite number of steps, the existence of at least one solution is guaranteed.

Theorem 2 may be demonstrated directly by an examination of the two-person nonzero-sum game in which one player has a payoff matrix A and has the columns for his pure strategies. The second player, whose strategies are the rows, has a payoff matrix given by $B = (b_{ij})$ where $b_{ij} = -1/c_{ij}^{\eta}$ (we may assume, without loss of generality, that $c_{ij} > 0$, since the theorem is unchanged if each entry in C is increased by the same amount). Theorem 2 may then be obtained by letting the exponent η tend to infinity and considering a convergent subsequence of equilibrium points for these games. The proof is quite simple, but since it involves the selection of a convergent subsequence it is basically non-constructive, even though the Lemke-Howson argument may be used to calculate an equilibrium point for each value of η. The remainder of this section will be devoted to a modification of this algorithm which is applicable to the limiting case directly.

The terminology introduced in the following definition will be useful in discussing the limiting algorithm.

DEFINITION: An *ordinal basis* for the matrix C consists of a set of n columns $j_1, j_2, \ldots, j_n$ so that if $u_i = \min(c_{ij_1}, c_{ij_2}, \ldots, c_{ij_n})$, then for every column k, there is at least one i with $u_i \geq c_{ik}$.

The term *ordinal basis* used in this definition is meant to be suggestive of an analogy with linear programming, as we shall see from some of the properties described below. But first of all it should be noticed that our theorem will be demonstrated if we can exhibit a feasible basis for A which is simultaneously an ordinal basis for C.

The algorithm for the determination of such a set of columns alternates between pivot steps for the linear equations, and a related operation on the matrix C. We make the standard nondegeneracy assumption that all of the variables associated with the n columns of a feasible basis for the equations $Ax = b$ are strictly positive. The nondegeneracy assumption for C takes the rather novel form that no two elements in the same row are equal. Both of these assumptions can be brought about by perturbations of the corresponding matrices.

Before turning to a discussion of pivot steps it is useful to note that the nondegeneracy assumption for C implies that each column of an ordinal basis has precisely one row minimizer to be used in forming the vector u. This may be seen if the column k, in the definition of an ordinal basis, is taken to be a column in the basis.

LEMMA 1: *Let $j_1, j_2, \ldots, j_n$ be the columns of a feasible basis for the equations $Ax = b$, and let j^* be an arbitrary column not in this collection. Then, if the problem is nondegenerate and the convex set $\{x \mid x \geq 0, Ax = b\}$ is bounded, there is a unique feasible basis consisting of column j^* and $n - 1$ columns of the original feasible basis.*

This is, of course, a standard result in linear programming, which says that if the constraint set is bounded and if the problem is nondegenerate, then any column outside of a basis may be introduced and as a result of a pivot step precisely one column will be eliminated. Something very much like taking a pivot step may be applied to an ordinal basis of the matrix C. With one exception, a specific column in an ordinal basis may be removed and a unique column introduced from outside so that the new set of columns is also an ordinal basis.

LEMMA 2: *Let $j_1, j_2, \ldots, j_n$ be an ordinal basis of C, and j_1 an arbitrary one of these columns. Assume that $j_2, \ldots, j_n$ are not all selected form the first n columns of C. Then if no two elements in the same row of C are equal and if C is in standard form, there is a unique column $j^* \neq j_1$ such that $j^*, j_2, \ldots, j_n$ is an ordinal basis.*

The steps involved in removing a specific column from an ordinal basis and replacing it by a column outside the basis will be called an ordinal pivot step. We shall first give a definition of an ordinal pivot step, then show that the new set of columns is an ordinal basis, and finally demonstrate that the introduction of no other column will lead to an ordinal basis.

DEFINITION: Consider an ordinal basis for C and a specific column to be removed. In the $n \times n - 1$ matrix of remaining columns precisely one column will contain two row minimizers, one of which is new and the other a row minimizer for the original basis. Let the row associated with the latter have an index i^*. Examine all columns in C for which

$c_{ik} > \min\{c_{ij}|j$ remains in the basis$\}$ holds for all i not equal to i^*. Of these columns, select the one which maximizes c_{i^*k}. An ordinal pivot step introduces this columns into the basis.

It is possible to carry out an ordinal pivot step if there is a column in C for which $c_{ik} > \min\{c_{ij}|j$ remains in the basis$\}$ for all $i \neq i^*$. But since the matrix C is in *standard form*, the column

$$\begin{bmatrix} c_{1i^*} \\ \vdots \\ c_{ni^*} \end{bmatrix}$$

will surely satisfy this condition unless the $n - 1$ columns remaining in the basis come from the first n columns of the matrix C.

If j^* is the column brought into the basis, then the new u vector is given by $u'_i = \min\{c_{ij}|j$ remains in the basis$\}$ for $i \neq i^*$, and $u'_{i^*} = c_{i^*j^*}$. The way in which j^* is selected implies that there is no column in C all of whose components are strictly larger than those of u', so that the new collection of columns is an ordinal basis.

In order to see that the introduction of no column other than j^* will lead to an ordinal basis, let us consider the square submatrix obtained from the original ordinal basis:

$$\begin{bmatrix} c_{1j_1} & c_{1j_2} & \cdots & c_{1j_n} \\ c_{2j_1} & c_{2j_2} & \cdots & c_{2j_n} \\ \vdots & & & \\ c_{nj_1} & c_{nj_2} & \cdots & c_{nj_n} \end{bmatrix}.$$

To be specific let us assume that the row minimizers occur along the diagonal, that column j_1 is to be removed from the basis, and that the second smallest element in the first row is c_{1j_2}. This implies $i^* = 2$.

When a new column is brought in to replace the first column, c_{3j_3} must still be the row minimum for the third row, since otherwise there would be no row minimum in the third column, and similarly for $c_{4j_4}, \ldots, c_{nj_n}$. We know therefore that if column j^* is brought into the basis, then $c_{3j^*} > c_{3j_3}, c_{4j^*} > c_{4j_4}, \ldots, c_{nj^*} > c_{nj_n}$.

Two cases occur for the row minima of the first two rows, either $c_{1j^*} < c_{1j_2}$ and $c_{2j^*} > c_{2j_2}$ or the reverse. The first case leads back to the original basis. To see this we notice that if the first case does take place, the new u vector will be given by

$$\begin{bmatrix} c_{1j^*} \\ c_{2j_2} \\ \vdots \\ c_{nj_n} \end{bmatrix},$$

and if the new set of columns are to be an ordinal basis, then for any column k we must have $u_i \geq c_{ik}$ for at least one i. But if k is the column j_1, this means $c_{ij^*} \geq c_{1j_1}$. On the other hand the old set of columns was assumed to be an ordinal basis, and so for any k, $c_{ij_i} \geq c_{ik}$ for some i. But here we may take $k = j^*$ and we see that $c_{1j_1} \geq c_{1j^*}$, so that $c_{1j_1} = c_{1j^*}$. By the nondegeneracy assumption, no two elements of the same row are equal and therefore $j^* = j_1$ and we are back to the original set of columns.

It is in the second variant, in which the minimizing elements in the first two rows are reversed, that we move to a new basis. In this case we look for a column j^* in which $c_{1j^*} > c_{1j_2}, c_{3j^*} > c_{3j_3}, \ldots, c_{nj^*} > c_{nj_n}$, or $c_{ij^*} > \min \{c_{ij} | j \text{ remains in the basis}\}$ for all $i \neq 2$, and in order for the new basis to be feasible we must select from these columns so as to maximize c_{2j^*}. But this is the column described in the definition of an ordinal pivot step, and Lemma 2 has therefore been demonstrated.

It is useful to note that ordinal pivot steps are reversible; if j_1 is eliminated from a basis and j^* brought in, then j^* may be eliminated from the new basis and the original basis will be obtained. Ordinal pivot steps are remarkably easy to carry out. They involve only ordinal comparisons of elements in the same row, and therefore the entries in the matrix C can be selected from arbitrary ordered sets, one for each row, rather than being real numbers. For example the entries in a row may be commodity bundles ordered by a preference ordering.

We are now ready to discuss the algorithm for determining a set of columns that is simultaneously a feasible basis for the equations $Ax = b$, and an ordinal basis for the matrix C. It is quite easy to find a pair of bases, one a feasible basis for the matrix A, and the other an ordinal basis for the matrix C which, while not identical, are quite close, and we shall use such a pair of bases as a starting point in the algorithm. The columns $(1, 2, \ldots, n)$ form a feasible basis for the matrix A, and the columns

$$\begin{bmatrix} c_{1j} & c_{12} & \cdots & c_{1n} \\ c_{2j} & c_{22} & & c_{2n} \\ \vdots & \vdots & & \vdots \\ c_{nj} & c_{n2} & \cdots & c_{nn} \end{bmatrix}$$

form an ordinal basis for the matrix C if j is selected from all of the columns $k > n$ so as to maximize c_{1k}. The columns in the C basis are given by $(j, 2, \ldots, n)$. The relationship between the two bases may be described by saying that the A basis contains column 1 and $n - 1$ remaining columns. The $n - 1$ remaining columns are also contained in the C basis along with one additional column other than the first. The ingenious idea introduced by Lemke and Howson is to insist that this relationship be maintained between the two bases.

In other words we will always be in a position where the feasible A basis can be described by the columns $(1, j_2, j_3, \ldots, j_n)$ and the ordinal basis for C by $(j_1, j_2, j_3, \ldots, j_n)$, with $j_1 \neq 1$. What steps can be taken so as to preserve this property? There are only two possible steps, one a pivot step for the matrix A and the other an ordinal pivot step for the matrix C.

A pivot step on the matrix A will leave this relationship unchanged only if column j_1 is introduced into the A basis. It is of course possible that column 1 will be eliminated from the A basis when column j_1 is brought in; the problem would be solved if this were to occur since the same basis $(j_1, j_2, \ldots, j_n)$ would then be obtained for both matrices. If column 1 is not eliminated by the pivot step then some other column, say j_i, will be. The two bases will still stand in the same relation with j_i being the column in the C basis which does not appear in the A basis.

The other possible continuation is to do an ordinal pivot step on the C matrix, eliminating one of its columns. The mutual relation between the A and C bases will be retained only if column j_1 is eliminated from the C basis. If j_1 is eliminated and column 1 is introduced into the C basis, the problem is solved since the same basis $(1, j_2, \ldots, j_n)$ will then be obtained for both matrices. On the other hand if column $j^* \neq 1$ is brought into the C basis when j_1 is eliminated, the two bases again stand in the same relationship with j^* being the column in the C basis which does not appear in the A basis.

As we have seen, the ordinal pivot step on the matrix C can always be carried out, except in the case where the columns other than j_1 in the ordinal basis for C are selected from the first n columns in the C matrix. For this case to occur the A basis must be given by $(1, 2, \ldots, n)$ and the C basis by $(j, 2, \ldots, n)$ so that we are in the starting position described above. From the starting position there is only one pivot step to be taken, namely to introduce column j into the A basis. From all other positions in which the A and C bases stand in the correct relationship, two pivot steps are available.

These considerations suggest the following algorithm. Start with the bases described above and take the one pivot step that is available. At any other point one of the two possible pivot steps will have been used in reaching that position, so that the only continuation is by means of the one remaining pivot step. There is a unique continuation at each step and the process can only terminate when we pivot into a solution of the problem. There are a finite number of possible positions, and if we never return to the same position the process will inevitably terminate at a set of columns which is simultaneously a feasible basis for $Ax = b$, and an ordinal basis for C.

Cycling is impossible, for if the first position to be repeated is the initial position, this would imply that there are two possible pivot steps from the initial position, which we have seen to be false. On the other hand if the process first repeats at some point other than the initial position, then there would be *three* possible pivot steps proceeding from that point, which is again impossible. The algorithm must therefore terminate in a finite number of steps with a solution to the problem.

This concludes the proof of Theorem 2, and as a consequence, Theorem 1, in the case in which the sets V_S have a finite number of corners for each proper subset S.

5. AN EXAMPLE OF THE ALGORITHM

Let us consider an example of the algorithm applied to a balanced three-person game. I will assume that each player by himself can achieve a zero utility level and nothing larger. Each two-player V_S will be assumed to have three corners, and since the game is balance $V_{(123)}$ need not be explicitly considered. The matrix C will consist of twelve columns, three for each two-player coalition, and one column for each single player. Each column will have a large arbitrary number entered in those rows corresponding to players not in the coalition, and in order to avoid degeneracy these numbers will be different for different columns:

$$C = \begin{bmatrix} 0 & M_2 & M_3 & 12 & 3 & 2 & 9 & 5 & 4 & M_{10} & M_{11} & M_{12} \\ M_1 & 0 & M_3 & 6 & 7 & 9 & M_7 & M_8 & M_9 & 5 & 2 & 8 \\ M_1 & M_2 & 0 & M_4 & M_5 & M_6 & 3 & 8 & 10 & 6 & 9 & 4 \end{bmatrix}.$$

The M's are arbitrary other than satisfying the inequalities $M_1 > M_2 > \cdots > M_{12} > 12$. The matrix A is the incidence matrix of players versus sets suitably repeated.

In order to avoid degeneracy in the matrix A, the last column has been perturbed by small ε's, subject to $0 < \varepsilon_1 < \varepsilon_2 < \varepsilon_3$.

Step 1. We begin with a basis for the matrix A consisting of columns $(1, 2, 3)$ and for C columns $(10, 2, 3)$, so that the u vector associated with this C basis is $u = (M_{10}, 0, 0)$. The first step is to bring the tenth column into the A basis, and the pivot element for this step is printed in bold:

$$A = \begin{bmatrix} 1 & 0 & 0 & 1 & 1 & 1 & 1 & 1 & 1 & 0 & 0 & 0 & 1 + \varepsilon_1 \\ 0 & 1 & 0 & 1 & 1 & 1 & 0 & 0 & 0 & 1 & 1 & 1 & 1 + \varepsilon_2 \\ 0 & 0 & 1 & 0 & 0 & 0 & 1 & 1 & 1 & 1 & 1 & 1 & 1 + \varepsilon_3 \end{bmatrix}.$$

Column two has been removed from the A basis, and therefore it must be removed from the C basis. In the remaining two columns, column ten has two row minimizers, with the new one in the second row. We therefore examine all columns k, with $c_{2k} > 5$, $c_{3k} > 0$ and select the one which maximizes c_{1k}. This is column twelve.

Step 2. The A matrix which appears after the first pivot step is

$$\begin{bmatrix} 1 & 0 & 0 & 1 & 1 & 1 & 1 & 1 & 1 & 0 & 0 & 0 & 1 + \varepsilon_1 \\ 0 & 1 & 0 & 1 & 1 & 1 & 0 & 0 & 0 & 1 & 1 & 1 & 1 + \varepsilon_2 \\ 0 & -1 & 1 & -1 & -1 & -1 & 1 & 1 & 1 & 0 & 0 & 0 & \varepsilon_3 - \varepsilon_2 \end{bmatrix}.$$

The A basis is $(1, 3, 10)$ and the C basis $(12, 3, 10)$, with a u vector given by $u = (M_{12}, 5, 0)$. We continue by bringing column twelve in the A basis. No calculation is required since column ten, which is identical with column twelve, will be eliminated. Column ten must therefore be eliminated from the C basis. If we consider the submatrix of columns

$(12, 3, 10)$,

$$\begin{bmatrix} M_{12} & M_3 & M_{10} \\ 8 & M_3 & 5 \\ 4 & 0 & 6 \end{bmatrix},$$

we see that when column ten is eliminated, column twelve has two row minima, with the new one appearing in the second row. We therefore maximize c_{1k} subject to $c_{2k} > 8$, $c_{3k} > 0$, and obtain column seven.

The algorithm then proceeds through the following pivot steps.

Step 3. The A basis is $(1, 3, 12)$, the C basis $(7, 3, 12)$ and $u = (9, 8, 0)$. Column seven is brought into the A basis and column three removed. Column three is then removed from the C basis and column eight is introduced.

Step 4. The A basis is $(1, 7, 12)$, the C basis $(8, 7, 12)$ and $u = (5, 8, 3)$. Column eight is brought into the A basis and column seven removed. Column seven is then removed from the C basis and column four is introduced.

Step 5. The A basis is $(1, 8, 12)$, the C basis $(4, 8, 12)$ and $u = (5, 6, 4)$. Column four is introduced into the A basis, and column one eliminated, so that the solution is obtained. Columns $(4, 8, 12)$ form a feasible basis for A and an ordinal basis for C. The utility vector $u = (5, 6, 4)$ cannot be blocked by any two- or one-player set, and if the game is balanced this vector must be in $V_{(123)}$. Any Pareto optimum point in $V_{(123)}$ which is greater than or equal to $(5, 6, 4)$ must be in the core.

There is a substantial amount of arbitrariness in initiating the algorithm: in the ordering of the M's, in the lexicographical ordering used in the A matrix, and in the determination of a pair of bases which differ in at most one column. The question of which points in the core will be determined by these variations is an intriguing one but one which I would prefer to postpone.

6. SOME GENERAL REMARKS

We have completed the proof of Theorem 1 in the case in which each V_S has a finite number of corners. The general case may be studied by selecting a finite list of vectors from each set V_S, and applying the algorithm to obtain a vector in V_N which cannot be blocked by any vector in the list. If the number of vectors in each V_S is systematically increased, becoming everywhere dense in the limit, we will obtain a sequence of vectors in V_N, and any limit point of this sequence will be blocked by no coalition, thereby demonstrating Theorem 1.

In Section 2 it was shown that in an exchange economy, convexity of preferences implies that the game is balanced and therefore has a nonempty core. If the number of players tends to infinity in an appropriate way, and some additional assumptions are made, the core approaches the set of competitive equilibria. The existence of competitive equilibria has therefore been demonstrated by an argument based on an algorithm rather than the use of fixed point theorems. An alternative procedure, discovered by Mr. Rolf Mantel, is to apply Theorem 2 directly to the existence of equilibrium prices without any reference to *n*-person game theory. Mr. Mantel's work appears in his thesis [6].

I would like to suggest two motives for attempting to avoid the abstract proofs of the existence of equilibria that have previously been given. First, the modern treatment of competitive models has tended to focus on problems of existence rather than problems of computation, and this second aspect of the problem is worth examination. Of course a prohibitive amount of information would be required to calculate equilibrium prices for an actual economy, and this will surely continue to be true even if the capabilities of electronic computers were to increase at a fanciful rate in the future. On the other hand experimental calculations on small economic models will, I think, be quite useful, and it is now possible to perform such calculations on models involving the consumer side of the economy.

To be sure, fixed point theorems do provide a method of calculating equilibrium prices in the following sense. Let $f_j(\pi)$ be the excess demand for the *j*th commodity at prices π. These functions are assumed to be continuous, satisfy the Walras Law $\{\pi \cdot f(\pi) = 0\}$ and be homogeneous of degree zero. At an equilibrium system of prices each excess demand is less than or equal to zero, and the property of homogeneity permits us to restrict our attention to prices lying on the simplex $\pi_j \geq 0$, $\Sigma\pi_j = 1$. Let the price simplex be divided into a simplicial subdivision and label each vertex of the subdivision with some commodity whose excess demand is less than or equal to zero at that system of prices. Then Sperner's Lemma, the central argument in the proof of fixed point theorems, tell us that there will be at least one small simplex all of whose vertices are labeled differently. See Figure 3.

If the subdivision is sufficiently fine, the price system at the center of the distinguished simplex will have excess demands which are less than

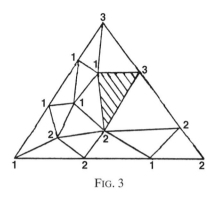

FIG. 3

or equal to zero or if positive the excess demands will be quite small. The price obtained in this way will be close to an equilibrium price in a functional sense though not necessarily in terms of Euclidean distance. The functional sense of distance in which each market is approximately in equilibrium is probably the relevant one, and since the excess demands for a finite number of price vectors can be calculated if consumer preference and the production set are known, we do have a computing procedure based on fixed point theorems for calculating equilibrium prices with a given degree of accuracy.

When the sets V_S are approximated by sets with a finite number of corners our algorithm will calculate, in a similar functional sense, a point arbitrary close to the core. Its advantage over the procedure based on Sperner's Lemma seems to me to be that the latter technique is essentially an exhaustive search, and not a systematic algorithm. This is admittedly a vague distinction which can only be clarified by computational experience, but the reader should compare the situation with that arising in linear programming. An exhaustive search of all vertices of a convex polyhedron in order to determine the maximum of a linear function is less efficient as a computational device than the orderly sequence of steps prescribed by the simplex method.

The second advantage of the method discussed in this paper, is in the possibility of its economic interpretation. Since the conventional Walrasian adjustment of prices does not always converge to equilibrium, we have at present no uniformly valid computational procedure for problems involving the consumer side of an economy which is at the same time economically suggestive. An exhaustive search based on Sperner's Lemma has no economic interpretation as a possible adjustment mechanism for arriving at an equilibrium state. On the other hand the method of this paper may be capable of such an interpretation. It proposes at each iteration a utility vector which cannot be blocked by any set of players. If the utility vector is feasible for the set of all players the problem is finished. If not a vector is proposed which differs from the old vector for only two players, one of whom receives more and the other less. The reader will have to judge for himself whether the algorithm has any economic interpretation as an adjustment mechanism, but my feeling is that the possibility of interpretation is distinctly higher than that arising from an exhaustive search.

Brouwer's fixed point theorem itself may be demonstrated quite simply by taking the matrix A as a repeated unit matrix in Theorem 2. Our algorithm can therefore replace Sperner's Lemma in providing the combinatorial basis for fixed point theorems. There is at present, however, insufficient evidence to know whether the algorithm seriously avoids an exhaustive search in approximating the fixed points of continuous mappings.

Cowles Foundation for Research in Economics, Yale University

REFERENCES

[1] Aumann, R. J., and B. Peleg: "Von Neumann-Morgenstern Solutions to Cooperative Games Without Side Payments," *Bulletin of the American Mathematical Society*, 66, (1960) pp. 173–179.
[2] Bondareva, O.: "The Core of an N Person Game," *Vestnik Leningrad University*, 17, (1962) No. 13, pp. 141–142.
[3] Debreu, G., and H. Scarf: "A Limit Theorem on the Core of an Economy," *International Economic Review*, Vol. 4, No. 3, Sept., 1963.
[4] Lemke, C. E.: "Bimatrix Equilibrium Points and Mathematical Programming," *Management Science*, Vol. 11, No. 7, May, 1965.
[5] Lemke, C. E., and J. T. Howson, "Equilibrium Points of Bi-matrix Games," *SIAM Journal*, Vol. 12, July, 1964.
[6] Mantel, Rolf: "Toward a Constructive Proof of the Existence of Equilibrium in a Competitive Economy," submitted as a thesis to the Department of Economics, Yale University, 1965.
[7] Peleg, B.: *An Inductive Method for Constructing Minimal Balanced Collection of Finite Sets*, Research Program in Game Theory and Mathematical Economics, Memorandum No. 3, February, 1964, Department of Mathematics, Hebrew University, Jerusalem.
[8] Shapley, L. S.: *On Balanced Sets and Cores*, RAND Corp. Memorandum, RM-4601-PR, June, 1965.

GAMES WITH INCOMPLETE INFORMATION PLAYED BY "BAYESIAN" PLAYERS

Part I. The Basic Model*†1

JOHN C. HARSANYI

UNIVERSITY OF CALIFORNIA, BERKELEY

The paper develops a new theory for the analysis of games with incomplete information where the players are uncertain about some important parameters of the game situation, such as the payoff functions, the strategies available to various players, the information other players have about the game, etc. However, each player has a subjective probability distribution over the alternative possibilities.

In most of the paper it is assumed that these probability distributions entertained by the different players are mutually "consistent," in the sense that they can be regarded as conditional probability distributions derived from a certain "basic probability distribution" over the parameters unknown to the various players. But later the theory is extended also to cases where the different players' subjective probability distributions fail to satisfy this consistency assumption.

In cases where the consistency assumption holds, the original game can be replaced by a game where nature first conducts a lottery in accordance

* Received June 1965, revised June 1966, accepted August 1966, and revised June 1967.

† Parts II and III of "Games with Incomplete Information Played by 'Bayesian' Players" will appear in subsequent issues of *Management Science: Theory*.

1 The original version of this paper was read at the Fifth Princeton Conference on Game Theory, in April, 1965. The present revised version has greatly benefitted from personal discussions with Professors Michael Maschler and Robert J. Aumann, of the Hebrew University, Jerusalem; with Dr. Reinhard Selten, of the Johann Wolfgang Goethe University, Frankfurt am Main; and with the other participants of the International Game Theory Workshop held at the Hebrew University in Jerusalem, in October-November 1965. I am indebted to Dr. Maschler also for very helpful detailed comments on my manuscript.

This research was supported by Grant No. GS-722 of the National Science Foundation as well as by a grant from the Ford Foundation to the Graduate School of Business Administration, University of California. Both of these grants were administered through the Center for Research in Management Science, University of California, Berkeley. Further support has been received from the Center for Advanced Study in the Behavioral Sciences, Stanford.

with the basic probability distribution, and the outcome of this lottery will decide which particular subgame will be played, i.e., what the actual values of the relevant parameters will be in the game. Yet, each player will receive only partial information about the outcome of the lottery, and about the values of these parameters. However, every player will know the "basic probability distribution" governing the lottery. Thus, technically, the resulting game will be a game with complete information. It is called the Bayes-equivalent of the original game. Part I of the paper describes the basic model and discusses various intuitive interpretations for the latter. Part II shows that the Nash equilibrium points of the Bayes-equivalent game yield "Bayesian equilibrium points" for the original game. Finally, Part III considers the main properties of the "basic probability distribution."

TABLE OF CONTENTS

Section 13. The proof of Theorem III.

Section 14. The assumption of mutual consistency among the different players' subjective probability distributions.

Section 15. Games with "inconsistent" subjective probability distributions.

Section 16. The possibility of spurious inconsistencies among the different players' subjective probability distributions.

Section 17. A suggested change in the formal definition of games with mutually consistent probability distributions.

GLOSSARY OF MATHEMATICAL NOTATION

I-game $\cdots$ A game with *incomplete* information.

C-game $\cdots$ A game with *complete* information.

G $\cdots$ The *I*-game originally given to us.

G^* $\cdots$ The Bayesian game equivalent to G. (G^* is a *C*-game.)

G^{**} $\cdots$ The Selten game equivalent to G and to G^*. (G^{**} is likewise a *C*-game.)

$\mathcal{N}(G), \mathcal{N}(G^*), \mathcal{N}(G^{**})$ $\cdots$ The normal form of G, G^*, and G^{**} respectively.

$\mathcal{S}(G), \mathcal{S}(G^*)$ $\cdots$ The semi-normal form of G and G^* respectively.

s_i $\cdots$ Some strategy (pure or mixed) of player i, with $i = 1, \ldots, n$.

$S_i = \{s_i\}$ $\cdots$ Player i's strategy space.

c_i $\cdots$ Player i's attribute vector (or information vector).

$C_i = \{c_i\}$ $\cdots$ The range space of vector c_i.

$c = (c_1, \ldots, c_n)$ $\cdots$ The vector obtained by combining the n vectors $c_1, \ldots, c_n$ into one vector.

$C = \{c\}$ $\cdots$ The range space of vector c.

$c^i = (c_1, \ldots, c_{i-1}, c_{i+1}, \ldots, c_n)$ $\cdots$ The vector obtained from vector c by omitting subvector c_i.

$C^i = \{c^i\}$ $\cdots$ The range space of vector c^i.

x_i $\cdots$ Player i's payoff (expressed in utility units).

$x_i = U_i(s_1, \ldots, s_n) = V_i(s_1, \ldots, s_n; c_1, \ldots, c_n)$ $\cdots$ Player i's payoff function.

$P_i(c_1, \ldots, c_{i-1}, c_{i+1}, \ldots, c_n) = P_i(c^i) = R_i(c^i | c_i)$ $\cdots$ The subjective probability distribution entertained by player i.

$R^* = R^*(c_1, \ldots, c_n) = R^*(c)$ $\cdots$ The basic probability distribution of the game.

$R_i^* = R^*(c_1, \ldots, c_{i-1}, c_{i+1}, \ldots, c_n | c_i) = R^*(c^i | c_i)$ $\cdots$ The conditional probability distribution obtained from R^* for a given value of vector c_i.

k_i $\cdots$ The number of different values that player i's attribute vector c_i can take in the game (in cases where this number is finite).

$K = \sum_{i=1}^{n} k_i$ $\cdots$ The number of players in the Selten game G^{**} (when this number is finite).

s_i^* $\cdots$ A normalized strategy of player i. (It is a function from the range space C_i of player i's attribute vector c_i, to his strategy space S_i.)

$S_i^* = \{s_i^*\}$ $\cdots$ The set of all normalized strategies s_i^* available to player i.

$\mathscr{E}$ $\cdots$ the expected-value operator.

$\mathscr{E}(x_i) = W_i(s_1^*, \ldots, s_n^*)$ $\cdots$ Player i's normalized payoff function, stating his *unconditional* payoff expectation.

$\mathscr{E}(x_i|c_i) = Z_i(s_1^*, \ldots, s_n^*|c_i)$ $\cdots$ Player i's semi-normalized payoff function, stating his *conditional* payoff expectation for a given value of his attribute vector c_i.

D $\cdots$ A cylinder set, defined by the condition $D = D_1 \times \cdots \times D_n$, where $D_1 \subseteq C_1, \ldots, D_n \subseteq C_n$.

$G(D)$ $\cdots$ For a given decomposable game G or G^*, $G(D)$ denotes the component game played in all cases where the vector c lies in cylinder D. D is called the defining cylinder of the component game $G(D)$.

Special Notation in Certain Sections

In section 3 (Part I):

a_{0i} denotes a vector consisting of those parameters of player i's payoff function U_i which (in player j's opinion) are *unknown* to all n players.

a_{ki} denotes a vector consisting of those parameters of the function U_i which (in j's opinion) are unknown to some of the players but are *known* to player k.

$a_0 = (a_{01}, \ldots, a_{0n})$ is a vector summarizing all information that (in j's opinion) none of the players have about the functions $U_1, \ldots, U_n$.

$a_k = (a_{k1}, \ldots, a_{kn})$ is a vector summarizing all information that (in j's opinion) player k has about the functions $U_1, \ldots, U_n$, except for the information that (in j's opinion) *all* n players have about these functions.

b_i is a vector consisting of all those parameters of player i's subjective probability distribution P_i which (in player j's opinion) are unknown to some or all of the players $k \neq i$.

In terms of these notations, players i's information vector (or attribute vector) c_i can be defined as

$c_i = (a_i, b_i)$.

V_i^* denotes player i's payoff function *before* vector a_0 has been integrated out. *After* elimination of vector a_0 the symbol V_i is used to denote player i's payoff function.

In sections 9–10 (Part II):

a^1 and a^2 denote the two possible values of player 1's attribute vector c_1.

b^1 and b^2 denote the two possible values of player 2's attribute vector c_2.

$r_{km} = R^*(c_1 = a^k$ and $c_2 = b^m)$ denotes the probability mass function corresponding to the basic probability distribution R^*.

$p_{km} = r_{km}/(r_{k1} + r_{k2})$ and $q_{km} = r_{km}/(r_{1m} + r_{2m})$ denote the corresponding *conditional* probability mass functions.

y^1 and y^2 denote player 1's two pure strategies.

z^1 and z^2 denote player 2's two pure strategies.

$y^{nt} = (y^n, y^t)$ denotes a normalized pure strategy for player 1, requiring the use of strategy y^n if $c_1 = a^1$, and requiring the use of strategy y^t if $c_1 = a^2$.

$z^{uv} = (z^u, z^v)$ denotes a normalized pure strategy for player 2, requiring the use of strategy z^u if $c_2 = b^1$, and requiring the use of strategy z^v if $c_2 = b^2$.

In section 11 (Part II):

a^1 and a^2 denote the two possible values that *either* player's attribute vector c_i can take.

$r_{km} = R^*(c_1 = a^k$ and $c_2 = a^m)$.

p_{km} and q_{km} have the same meaning as in sections 9–10.

y_i^* denotes player i's payoff demand.

y_i denotes player i's *gross* payoff.

x_i denotes player i's *net* payoff.

x_i^* denotes player i's net payoff in the case $(c_1 = a^1, c_2 = a^2)$.

x_i^{**} denotes player i's net payoff in the case $(c_1 = a^2, c_2 = a^1)$.

In section 13 (Part III):

$\alpha, \beta, \gamma, \delta$ denote specific values of vector c.

$\alpha_i, \beta_i, \gamma_i, \delta_i$ denote specific values of vector c_i.

$\alpha^i, \beta^i, \gamma^i, \delta^i$ denote specific values of vector c^i, etc.

$r_i(\gamma^i|\gamma_i) = R_i(c^i = \gamma^i|c_i = \gamma_i)$ denotes the probability mass function corresponding to player i's subjective probability distribution R_i (when R_i is a discrete distribution).

$r^*(\gamma) = R^*(c = \gamma)$ denotes the probability mass function corresponding to the basic probability distribution R^* (when R^* is a discrete distribution).

$\mathscr{R} = \{r^*\}$ denotes the set of all admissible probability mass functions r^*.

E denotes a similarity class, i.e., set of nonnull points $c = \alpha, c = \beta, \cdots$ similar to one another (in the sense defined in Section 13).

In section 16 (Part III):

$R^{(i)}$ denotes the basic probability distribution R^* as assessed by player i ($i = 1, \ldots, n$).

$R^{*\prime}$ denotes a given player's (player j's) *revised* estimate of the basic probability distribution R^*.

$c_i' = (c_i, d_i)$ denotes player j's *revised* definition of player i's attribute vector c_i. (It is in general a larger vector than the vector c_i originally assumed by player j.)

R_i' denotes player j's *revised* estimate of player i's subjective probability distribution R_i.

1.

Following von Neumann and Morgenstern [7, p. 30], we distinguish between games with *complete information*, to be sometimes briefly called *C*-games in this paper, and games with *incomplete information*, to be called *I*-games. The latter differ from the former in the fact that some or all of the players lack full information about the "rules" of the game, or equivalently about its normal form (or about its extensive form). For example, they may lack full information about other players' or even their own payoff functions, about the physical facilities and strategies available to other players or even to themselves, about the amount of information the other players have about various aspects of the game situation, etc.

In our own view it has been a major analytical deficiency of existing game theory that it has been almost completely restricted to *C*-games, in spite of the fact that in many real-life economic, political, military, and other social situations the participants often lack full information about some important aspects of the "game" they are playing.[2]

It seems to me that the basic reason why the theory of games with incomplete information has made so little progress so far lies in the fact that these games give rise, or at least appear to give rise, to an infinite regress in reciprocal expectations on the part of the players, [3, pp. 30–32]. For example, let us consider any two-person game in which the players do not know each other's payoff functions. (To simplify our discussion I shall assume that each player knows his own payoff function. If we made the opposite assumption, then we would have to introduce even more complicated sequences of reciprocal expectations.)

In such a game player 1's strategy choice will depend on what he expects (or believes) to be player 2's payoff function U_2, as the nature of the latter will be an important determinant of player 2's behavior in the game. This expectation about U_2 may be called player 1's *first-order* expectation. But his strategy choice will also depend on what he expects to be player 2's first-order expectation about his own (player 1's) payoff function U_1. This may be called player 1's *second order* expectation, as it is an expectation concerning a first-order expectation. Indeed, player 1's strategy choice will also depend on what he expects to be player 2's second-order expectation—that is, on what player 1 thinks that player 2 thinks that player 1 thinks about player 2's payoff function U_2. This we may call player 1's *third-order* expectation—and so on *ad infinitum*. Likewise, player 2's strategy choice will depend on an infinite sequence consisting of his first-order, second-order, third-order, etc., expectations concerning the payoff functions U_1 and U_2. We shall call any model of this kind a *sequential-expectations* model for games with incomplete information.

If we follow the Bayesian approach and represent the players expectations or beliefs by subjective probability distributions, then player 1's *first-order* expectation will have the nature of a subjective probability distribution $P_1^1(U_2)$ over all alternative payoff functions U_2 that player 2

---

[2] The distinction between games with *complete* and *incomplete* information (between *C*-games and *I*-games) must not be confused with that between games with *perfect* and *imperfect* information. By common terminological convention, the first distinction always refers to the amount of information the players have about the *rules* of the game, while the second refers to the amount of information they have about the other players' and their own previous *moves* (and about previous chance moves). Unlike games with *incomplete* information, those with *imperfect* information have been extensively discussed in the literature.

may possibly have. Likewise, player 2's first-order expectation will be a subjective probability distribution $P_2^1(U_1)$ over all alternative payoff functions U_1 that player 1 may possibly have. On the other hand, player 1's *second-order* expectation will be a subjective probability distribution $P_1^2(P_2^1)$ over all alternative first-order subjective probability distributions P_2^1 that player 2 may possibly choose, etc. More generally, the kth-order expectation ($k > 1$) of either player i will be a subjective probability distribution $P_i^k(P_j^{k-1})$ over all alternative $(k-1)$th-order subjective probability distributions P_j^{k-1} that the other player j may possibly entertain.[3]

In the case of n-person I-games the situation is, of course, even more complicated. Even if we take the simpler case in which the players know at least their own payoff functions, each player in general will have to form expectations about the payoff functions of the other $(n-1)$ players, which means forming $(n-1)$ different first-order expectations. He will also have to form expectations about the $(n-1)$ first-order expectations entertained by each of the other $(n-1)$ players, which means forming $(n-1)^2$ second-order expectations, etc.

The purpose of this paper is to suggest an alternative approach to the analysis of games with incomplete information. This approach will be based on constructing, for any given I-game G, some C-game G^* (or possibly several different C-games G^*) game-theoretically *equivalent* to G. By this means we shall reduce the analysis of I-games to the analysis of certain C-games G^*; so that the problem of such sequences of higher and higher-order reciprocal expectations will simply not arise.

As we have seen, if we use the Bayesian approach, then the sequential-expectations model for any given I-game G will have to be analyzed in terms of infinite sequences of higher and higher-order subjective probability distributions, i.e., subjective probability distributions over subjective probability distributions. In contrast, under our own model, it will be possible to analyze any given I-game G in terms of *one* unique

---

[3] Probability distributions over some space of payoff functions or of probability distributions, and more generally probability distributions over function spaces, involve certain technical mathematical difficulties [5, pp. 355–357]. However, as Aumann has shown [1] and [2], these difficulties can be overcome. But even if we succeed in defining the relevant higher-order probability distributions in a mathematically admissible way, the fact remains that the resulting model—like *all* models based on the sequential-expectations approach —will be extremely complicated and cumbersome. The main purpose of this paper is to describe an alternative approach to the analysis of games with incomplete information, which completely avoids the difficulties associated with sequences of higher and higher-order reciprocal expectations.

probability distribution R^* (as well as certain conditional probability distributions derived from R^*).

For example, consider a two-person non-zero-sum game G representing price competition between two duopolist competitors where neither player has precise information about the cost functions and the financial resources of the other player. This, of course, means that neither player i will know the true payoff function U_j of the other player j, because he will be unable to predict the profit (or the loss) that the other player will make with any given choice of strategies (i.e., price and output policies) s_1 and s_2 by the two players.

To make this example more realistic, we may also assume that each player has *some* information about the other player's cost functions and financial resources (which may be represented, e.g., by a subjective probability distribution over the relevant variables); but that each player i lacks exact information about *how much* the other player j actually knows about player i's cost structure and financial position.

Under these assumptions this game G will be obviously an I-game, and it is easy to visualize the complicated sequences of reciprocal expectations (or of subjective probability distributions) we would have to postulate if we tried to analyze this game in terms of the sequential-expectations approach.

In contrast, the new approach we shall describe in this paper will enable us to reduce this I-game G to an equivalent C-game G^* involving four random events (i.e., chance moves) e_1, e_2, f_1, and f_2, assumed to occur *before* to the two players choose their strategies s_1 and s_2. The random event e_i $(i = 1, 2)$ will determine player i's cost functions and the size of his financial resources; and so will completely determine his payoff function U_1 in the game. On the other hand, the random event f_i will determine the amount of information that player i will obtain about the cost functions and the financial resources of the other player j $(j = 1, 2$ and $\neq i)$, and will thereby determine the actual amount of information[4] that player i will have about player j's payoff function U_j.

---

[4] In terms of the terminology we shall later introduce, the variables determined by the random events e_i and f_i will constitute the random vector c_i $(i = 1, 2)$, which will be called player i's information vector or attribute vector, and which will be assumed to determine player i's "type" in the game (cf. the third paragraph below).

Both players will be assumed to know the joint probability distribution $R^*(e_1, e_2, f_1, f_2)$ of these four random events.[5] But, e.g., player 1 will know the actual outcomes of these random events only in the case of e_1 and f_1, whereas player 2 will know the actual outcomes only in the case of e_2 and f_2. (In our model this last assumption will represent the facts that each player will know only his own cost functions and financial resources but will not know those of his opponent; and that he will, of course, know how much information he himself has about the opponent but will not know exactly how much information the opponent will have about him.)

As in this new game G^* the players are assumed to know the probability distribution $R^*(e_1, e_2, f_1, f_2)$, this game G^* will be a C-game. To be sure, player 1 will have no information about the outcomes of the chance moves e_2 and f_2, whereas player 2 will have no information about the outcomes of the chance moves e_1 and f_1. But these facts will not make G^* a game with "incomplete" information but will make it only a game with "imperfect" information (cf. Footnote 2 above). Thus, our approach will basically amount to replacing a game G involving *incomplete* information, by a new game G^* which involves *complete* but *imperfect* information, yet which is, as we shall argue, essentially equivalent to G from a game-theoretical point of view (see Section 5 below).

As we shall see, this C-game G^* which we shall use in the analysis of a given I-game G will also admit of an alternative intuitive interpretation. Instead of assuming that certain important *attributes* of the players are determined by some hypothetical random events at the beginning of the game, we may rather assume that the players *themselves* are drawn at random from certain hypothetical populations containing a mixture of individuals of different "types," characterized by different attribute vectors (i.e., by different combinations of the relevant attributes). For instance, in our duopoly example we may assume that each player i ($i = 1, 2$) is drawn from some hypothetical population Π_i containing individuals of different "types," each possible "type" of player i being characterized by a different attribute vector c_i, i.e., by a different combination of production costs, financial resources, and states of information. Each player i will know his own type or attribute vector c_i but will be, in general, ignorant of his opponent's. On the other hand,

---

[5] For justification of this assumption, see Sections 4 and 5 below, as well as Part III of this paper.

both players will again be assumed to know the joint probability distribution $R^*(c_i, c_2)$ governing the selection of players 1 and 2 of different possible types c_1 and c_2 from the two hypothetical populations Π_1 and Π_2.

It may be noted, however, that in analyzing a given I-game G, construction of an equivalent C-game G^* is only a partial answer to our analytical problem, because we are still left with the task of defining a suitable solution concept for this C-game G^* itself, which may be a matter of some difficulty. This is so because in many cases the C-game G^* we shall obtain in this way will be a C-game of unfamiliar form, for which no solution concept has been suggested yet in the game-theoretical literature.[6] Yet, since G^* will always be a game with complete information, its analysis and the problem of defining a suitable solution concept for it, will be at least amenable to the standard methods of modern game theory. We shall show in some examples how one actually can define appropriate solution concepts for such C-games G^*.

2.

Our analysis of I-games will be based on the assumption that, in dealing with incomplete information, every player i will use the Bayesian approach. That is, he will assign a *subjective* joint probability distribution P_i to all variables unknown to him—or at least to all unknown *independent* variables, i.e., to all variables not depending on the players' own strategy choices. Once this has been done he will try to maximize the mathematical expectation of his own payoff x_i in terms of this probability distribution P_i.[7] This assumption will be called the *Bayesian hypothesis*.

If incomplete information is interpreted as lack of full information by the players about the *normal form* of the game, then such incomplete information can arise in three main ways.

---

[6] More particularly, this game G^* will have the nature of a game with *delayed commitment* (see Section 11 in Part II of this paper).

[7] A *subjective* probability distribution P_i entertained by a given player i is defined in terms of his own choice behavior, cf. [6]. In contrast, an *objective* probability distribution P^* is defined in terms of the long-run frequencies of the relevant events (presumably as established by an independent observer, say, the umpire of the game). It is often convenient to regard the subjective probabilities used by a given player i as being his personal *estimates* of the corresponding objective probabilities or frequencies unknown to him.

1. The players may not know the *physical outcome function* Y of the game, which specifies the physical outcome $y = Y(s_1, \ldots, s_n)$ produced by each strategy n-tuple $s = (s_1, \ldots, s_n)$ available to the players.

2. The players may not know their own or some other players' *utility functions* X_i, which specify the utility payoff $x_i = X_i(y)$ that a given player i derives from every possible physical outcome y.[8]

3. The players may not know their own or some other players' *strategy spaces* S_i, i.e., the set of all strategies s_i (both pure and mixed) available to various players i.

All other cases of incomplete information can be reduced to these three basic cases—indeed sometimes this can be done in two or more different (but essentially equivalent) ways. For example, incomplete information may arise by some players' ignorance about the amount or the quality of physical resources (equipment, raw materials, etc.) available to some other players (or to themselves). This situation can be equally interpreted *either* as ignorance about the physical outcome function of the game (case 1), *or* as ignorance about the strategies available to various players (case 3). Which of the two interpretations we have to use will depend on how we choose to define the "strategies" of the players in question. For instance, suppose that in a military engagement our own side does not know the number of firearms of a given quality available to the other side. This can be interpreted as inability on our part to predict the *physical outcome* (i.e., the amount of destruction) resulting from alternative strategies of the opponent, where any given "strategy" of his is defined as firing a given *percentage* of his firearms (case 1). But is can also be interpreted as inability to decide whether certain strategies are *available* to the opponent at all, where now any given "strategy" of his is defined as firing a specified *number* of firearms (case 3).

Incomplete information can also take the form that a given player i does not know whether another player j does or does not have information about the occurrence or nonoccurrence of some specified event e.

---

[8] If the physical outcome y is simply a vector of money payoffs $y_1, \ldots, y_n$ to the n players then we can usually assume that any player i's utility payoff $x_i = X_i(y_i)$ is a (strictly increasing) function of his money payoff y_i and that all players will know this. However, the other players j may not know the specific mathematical form of player i's utility function for money, X_i. In other words, even though they may know player i's *ordinal* utility function, they may not know his *cardinal* utility function. That is to say, they may not know how much *risk* he would be willing to take in order to increase his money payoff y_i by given amounts.

Such a situation will always come under case 3. This is so because in a situation of this kind, from a game-theoretical point of view, the crucial fact is player i's inability to decide whether player j is in a position to use any strategy s_j^0 involving *one* course of action in case event e does occur, and *another* course of action in case event e does not occur. That is, the situation will essentially amount to ignorance by player i about the availability of certain strategies s_j^0 to player j.

Going back to the three main cases listed above, cases 1 and 2 are both special cases of ignorance by the players about their own or some other players' *payoff functions* $U_i = X_i(Y)$ specifying the utility payoff $x_i = U_i(s_1, \ldots, s_n)$ a given player i obtains if the n players use alternative strategy n-tuples $s = (s_1, \ldots, s_n)$.

Indeed, case 3 can also be reduced to this general case. This is so because the assumption that a given strategy $s_i = s_i^0$ is not *available* to player i is equivalent, from a game-theoretical point of view, to the assumption that player i will never actually *use* strategy s_i^0 (even though it would be physically available to him) because by using s_i^0 he would always obtain some extremely low (i.e., highly negative) payoffs $x_i = U_i(s_1, \ldots, s_i^0, \ldots, s_n)$, whatever strategies $s_1, \ldots, s_{i-1}, s_{i+1}, \ldots, s_n$ the other players $1, \ldots, i-1, i+1, \ldots, n$ may be using.

Accordingly, let $S_i^{(j)}$ ($j = 1$ or $j \neq 1$) denote the *largest* set of strategies s_i which in player j's opinion may be *conceivably* included in player i's strategy space S_i. Let $S_i^{(0)}$ denote player i's "true" strategy space. Then, for the purposes of our analysis, we shall define player i's strategy space S_i as

$$S_i = \bigcup_{k=0}^{n} S_i^{(k)}. \tag{2.1}$$

We lose no generality by assuming that this set S_i as defined by (2.1) is *known* to all players because any lack of information on the part of some player j about this set S_i can be represented within our model as lack of information about the numerical values that player i's payoff function $x_i = U_i(s_1, \ldots, s_i, \ldots, s_n)$ takes for some specific choices of s_i, and in particular whether these values are so low as completely to discourage player i from using these strategies s_i.[9]

---

[9] Likewise, instead of assuming that player j assigns subjective probabilities to events of the form $E = \{s_i^0 \notin S_i\}$, we can always assume that he assigns these probabilities to events of the form $E = \{U_i(s_1, \ldots, s_i, \ldots, s_n) < x_i^0$ whenever $s_i = s_i^0\}$, etc.

Accordingly, we define an *I*-game *G* as a game where every player *j* *knows* the strategy spaces S_i of all players $i = 1, \ldots, j, \ldots, n$ but where, in general, he does *not* know the payoff functions U_i of these players $i = 1, \ldots, j, \ldots, n$.

3.

In terms of this definition, let us consider a given *I*-game *G* from the point of view of a particular player *j*. He can write the payoff function U_i of each player *i* (including his own payoff function U_j for $i = j$) in a more explicit form as

$$x_i = U_i(s_1, \ldots, s_n) = V_i^*(s_1, \ldots, s_n; a_{0i}, a_{1i}, \ldots, a_{ii}, \ldots, a_{ni}), \quad (3.1)$$

where V_i^*, unlike U_i, is a function whose mathematical form is (in player *j*'s opinion) *known* to all *n* players; whereas a_{0i} is a vector consisting of those parameters of function U_i which (in *j*'s opinion) are *unknown* to *all* players; and where each a_{ki} for $k = 1, \ldots, n$ is a vector consisting of those parameters of function U_i which (in *j*'s opinion) are *unknown* to some of the players but are *known* to player *k*. If a given parameter α is known *both* to players *k* and *m* (without being known to all players), then this fact can be represented by introducing two variables α_{ki} and α_{mi} with $\alpha_{ki} = \alpha_{mi} = \alpha$, and then making α_{ki} a component of vector a_{ki} while making α_{mi} a component of vector a_{mi}.

For each vector a_{ki} ($k = 0, 1, \ldots, n$), we shall assume that its *range space* $A_{ki} = \{a_{ki}\}$, i.e., the set of all possible values it can take, is the whole Euclidean space of the required number of dimensions. Then V_i^* will be a function from the Cartesian product $S_1 \times \cdots \times S_n \times A_{0i} \times \cdots \times A_{ni}$ to player *i*'s utility line Ξ_i, which is itself a copy of the real line *R*.

Let us define a_k as the vector combining the components of all *n* vectors $a_{k1}, \ldots, a_{kn}$. Thus we write

$$a_k = (a_{k1}, \ldots, a_{kn}), \quad (3.2)$$

for $k = 0, 1, \ldots, i, \ldots, n$. Clearly, vector a_0 summarizes the information that (in player *j*'s opinion) none of the players has about the *n* functions $U_1, \ldots, U_n$, whereas vector a_k ($k = 1, \ldots, n$) summarizes the information that (in *j*'s opinion) player *k* has about these functions, except for the information that (in *j*'s opinion) *all n* players share about them. For each vector a_k, its range space will be the set $A_k = \{a_k\} = A_{k1} \times \cdots \times A_{kn}$.

In equation (3.1) we are free to replace each vector a_{ki} ($k = 0, \ldots, n$) by the larger vector $a_k = (a_{k1}, \ldots, a_{ki}, \ldots, a_{kn})$, even though this will mean that in each case the $(n - 1)$ subvectors $a_{k1}, \ldots, a_{k(i-1)}, a_{k(i+1)}, \ldots, a_{kn}$ will occur vacuously in the resulting new equation. Thus, we can write

$$x_i = V_i^*(s_1, \ldots, s_n; a_0, a_1, \ldots, a_i, \ldots, a_n). \tag{3.3}$$

For any given player i the n vectors $a_0, a_1, \ldots, a_{i-1}, a_{i+1}, \ldots, a_n$ in general will represent *unknown* variables; and the same will be true for the $(n - 1)$ vectors $b_i, \ldots, b_{i-1}, b_{i+1}, \ldots, b_n$ to be defined below. Therefore, under the Bayesian hypothesis, player i will assign a subjective joint probability distribution

$$P_i = P_i(a_0, a_1, \ldots, a_{i-1}, a_{i+1}, \ldots, a_n; b_1, \ldots, b_{i-1}b_{i+1}, \ldots, b_n) \tag{3.4}$$

to all these unknown vectors.

For convenience we introduce the shorter notations $a = (a_1, \ldots, a_n)$ and $b = (b_1, \ldots, b_n)$. The vectors obtained from a and b by omitting the subvector a_i and b_i, respectively, will be denoted by a^i and b^i. The corresponding range spaces can be written as $A = A_1 \times \cdots \times A_n$; $B = B_1 \times \cdots \times B_n$; $A^i = A_1 \times \cdots \times A_{i-1} \times A_{i+1} \times \cdots \times A_n$; $B^i = B_1 \times \cdots B_{i-1} \times B_{i+1} \times \cdots \times B_n$.

Now we can write equations (3.3) and (3.4) as

$$x_i = V_i^*(s_1, \ldots, s_n; a_0, a), \tag{3.5}$$

$$P_i = P_i(a_0, a^i; b^i), \tag{3.6}$$

where P_i is a probability distribution over the vector space $A_0 \times A^i \times B^i$.

The other $(n - 1)$ players in general will not know the subjective probability distribution P_i used by player i. But player j (from whose point of view we are analyzing the game) will be able to write P_i for each player i (both $i = j$ and $i \neq j$) in the form

$$P_i(a_0, a^i; b^i) = R_i(a_0, a^i; b^i|b_i), \tag{3.7}$$

where R_i, unlike P_i, is a function whose mathematical form is (in player

j's opinion) *known* to all n players; whereas b_i is a vector consisting of those parameters of function P_i which (in j's opinion) are *unknown* to some or all of the players $k \neq i$. Of course, player j will realize that player i himself will *know* vector b_i since b_i consists of parameters of player i's own subjective probability distribution P_i.

The vectors $b_1, \ldots, b_{i-1}, b_{i+1}, \ldots, b_n$ occurring in equation (3.4), which so far have been left undefined, are the parameter vectors of the subjective probability distributions $P_1, \ldots, P_{i-1}, P_{i+1}, \ldots, P_n$, unknown to player i. The vector b^i occurring in equations (3.6) and (3.7) is a combination of all these vectors $b_1, \ldots, b_{i-1}, b_{i+1}, \ldots, b_n$, and summarizes the information that (in player j's opinion) player i lacks about the other $(n-1)$ players' subjective probability distribution $P_1, \ldots, P_{i-1},$ $P_{i+1}, \ldots, P_n$.

Clearly, function R_i is a function yielding, for each specific value of vector b_i, a probability distribution over the vector space $A^i \times B^i$.

We now propose to eliminate the vector a_0, unknown to all players, from equations (3.5) and (3.7). In the case of equation (3.5) this can be done by taking *expected values* with respect to a_0 in terms of player i's own subjective probability distribution $P_i(a_0, a^i; b^i) = R_i(a_0, a^i; b^i|b_i)$. We define

$$V_i(s_1, \ldots, s_n; a|b_i) = V_i(s_1, \ldots, s_n; a, b_i)$$

$$= \int_{A_0} V_i^*(s_1, \ldots, s_n; a_0, a) d_{(a_0)} R_i(a_0, a^i; b^i|b_i).$$

$$(3.8)$$

Then we write

$$x_i = V_i(s_1, \ldots, s_n; a, b_i), \qquad (3.9)$$

where x_i now denotes the *expected value* of player i's payoff in terms of his own subjective probability distribution.

In the case of equation (3.7) we can eliminate a_0 by taking the appropriate *marginal* probability distributions. We define

$$P_i(a^i, b^i) = \int_{A_0} d_{(a_0)} P_i(a_0, a^i; b^i), \qquad (3.10)$$

and

$$R_i(a^i, b^i | b_i) = \int_{A_0} d_{(a_0)} R_i(a_0, a^i; b^i | b_i). \qquad (3.11)$$

Then we write

$$P_i(a^i, b^i) = R_i(a^i, b^i | b_i). \qquad (3.12)$$

We now rewrite equation (3.9) as

$$x_i = V_i(s_1, \ldots, s_n; a, b_i, b^i) = V_i(s_1, \ldots, s_n; a, b), \qquad (3.13)$$

where vector b^i occurs only vacuously. Likewise we rewrite equation (3.12) as

$$P_i(a^i, b^i) = R_i(a^i, b^i | a_i, b_i), \qquad (3.14)$$

where on the right-hand side vector a_i occurs only vacuously.

Finally, we introduce the definitions $c_i = (a_i, b_i)$; $c = (a, b)$; and $c^i = (a^i, b^i)$. Moreover, we write $C_i = A_i \times B_i$; $C = A \times B$; and $C^i = A^i \times B^i$. Clearly, vector c_i represents the *total information* available to player i in the game (if we disregard the information available to all n players). Thus, we may call c_i player i's *information vector*.

From another point of view, we can regard vector c_i as representing certain physical, social, and psychological *attributes* of player i *himself*, in that is summarizes some crucial parameters of player i's own payoff function U_i as well as the main parameters of his beliefs about his social and physical environment. (The relevant parameters of player i's payoff function U_i again partly represent parameters of his subjective utility function X_i and partly represent parameters of his environment, e.g., the amounts of various physical or human resources available to him, etc.) From this point of view, vector c_i may be called player i's *attribute vector*.

Thus, under this model, the players' incomplete information about the true nature of the game situation is represented by the assumption that in general the actual value of the attribute vector (or information vector) c_i of any given player i will be known only to player i himself, but will be *unknown* to the other $(n - 1)$ players. That is, as far as these other players are concerned, c_i could have any one of a number—possibly even of an infinite number—of alternative values (which together form the range space $C_i = \{c_i\}$ of vector c_i). We may also express this assumption by saying that in an I-game G, in general, the rules of the game as such allow any given player i to belong to any one of a number of possible "*types*," corresponding to the alternative values his attribute vector c_i could take (and so representing the alternative payoff functions U_i and the alternative subjective probability distributions P_i that player i might have in the game.) Each player is always assumed to know his own actual type but to be in general ignorant about the other players' actual types.

Equations (3.13) and (3.14) now can be written as

$$x_i = V_i(s_1, \ldots, s_n; c) = V_i(s_1, \ldots, s_n; c_1, \ldots, c_n) \qquad (3.15)$$

$$P_i(c^i) = R_i(c^i|c_i) \qquad (3.16)$$

or

$$P_i(c_1, \ldots, c_{i-1}, c_{i+1}, \ldots, c_n) = R_i(c_1, \ldots, c_{i-1}, c_{i+1}, \ldots, c_n|c_i). \quad (3.17)$$

We shall regard equations (3.15) and (3.17) [or (3.16)] as the standard forms of the equations defining an I-game G, considered from the point of view of some particular player j.

Formally we define the *standard form* of a given I-game G for some particular player j as an ordered set G such that

$$G = \{S_1, \ldots, S_n; C_1, \ldots, C_n; V_1, \ldots, V_n; R_1, \ldots, R_n\} \qquad (3.18)$$

where for $i = 1, \ldots, n$ we write $S_i = \{s_i\}$; $C_i = \{c_i\}$; moreover, where V_i is a function from the set $S_1 \times \cdots \times S_n \times C_1 \times \cdots \times C_n$ to player i's utility line Ξ_i (which is itself a copy of the real line R); and where, for any specific value of the vector c_i, the function $R_i = R_i(c^i|c_i)$ is a probability distribution over the set $C^i = C_1 \times \cdots \times C_{i-1} \times C_{i+1} \times \cdots \times C_n$.

4.

Among C-games the natural analogue of this I-game G will be a C-game G^* with the same payoff functions V_i and the same strategy spaces S_i. However, in G^* the vectors c_i will have to be reinterpreted as being *random vectors* (chance moves) with an *objective* joint probability distribution

$$R^* = R^*(c_1, \ldots, c_n) = R^*(c) \qquad (4.1)$$

known to all n players.[10] (If some players did not know R^*, then G^* would not be a C-game.) To make G^* as similar to G as possible, we shall assume that each vector c_i will take its values from the same range space C_i in either game. Moreover, we shall assume that in game G^*, just as in game G, when player i chooses his strategy s_i, he will know only the value of his *own* random vector c_i but will not know the random vectors $c_1, \ldots, c_{i-1}, c_{i+1}, \ldots, c_n$ of the other $(n-1)$ players. Accordingly we may again call c_i the *information vector* of player i.

Alternatively, we may again interpret this random vector c_i as representing certain physical, social, and psychological attributes of player i himself. (But, of course, now we have to assume that for all n players these attributes are determined by some sort of random process, governed by the probability distribution R^*.) Under this interpretation we may again call c_i the *attribute vector* of player i.

We shall say that a given C-game G^* is in *standard form* if

1. the payoff functions V_i of G^* have the form indicated by equation (3.15);
2. the vectors $c_1, \ldots, c_n$ occurring in equation (3.15) are random vectors with a joint probability distribution R^* [equation (4.1)] known to all players;
3. each player i is assumed to know only his own vector c_i, and does not know the vectors $c_1, \ldots, c_{i-1}, c_{i+1}, \ldots, c_n$ of the other players when he chooses his strategy s_i.

Sometimes we shall again express these assumptions by saying that the rules of the game allow each player i to belong to any one of a number of alternative *types* (corresponding to alternative specific values that the random vector c_i can take); and that each player will always know his own actual type, but in general will not know those of the other players.

Formally we define a C-game G^* in standard form as an ordered set G^* such that

$$G^* = \{S_1, \ldots, S_n; C_1, \ldots, C_n; V_1, \ldots, V_n; R^*\}. \qquad (4.2)$$

Thus, the ordered set G^* differs from the ordered set G [defined by

---

[10] Assuming that a joint probability distribution R^* of the required mathematical form exists (see Section 5 below, as well as Part III of this paper).

equation (3.18)] only in the fact that the n-tuple $R_1, \ldots, R_n$ occurring in G is replaced in G^* by the singleton R^*.

If we consider the normal form of a game as a special limiting case of a standard form (viz. as the case where the random vectors $c_1, \ldots, c_n$ are empty vectors without components), then, of course, every C-game has a standard form. But only a C-game G^* containing random variables (chance moves) will have a standard form nontrivially different from its normal form.

Indeed, if G^* contains more than one random variable, then it will have several different standard forms. This is so because we can always obtain new standard forms G^{**}—intermediate between the original standard form G^* and the normal form G^{***}—if we suppress *some* of the random variables occurring in G^*, without suppressing *all* of them (as we would do if we wanted to obtain the normal form G^{***} itself). This procedure can be called *partial* normalization as distinguished from the *full* normalization, which would yield the normal form G^{***}.[11]

5.

Suppose that G is an I-game (considered from player j's point of view) while G^* is a C-game, both games being given in standard form. To obtain complete similarity between the two games, it is not enough if the strategy spaces $S_1, \ldots, S_n$, the range spaces $C_1, \ldots, C_n$, and the payoff functions $V_1, \ldots, V_n$ of the two games are the same. It is necessary also that each player i in either game should always assign the same *numerical* probability p to any given specific event E. Yet in game G player i will assess all probabilities in terms of his *subjective* probability distribution $R_i(c^i|c_i)$; whereas in game G^*—since vector c_i is known to him—he will assess all probabilities in terms of the objective *conditional* probability distribution $R^*(c^i|c_i)$ generated by the basic probability distribution $R^*(c)$ of the game G^*. Therefore, if the two games are to be equivalent, then numerically the distributions $R_i(c^i|c_i)$ and $R^*(c^i|c_i)$ must be identically equal.

---

[11] Partial normalization involves essentially the same operations as full normalization (see Section 7 below). It involves taking the expected values of the payoff functions V_i with respect to the random variables to be suppressed, and redefining the players' strategies where necessary. However, in the case of partial normalization we also have to replace the probability distribution R^* of the original standard form G^*, by a *marginal* probability distribution not containing the random variables to be suppressed. (In the case of full normalization no such marginal distribution has to be computed because the normal form G^{***} will not contain random variables at all.)

This leads to the following definition. Let G be an I-game (as considered by player j), and let G^* be a C-game, both games being given in standard form. We shall say that G and G^* are *Bayes-equivalent* for player j if the following conditions are fulfilled:

1. The two games must have the same strategy spaces $S_1, \ldots, S_n$ and the same range spaces $C_1, \ldots, C_n$.
2. They must have the same payoff functions $V_1, \ldots, V_n$.
3. The subjective probability distribution R_i of each player i in G must satisfy the relationship

$$R_i(c^i|c_i) = R^*(c^i|c_i), \tag{5.1}$$

where $R^*(c) = R^*(c_i, c^i)$ is the basic probability distribution of game G^* and where

$$R^*(c^i|c_i) = R^*(c_i, c^i) \Big/ \int_{C_i} d_{(c^i)} R^*(c_i, c^i). \tag{5.2}$$

In view of equations (5.1) and (5.2) we can write

$$R^*(c) = R^*(c_i, c^i) = R_i(c^i|c_i) \cdot \int_{C^i} d_{(c^i)} R^*(c_i, c^i). \tag{5.3}$$

In contrast to equation (5.2), which ceases to have a clear mathematical meaning when the denominator on its right-hand side becomes zero, equation (5.3) always retains a clear mathematical meaning.

We propose the following postulate.

Postulate 1. Bayes-equivalence. Suppose that some I-game G and some C-game G^* are Bayes-equivalent for player j. Then the two games will be completely equivalent for player j from a game-theoretical standpoint; and, in particular, player j's strategy choice will be governed by the same decision rule (the same solution concept) in either game.

This postulate follows from the Bayesian hypothesis, which implies that every player will use his *subjective* probabilities exactly in the same way as he would use known *objective* probabilities numerically equal to the former. Game G (as assessed by player j) and game G^* agree in all defining characteristics, including the numerical probability distributions used by the players. The only difference is that in G the probabilities used by each player are subjective probabilities whereas in G^* these probabilities are objective (conditional) probabilities. But by the Bayesian hypothesis this difference is immaterial.

Of course, under the assumptions of the postulate, all we can say is that *for player j himself* the two games are completely equivalent for game-theoretical purposes. We cannot conclude on the basis of the information assumed that the two games are likewise equivalent also for some *other* players $k \neq j$. In order to reach this latter conclusion we would have to know that G and G^* would preserve their Bayes-equivalence even if G were analyzed in terms of the functions $V_1, \ldots, V_n$ and $R_1, \ldots, R_n$ postulated by these *other* players k, instead of being analyzed in terms of the functions $V_1, \ldots, V_n$ and $R_1, \ldots, R_n$ postulated by player j himself. But so long as we are interested only in the decision rules that player j himself will follow in game G, all we have to know are the functions $V_1, \ldots, V_n$ and $R_1, \ldots, R_n$ that player j will be using.

Postulate 1 naturally gives rise to the following questions. Given any *I*-game G, is it always possible to construct a *C*-game G^* Bayes-equivalent to G? And, in cases where this is possible, is this *C*-game G^* always unique? These questions are tantamount to asking whether for any arbitrarily chosen n-tuple of subjective probability distributions $R_1(c^1|c_1), \ldots, R_n(c^n|c_n)$, there always *exists* a probability distribution $R^*(c_1, \ldots, c_n)$ satisfying the functional equation (5.3), and whether this distribution R^* is always *unique* in cases where it does exist. As these questions require an extended discussion, we shall answer them in Part III of this paper (see Theorem III and the subsequent heuristic discussion). We shall see that a given *I*-game G will have a *C*-game analogue G^* only if G itself satisfies certain consistency requirements. On the other hand, if such a *C*-game analogue G^* exists for G then it will be "essentially" unique (in the sense that, in cases where two different *C*-games G_1^*, and G_2^* are both Bayes-equivalent to a given *I*-game G, it will make no difference whether we use G_1^* or G_2^* for the analysis of G). In the rest of the present Part *I* of this paper, we shall restrict our analysis to *I*-games G for which a Bayes-equivalent *C*-game analogue G^* does exist.

As we shall make considerable use of Bayes-equivalence relationships between certain *I*-games G and certain *C*-games G^* given in standard form, it will be convenient to have a short designation for the latter. Therefore, we shall introduce the term *Bayesian games* as a shorter name for *C*-games G^* given in standard form. Depending on the nature of the *I*-game G we shall be dealing with in particular cases, we shall also speak of Bayesian two-person zero-sum games, Bayesian bargaining games, etc.

6.

In view of the important role that Bayesian games will play in our analysis, we shall now consider two alternative (but essentially equivalent) models for these games, which for some purposes will usefully supplement the model we have defined in Sections 4 and 5.

So far we have defined a Bayesian game G^* as a game where each player's payoff $x_i = V_i(s_1, \ldots, s_n; c_1, \ldots, c_n)$ depends, not only on the strategies $s_1, \ldots, s_n$ chosen by the n players, but also on some random vectors (information vectors or attribute vectors) $c_1, \ldots, c_n$. It has also been assumed that all players will know the joint probability distribution $R^*(c_1, \ldots, c_n)$ of these random vectors, but that in general the actual value of any given vector c_i will be known only to player i himself whose information vector (or attribute vector) it represents. This model will be called the *random-vector* model for Bayesian games.

An alternative model for Bayesian games can be described as follows. The actual individuals who will play the roles of players $1, \ldots, n$ in game G^* on any given occasion, will be selected *by lot* from certain populations $\Pi_1, \ldots, \Pi_n$ of potential players. Each population Π_i from which a given player i is to be selected will contain individuals with a variety of different attributes, so that *every* possible combination of attributes (i.e., every possible "type" of player i), corresponding to any specific value $c_i = c_i^0$ that the attribute vector c_i can take in the game, will be represented in this population Π_i. If in population Π_i a given individual's attribute vector c_i has the specific value $c_i = c_i^0$, then we shall say that he belongs to the *attribute class* c_i^0. Thus, each population Π_i will be partitioned into that many attribute classes as the number of different values that player i's attribute vector c_i can take in the game.

As to the random process selecting n players from the n populations $\Pi_1, \ldots, \Pi_n$, we shall assume that the probability of players $1, \ldots, n$ being selected from any specific n-tuple of attribute classes $c_1^0, \ldots, c_n^0$ will be governed[12] by the probability distribution $R^*(c_1, \ldots, c_n)$. We shall also retain the assumptions that this probability distribution R^* will be known to all n players, and that each player i will also know his own attribute class $c_i = c_i^0$ but, in general, will not know the other

---

[12] Under our assumptions in general the selection of players $1, \ldots, n$ from the respective populations $\Pi_1, \ldots, \Pi_n$ will not be statistically independent events because the probability distribution $R^*(c_1, \ldots, c_n)$ in general will not permit of factorization into n independent probability distributions $R_1^*(c_1), \ldots, R_n^*(c_n)$. Therefore, strictly speaking, our model postulates *simultaneous* random selection of a whole player n-tuple from a population Π of all possible player n-tuples, where Π is the Cartesian product $\Pi = \Pi_1 \times \cdots \times \Pi_n$.

players' attribute classes $c_1 = c_1^0, \ldots, c_{i-1} = c_{i-1}^0, c_{i+1} = c_{i+1}^0, \ldots, c_n = c_n^0$. As in this model the lottery by which the players are selected occurs prior to any other move in the game, it will be called the *prior-lottery model* for Bayesian games.

Let G be a real-life game situation where the players have incomplete information, and let G^* be a Bayesian game Bayes-equivalent to G (as assessed by a given player j). Then the Bayesian game G^*, interpreted in terms of the prior-lottery model, can be regarded as a possible representation (of course a highly schematic representation) of the real-life random social process which has actually created this game situation G. More particularly, the prior-lottery model pictures this social process as it would be seen by an *outside observer* having information about some aspects of the situation but lacking information about some other aspects. He could not have enough information to predict the attribute vectors $c_1 = c_1^0, \ldots, c_n = c_n^0$ of the n individuals to be selected by this social process to play the roles of players $1, \ldots, n$ in game situation G. But he would have to have enough information to predict the joint probability distribution R^* of the attribute vectors $c_1, \ldots, c_n$ of these n individuals, and, of course, also to predict the mathematical form of the payoff functions $V_1, \ldots, V_n$. (But he could not have enough information to predict the payoff functions $U_1, \ldots, U_n$ because this would require knowledge of the attribute vectors of all n players.)

In other words, the hypothetical observer must have exactly all the information *common* to the n players, but must not have access to any additional information private to any one player (or to any sectional group of players—and, of course, he must not have access to any information inaccessible to all of the n players). We shall call such an observer a *properly informed* observer. Thus, the prior-lottery model for Bayesian games can be regarded as a schematic representation of the relevant real-life social process as seen by a properly informed outside observer.

As an example, let us again consider the price-competition game G with incomplete information, and the corresponding Bayesian game G^*, discussed in Section 1 above. Here each player's attribute vector c_i will consist of the variables defining his cost functions, his financial resources, and his facilities to collect information about the other player.[13] Thus, the prior-lottery model of G^* will be a model where each player is chosen at random from some population of possible players with

[13] Cf. Footnote 4 above.

different cost functions, different financial resources, and different information-gathering facilities. We have argued that such a model can be regarded as a schematic representation of the real-life social process which has actually produced the assumed competitive situation, and has actually determined the cost functions, financial resources, and information-gathering facilities, of the two players.

Dr. Selten has suggested[14] a third model for Bayesian games, which we shall call the *Selten model* or the *posterior-lottery model*. Its basic difference from the prior-lottery model consists in the assumption that the lottery selecting the active participants of the game will take place only *after* each potential player has chosen the strategy he would use in case he were in fact selected for active participation in the game.

More particularly, suppose that the attribute vector c_i of player i $(i = 1, \ldots, n)$ can take k_i different values in the game. (We shall assume that all k_i's are finite but the model can be easily extended also to the infinite case.) Then, instead of having one randomly selected player i in the game, we shall assume that the role of player i will be played at the same time by k_i different players, each of them representing a different value of the attribute vector c_i. The set of all k_i individuals playing the role of player i in the game will be called the *role class i*. Different individuals in the same role class i will be distinguished by subscripts as players $i_1, i_2, \ldots$. Under these assumptions, obviously the total number of players in the game will not be n but rather will be the larger (usually much larger) number

$$K = \sum_{i=1}^{n} k_i. \tag{6.1}$$

It will be assumed that each player i_m from a given role class i will choose some strategy s_i from player i's strategy space S_i. Different members of the same role class i may (but need not) choose different strategies s_i from this strategy space S_i.

After all K players have chosen their strategies, one player i_m from each role class i will be randomly selected as an *active player*. Suppose that the attribute vectors of the n active players so selected will be $c_1 = c_1^0, \ldots, c_n = c_n^0$, and that these players, prior to their selection, have chosen the strategies $s_1 = s_1^0, \ldots, s_n = s_n^0$. Then each active player i_m, selected from role class i, will obtain a payoff

$$x_i = V_i(s_1^0, \ldots, s_n^0; c_1^0, \ldots, c_n^0). \tag{6.2}$$

---

[14] In private communication (cf. Footnote 1 above).

All other $(K - n)$ players not selected as active players will obtain *zero* payoffs.

It will be assumed that, when the n active players are randomly selected from the n role classes, the probability of selecting individuals with any specific combination of attribute vectors $c_1 = c_1^0, \ldots, c_n = c_n^0$ will be governed by the probability distribution $R^*(c_1, \ldots, c_n)$.[15]

It is easy to see that in all three models we have discussed—in the random-vector model, the prior-lottery model, and the posterior-lottery model—the players' payoff functions, the information available to them, and the probability of any specific even in the game, are all essentially the same.[16] Consequently, all three models can be considered to be essentially equivalent. But, of course, formally they represent quite

---

[15] In actual fact, we could just as well assume that each player would choose his strategy only *after* the lottery, and *after* being informed whether this lottery has selected him as an active player or not. (Of course if we made this assumption then players not selected as active players could simply forget about choosing a strategy at all.) From a game-theoretical point of view this assumption would make no real difference so long as each active player would have to choose his strategy without being told the names of the *other* players selected as active players, and in particular without being told the attribute classes to which these other active players would belong.

Thus the fundamental theoretical difference between our second and third models is not so much in the actual *timing* of the postulated lottery as such. It is not so much in the fact that in one case the lottery precedes, and in the other case it follows, the players' strategy choices. The fundamental difference rather lies in the fact that our second model (like our first) conceives of the game as an *n-person* game, in which only the n active players are formally "players of the game"; whereas our third model conceives of the game as a *K-person* game, in which both the active and the inactive players are formally regarded as "players." Yet, to make it easier to avoid confusion between the two models, it is convenient to assume also a difference in the actual timing of the assumed lottery.

[16] Technically speaking, the players' effective payoff functions under the posterior-lottery model are not quite identical with their payoff functions under the other two models, but this difference is immaterial for our purposes. Under the posterior-lottery model, let $r = r_i(c_i^0)$ be the probability (marginal probability) that a given player i_m with attribute vector $c_i = c_i^0$ will be selected as the active player from role class i. Then player i_m will have the probability r of obtaining a payoff corresponding to the payoff function V_i and will have the probability $(1 - r)$ of obtaining a zero payoff whereas under the other two models each player i will *always* obtain a payoff corresponding to the payoff function V_i. Consequently, under the posterior-lottery model player i_m's expected payoff will be only r times $(0 < r \leq 1)$ the expected payoff he could anticipate under the other two models. However, under most game-theoretical solution concepts (and in particular under all solution concepts we would ourselves choose for analyzing game situations), the solution of the game will remain invariant if the players' payoff functions are multiplied by positive constants r (even if different constants r are used for different players).

In any case, the posterior-lottery model can be made completely equivalent to the other two models if we assume that each active active player i_m will obtain a payoff corresponding to the payoff function $V_i/r_i(c_i^0)$, instead of obtaining a payoff corresponding to the payoff function V_i as such [as prescribed by equation (6.2)].

different game-theoretical models, as the random-vector model corresponds to an n-person game G^* with complete information, whereas the posterior-lottery model corresponds to a *K-person* game G^{**} with complete information. In what follows, unless the contrary is indicated, by the term "Bayesian game" we shall always mean the n-person game G^* corresponding to the random-vector model, whereas the K-person game G^{**} corresponding to the posterior-lottery model will be called the *Selten game*.

In contrast to the other two models, the prior-lottery model formally does not qualify as a true "game" at all because it assumes that the n players are selected by a chance move representing the first move of the game, whereas under the formal game-theoretical definition of a game the identity of the players must always be known from the very beginning, *before* any chance move or personal move has occurred in the game.

Thus, we may characterize the situation as follows. The real-life social process underlying the I-game G we are considering is best represented by the prior-lottery model. But the latter does not correspond to a true "game" in a game-theoretical sense. The other two models are two alternative ways of converting the prior-lottery model into a true "game." In both cases this conversion entails a price in the form of introducing some unrealistic assumptions. In the case of the posterior-lottery model corresponding to the Selten game G^{**}, the price consists in introducing $(K - n)$ fictitious players in addition to the n real players participating in the game.[17]

In the case of the random-vector model corresponding to the Bayesian game G^*, there are no fictitious players, but we have to pay the price of making the unrealistic assumption that the attribute vector c_i of each player i is determined by a chance move *after* the beginning of the game—which seems to imply that player i will be in existence for some period of time, however short, during which *he will not know yet* the specific value $c_i = c_i^0$ his attribute vector c_i will take. So long as the Bayesian game G^* corresponding to the random-vector model is being considered in its standard form, this unrealistic assumption makes very little difference. But, as we shall see, when we convert G^* into its normal form this unrealistic assumption implied by our model does

---

[17] This will be true even if we change the timing of the assumed lottery in Selten's model (see Footnote 15 above).

cause certain technical difficulties, because it seems to commit us to the assumption that each player can choose his normalized strategy (i.e., his strategy for the normal-form version of G^*) *before* he learns the value of his own attribute vector c_i. An important advantage of the Selten game G^{**} lies in the fact that it does not require this particular unrealistic assumption: we are free to assume that every player i_m will know his own attribute vector c_i from the very beginning of the game, and will always choose his own strategy in light of this information.[18]

Thus, as analytical tools used in the analysis of a given I-game G, both the Bayesian game G^* and the Selten game G^{**} have their own advantages and disadvantages.[19]

7.

Let G be an I-game given in standard form, and let G^* be a Bayesian game Bayes-equivalent to G. Then we define the normal form $\mathcal{N}(G)$ of this I-game G as being the normal form $\mathcal{N}(G^*)$ of the Bayesian game G^*.

To obtain this normal form we first have to replace the strategies s_i of each player i by *normalized strategies* s_i^*. A normalized strategy s_i^* can be regarded as a conditional statement specifying the strategy $s_i = s_i^*(c_i)$ that player i would use if his information vector (or attribute vector) c_i took any given specific value. Mathematically, a normalized strategy s_i^* is a function from the range space $C_i = \{c_i\}$ of vector c_i to player i's strategy space $S_i = \{s_i\}$. The set of all possible such functions s_i^* is called player i's normalized-strategy space $S_i^* = \{s_i^*\}$. In contrast to these normalized strategies s_i^*, the strategies s_i available to player i in the standard form of the game will be called his *ordinary strategies*.

If in a given game the information vector c_i of a certain player i can

---

[18] Moreover, as Selten has pointed out, his model also has the advantage that it can be extended to the case where the subjective probability distributions $R_1, \ldots, R_n$ of a given I-game G fail to satisfy the required consistency conditions, so that no probability distribution R^* satisfying equation (5.3) will exist, and therefore no Bayesian game G^* Bayes-equivalent to G can be constructed at all. In other words, for any I-game G we can always define an equivalent Selten game G^{**}, even in cases where we cannot define an equivalent Bayesian game G^*. (See Section 15, Part III.)

[19] We have given intuitive reasons why a Bayesian game G^* and the corresponding Selten game G^{**} are essentially equivalent. For a more detailed and more rigorous game-theoretical proof the reader is referred to a forthcoming paper by Reinhard Selten.

take only k different values (with k finite) so that we can write

$$c_i = c_i^1, \ldots, c_i^k, \tag{7.1}$$

then any normalized strategy s_i^* of this player can be defined simply as a k-tuple of ordinary strategies

$$s_i^* = (s_i^1, \ldots, s_i^k), \tag{7.2}$$

where $s_i^m = s_i^*(c_i^m)$, with $m = 1, \ldots, k$, denotes the strategy that player i would use in the standard form of the game if his information vector c_i took the specific value $c_i = c_i^m$. In this case player i's normalized strategy space $S_i^* = \{s_i^*\}$ will be the set of all such k-tuples s_i^*, that is, it will be the k-times repeated Cartesian product of player i's ordinary strategy space S_i by itself. Thus we can write $S_i^* = S_i^1 \times \cdots \times S_i^k$ with $S_i^1 = \cdots = S_i^k = S_i$.

Under either of these definitions, the normalized strategies s_i^* will not have the nature of *mixed* strategies but rather that of *behavioral* strategies. Nevertheless, these definitions are admissible because any game G^* in standard form is a game of perfect recall, and so it will make no difference whether the players are assumed to use behavioral strategies or mixed strategies [4].

Equation (3.15) can now be written as

$$x_i = V_i(s_1^*(c_1), \ldots, s_n^*(c_n); \quad c_1, \ldots, c_n) = V_i(s_1^*, \ldots, s_n^*; c). \tag{7.3}$$

In order to obtain the normal form $\mathcal{N}(G) = \mathcal{N}(G^*)$, all we have to do now is to take expected values in equation (7.3) with respect to the whole random vector c, in terms of the basic probability distribution $R^*(c)$ of the game. We define

$$\mathcal{E}(x_i) = W_i(s_1^*, \ldots, s_n^*) = \int_C V_i(s_1^*, \ldots, s_n^*; c) \, d_c R^*(c). \tag{7.4}$$

Since each player will treat his expected payoff as his effective payoff from the game, we can replace $\mathcal{E}(x_i)$ simply by x_i and write

$$x_i = W_i(x_1^*, \ldots, s_n^*). \tag{7.5}$$

We can now define the normal form of games G and G^* as the ordered set

$$\mathcal{N}(G) = \mathcal{N}(G^*) = \{S_1^*, \ldots, S_n^*; W_1, \ldots, W_n\}. \tag{7.6}$$

Compared with equations (3.18) and (4.2) defining the standard forms of these two games, in equation (7.6) the ordinary strategy spaces S_i have been replaced by the normalized strategy spaces S_i^*, and the ordinary payoff functions V_i have been replaced by the normalized payoff function W_i. On the other hand, the range spaces C_i as well as the probability distributions R_i or R^* have been omitted because the normal form $\mathcal{N}(G) = \mathcal{N}(G^*)$ of games G and G^* does not any more involve the random vectors $c_1, \ldots, c_n$.

This normal form, however, has the disadvantage that it is defined in terms of the players' *unconditional* payoff expectations $\mathscr{E}(x_i) = W_i(s_1^*, \ldots, s_n^*)$, though in actual fact each player's strategy choice will be governed by his *conditional* payoff expectation $\mathscr{E}(x_i|c_i)$, because he will always know his own information vector c_i at the time of making his strategy choice. This conditional expectation can be defined as

$$\mathscr{E}(x_i|c_i) = Z_i(s_1^*, \ldots, s_n^*|c_i) = \int_{C^i} V_i(s_1^*, \ldots, s_n^*; c_i, c^i) \, d_{(c^i)} R^*(c^i|c_i). \tag{7.7}$$

To be sure, it can be shown (see Theorem I of Section 8, Part II) that if any given player i maximizes his *unconditional* payoff expectation W_i, then he will also be maximizing his *conditional* payoff expectation $Z_i(\cdot|c_i)$ for each specific value of c_i, with the possible exception of a small set of c_i values which can occur only with probability zero. In this respect our analysis bears out von Neumann and Morgenstern's Normalization Principle [7, pp. 79–84], according to which the players can safely restrict their attention to the normal form of the game when they are making their strategy choices.

However, owing to the special nature of Bayesian games, the Normalization Principle has only restricted validity for them, and their normal form $\mathcal{N}(G^*)$ must be used with special care, because solution concepts based on uncritical use of the normal form may give counterintuitive results (see Section 11 of Part II of this paper). In view of this fact, we shall introduce the concept of a semi-normal form. The *semi-normal* *form* $\mathscr{S}(G) = \mathscr{S}(G^*)$ of games G and G^* will be defined as a game where the players' strategies are the normalized strategies s_i^* described above, but where their payoff functions are the *conditional* payoff-expectation functions $Z_i(\cdot|c_i)$ defined by equation (7.7). Formally we define the semi-normal form of the games G and G^* as the ordered set

$$\mathscr{S}(G) = \mathscr{S}(G^*) = \{S_1^*, \ldots, S_n^*; C_1, \ldots, C_n; Z_1, \ldots, Z_n; R^*\}. \quad (7.8)$$

As the semi-normal form, unlike the normal form, does involve the random vectors $c_i, \ldots, c_n$, now the range spaces $C_1, \ldots, C_n$, and the probability distribution R^*, which have been omitted from equation (7.6), reappear in equation (7.8).

Instead of von Neumann and Morgenstern's Normalization Principle, we shall use only the weaker Semi-normalization Principle (Postulate 2 below), which is implied by the Normalization Principle but which does not itself imply the latter:

Postulate 2. Sufficiency of the Semi-normal Form. The solution of any Bayesian game G^*, and of the Bayes-equivalent I-game G, can be defined in terms of the semi-normal form $\mathscr{S}(G^*) = \mathscr{S}(G)$, without going back to the standard form of G^* or of G.

References

1. Robert J. Aumann, "On Choosing a Function at Random," in Fred B. Wright (editor), *Symposium on Ergodic Theory*, New Orleans: Academic Press, 1963, pp. 1–20.
2. ———, "Mixed and Behavior Strategies in Infinite Extensive Games," in M. Dresher, L. S. Shapley, and A. W. Tucker (editors), *Advances in Game Theory*, Princeton: Princeton University Press, 1964, pp. 627–650.
3. John C. Harsanyi, "Bargaining in Ignorance of the Opponent's Utility Function," *Journal of Conflict Resolution*, 6 (1962), pp. 29–38.
4. H. W. Kuhn, "Extensive Games and the Problem of Information," in H. W. Kuhn and A. W. Tucker (editors), *Contributions to the Theory Games*, Vol. II. Princeton: Princeton University Press, 1953, pp. 193–216.
5. J. C. C. McKinsey, *Introduction to the Theory of Games*, New York: McGraw-Hill, 1952.
6. Leonard J. Savage, *The Foundation of Statistics*, New York: John Wiley and Sons, 1954.
7. John von Neumann and Oskar Morgenstern, *Theory of Games and Economic Behavior*, Princeton: Princeton University Press, 1953.

Part II. Bayesian Equilibrium Points*[†][1]

John C. Harsanyi

University of California, Berkeley

Part I of this paper has describe a new theory for the analysis of games with incomplete information. It has been shown that, if the various players' subjective probability distributions satisfy a certain mutual-consistency requirement, then any given game with incomplete information will be equivalent to a certain game with complete information, called the "Bayes-equivalent" of the original game, or briefly a "Bayesian game."

Part II of the paper will now show that any Nash equilibrium point of this Bayesian game yields a "Bayesian equilibrium point" for the original game and conversely. This result will then be illustrated by numerical examples, representing two-person zero-sum games with incomplete information. We shall also show how our theory enables us to analyze the problem of exploiting the opponent's erroneous beliefs.

However, apart from its indubitable usefulness in locating Bayesian equilibrium points, we shall show it on a numerical example (the Bayes-equivalent of a two-person cooperative game) that the normal form of a Bayesian game is in many cases a highly unsatisfactory representation of the game situation and has to be replaced by other representations (e.g., by the semi-normal form). We shall argue that this rather unexpected result is due to the fact that Bayesian games must be interpreted as games with "delayed commitment" whereas the normal-form representation always envisages a game with "immediate commitment."

8.

Let G be an I-game (as assessed by player j), and let G^* be a Bayesian game Bayes-equivalent to G and having the function $R^*(c_1, \ldots, c_n)$ as its basic probability distribution. Let s_i^* be a normalized strategy of

* Received June 1965 and revised May 1966.

† Part I of "Games with Incomplete Information Played by 'Bayesian' Players" appeared in the preceding issue of *Management Science*. Part III will appear in the next issue. A "Table of Contents" and "Glossary" appeared with Part I.

[1] The numbering of sections and theorems will be consecutive in Parts I, II, and III of this paper. The author wishes again to express his indebtedness to the persons and institutions listed in Footnote 1 of Part I.

player i in game G. Suppose that, for some specific value $c_i = c_i^0$ of player i's attribute vector c_i, the (ordinary) strategy $s_i = s_i^*(c_i)$ selected by this normalized strategy s_i^* maximizes player i's conditional payoff expectation

$$\mathscr{E}(x_i|c_i^0) = Z_i(s_1^*, \ldots, s_i^*, \ldots, s_n^*|c_i^0) \tag{8.1}$$

if the normalized strategies $s_1^*, \ldots, s_{i-1}^*, s_{i+1}^*, \ldots, s_n^*$ of the other $(n-1)$ players are kept constant. Then we shall say that this normalized strategy s_i^* of player i is a *best reply at the point* $c_i = c_i^0$ to the other players' normalized strategies $s_1^*, \ldots, s_{i-1}^*, s_{i+1}^*, \ldots, s_n^*$.

Now suppose that s_i^* actually possesses this best-reply property at *all* possible values of the attribute vector c_i, with the possible exception of a small set C_i^* of c_i values, having a total probability mass zero. That is, we are assuming that the event $E = \{c_i \in C_i^*\}$ is being assigned zero probability by the marginal probability distribution

$$R^*(c_i) = \int_{C^i} d_{(c^i)} R^*(c_i, c^i) \tag{8.2}$$

derived from the basic probability distribution $R^*(c_1, \ldots, c_i, \ldots, c_n) = R^*(c_i, c^i)$. Then we shall say that s_i^* is *almost uniformly* a best reply to $s_1^*, \ldots, s_{i-1}^*, s_{i+1}^*, \ldots, s_n^*$.

Finally suppose that in a given normalized strategy n-tuple $s^* = (s_1^*, \ldots, s_n^*)$ *every* component s_i^* $(i = 1, \ldots, n)$ is almost uniformly a best reply to the other $(n-1)$ components $s_1^*, \ldots, s_{i-1}^*, s_{i+1}^*, \ldots, s_n^*$. Then we shall say that s^* is a *Bayesian equilibrium point in game G.*

We can now state the following theorem.

THEOREM I. *Let G be an I-game, and let G^* be a Bayesian game Bayes-equivalent to G (as assessed by player j). In order that any given n-tuple of normalized strategies $s^* = (s_1^*, \ldots, s_n^*)$ be a Bayesian equilibrium point in game G, it is both* sufficient *and* necessary *that in the normal form $\mathcal{N}(G^*)$ of game G^* this n-tuple s^* be an equilibrium point in Nash's sense* [3, 5].

PROOF. In view of equations (5.3), (7.4), and (7.7) we can write

$$W_i(s_1^*, \ldots, s_i^*, \ldots, s_n^*) = \int_{C_i} Z_i(s_1^*, \ldots, s_i^*, \ldots, s_n^*|c_i) \, d_{(c_i)} R^*(c_i) \tag{8.3}$$

where $R^*(c_i)$ is again the marginal probability distribution defined by equation (8.2).

To prove the sufficiency part of the theorem, suppose that s^* is not a Bayesian equilibrium point in G. This means that at least one of the components of s^*, say, s_i^*, is not an almost uniformly best reply to the other components of s^*. Consequently, there must be some set C_i^* of possible c_i values at which the function $Z_i(\cdot|c_i)$ could be increased if we replaced the ordinary strategies $s_i = s_i^*(c_i)$ selected by the normalized strategy s_i^*, with some alternative ordinary strategies $s_i' = s_i^{**}(c_i)$. That is, for all $c_i \in C_i^*$ we must have

$$Z_i\left[s_i^{**}(c_i),(s^*)^i\right] > Z_i\left[s_i^*(c_i),(s^*)^i\right], \tag{8.4}$$

where $(s^*)^i$ denotes the $(n-1)$-tuple

$$(s^*)^i = (s_1^*,\ldots,s_{i-1}^*,s_{i+1}^*,\ldots,s_n^*). \tag{8.5}$$

Moreover, the probability distribution $R^*(c_i)$ must associate a nonzero total probability mass with this set C_i^*.

Now let s_i^{***} be a normalized strategy of player i, selecting the strategies $s_i^{**}(c_i)$ for all points c_i in set C_i^*, but coinciding with s_i^* everywhere else. That is,

$$s_i^{***}(c_i) = s_i^{**}(c_i) \quad \text{for all } c_i \in C_i^*,$$

$$s_i^{***}(c_i) = s_i^*(c_i) \quad \text{for all } c_i \notin C_i^*. \tag{8.6}$$

Then, in view of equations (8.3) and (8.4), we must have

$$W_i\left[s_i^{***},(s^*)^i\right] > W_i\left[s_i^*,(s^*)^i\right]. \tag{8.7}$$

Consequently, in game $\mathcal{N}(G^*)$ the n-tuple s^* is not an equilibrium point in Nash's sense. Thus, if s^* is not a Bayesian equilibrium point in G, then it cannot be a Nashian equilibrium point in $\mathcal{N}(G^*)$, contrary to our hypothesis.

To prove the necessity part of the theorem, suppose that s^* is not an equilibrium point in Nash's sense in game $\mathcal{N}(G^*)$. This means that at least one component of s^*, say, s_i^*, can be replaced by some alternative normalized strategy s_i^{***} in such a way as to increase the numerical value of the function W_i. That is, there must exist some normalized strategy s_i^{***} satisfying (8.4). But in view of equation (8.3) this implies that, for some set C_i^* of possible c_i values, we must have

$$Z_i\left[s_i^{***}(c_i),(s^*)^i\right] > Z_i\left[s_i^*(c_i),(s^*)^i\right]. \qquad (8.8)$$

Moreover, this set C_i^* must have a nonzero total probability mass associated with it under the probability distribution $R^*(c_i)$. Consequently, s^* will not be a Bayesian equilibrium point in game G. Thus, if s^* is not a Nashian equilibrium point in $\mathcal{N}(G^*)$ then it cannot be a Bayesian equilibrium point in game G, contrary to our hypothesis. This completes the proof.

In view of Nash's [3, 5] equilibrium-point theorem and its various generalizations, any I-game satisfying some rather mild regularity conditions will have at least one Bayesian equilibrium point. In particular:

THEOREM II. *Let G be a standard-form I-game (as assessed by player j), for which a Bayes-equivalent Bayesian game G^* exists. Suppose that G is a finite game, in which each player i has only a finite number of (ordinary) pure strategies s_i. Then G will have at least one Bayesian equilibrium point.*

PROOF. Even if each player i has only a finite number of pure strategies s_i in the standard form of game G, he will have an infinity of pure normalized strategies s_i^* in the normal form $\mathcal{N}(G)$ of the game if his information vector c_i can take an infinity of different values. But it is easy to see that the normalized behavioral strategies of $\mathcal{N}(G)$ will even then always satisfy the continuity and contractibility requirements of Debreu's generalized equilibrium-point theorem [1]. Hence, $\mathcal{N}(G)$ will always contain an equilibrium point s^* in Nash's sense, and so by Theorem I s^* will be a Bayesian equilibrium point for game G itself as originally given in standard form.

9.

Now we propose to discuss two numerical examples, partly to illustrate the concept of Bayesian equilibrium points as defined in the last section, and partly to illustrate the nature of the solution concept our approach yields in the simplest class of I-games, viz., two-person zero-sum games with incomplete information.

Suppose that in a given two-person zero-sum game G player 1 can belong to either of two attribute classes $c_1 = a^1$ and $c_1 = a^2$, where his belonging to class a^1 may be intuitively interpreted that he is in a "weak" position (e.g., in terms of military equipment, or man power, or the economic resources available to him, etc.) whereas his belonging to class a^2 may be interpreted that he is in a "strong" position. Likewise, player 2 can belong to either of two attribute classes $c_2 = b_1$ and $c_2 = b^2$, where class b^1 again may be taken to indicate that he is in a "weak" position whereas class b^2 indicates that he is in a "strong" position. Thus, we have altogether four possible cases in that the two players may belong to classes (a^1, b^1) or (a^1, b^2) or (a^2, b^1) or (a^2, b^2).

In any one of these cases, each player has two pure strategies, called $s_1 = y^1$ and $s_1 = y^2$ in the case of player 1, and called $s_2 = z^1$ and $s_2 = z^2$ in the case of player 2. (In terms of an intuitive interpretation, y^1 and z^1 may represent "more aggressive" strategies while y^2 and z^2 may represent "less aggressive" ones.) These strategies will yield the following payoffs to player 1 in the four possible cases (player 2's payoff is always the negative of player 1's).

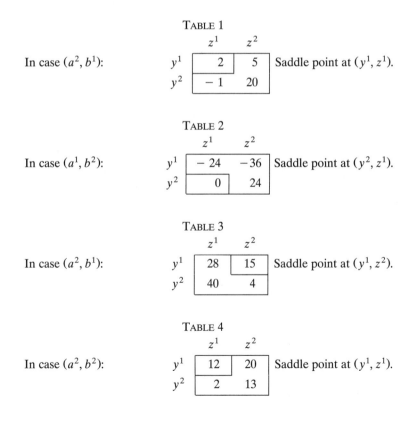

TABLE 1

In case (a^2, b^1):

	z^1	z^2
y^1	2	5
y^2	-1	20

Saddle point at (y^1, z^1).

TABLE 2

In case (a^1, b^2):

	z^1	z^2
y^1	-24	-36
y^2	0	24

Saddle point at (y^2, z^1).

TABLE 3

In case (a^2, b^1):

	z^1	z^2
y^1	28	15
y^2	40	4

Saddle point at (y^1, z^2).

TABLE 4

In case (a^2, b^2):

	z^1	z^2
y^1	12	20
y^2	2	13

Saddle point at (y^1, z^1).

(For the sake of simplicity, in all four cases we have chosen payoff matrices with saddle points in pure strategies.)

Since both players' attribute vectors c_1 and c_2 can take only a finite number (viz. two) alternative values, all probability distributions of the game can be represented by the corresponding probability-mass functions. In particular, the basic probability distribution $R^*(c_1, c_2)$ can be represented by the matrix of joint probabilities $r_{km} = \mathrm{Prob}(c_1 = a^k$ and $c_2 = b^m)$ with $k, m = 1, 2$. We shall assume that this matrix $\{r_{km}\}$ will be as follows.

TABLE 5

	$c_2 = b^1$	$c_2 = b^2$
$c_1 = a^1$	$r_{11} = 0.40$	$r_{12} = 0.10$
$c_1 = a^2$	$r_{21} = 0.20$	$r_{22} = 0.30$

TABLE 6

	$c_2 = b^1$	$c_2 = b^2$
$\mathrm{Prob}(c_2 = b^m \mid c_1 = a^1)$	$p_{11} = 0.8$	$p_{12} = 0.2$
$\mathrm{Prob}(c_2 = b^m \mid c_1 = a^2)$	$p_{21} = 0.4$	$p_{22} = 0.6$

Here $p_{km} = r_{km}/(r_{k1} + r_{k2})$. The first row of this matrix states the conditional probabilities (subjective probabilities) that player 1 will assign to the two alternative hypotheses about player 2's attribute class (viz. to $c_2 = b^1$ and to $c_2 = b^2$) when his own attribute class is $c_1 = a^1$. The second row states the probabilities he will assign to these two hypotheses when his own attribute class is $c_1 = a^2$.

In contrast, the conditional probabilities corresponding to the two *columns* of matrix $\{r_{km}\}$ are:

TABLE 7

	$\mathrm{Prob}(c_1 = a^k \mid c_2 = b^1)$	$\mathrm{Prob}(c_1 = a^k \mid c_2 = b^2)$
$c_1 = a^1$	$q_{11} = 0.67$	$q_{12} = 0.25$
$c_1 = a^2$	$q_{21} = 0.33$	$q_{22} = 0.75$

Here $q_{km} = r_{km}/(r_{1m} + r_{2m})$. The first column of this matrix states the conditional probabilities (subjective probabilities) that player 2 will assign to the two alternative hypotheses about player 1's attribute class (viz. to $c_1 = a^1$ and to $c_1 = a^2$) when his own attribute class is $c_2 = b^1$. The second column states the probabilities he will use when his own attribute class is $c_2 = b^2$.

All three probability matrices stated in Tables 5 to 7 show that in our examples the two players assume a positive correlation between their attribute classes, in the sense that it will tend to *increase* the probability a given player will assign to the hypothesis that his opponent is in a strong (or weak) position if he himself is in a strong (or weak) position. (Such a situation may arise if both player's power positions are determined largely by the same environmental influences; or if each player is making continual efforts to keep his own military, industrial, or other kind of strength more or less on a par with the other player's etc. Of course in other situations there may be a negative correlation, or none at all.)

In this game, player 1's normalized pure strategies will be of the form $s_1^* = y^{nt} = (y^n, y^t)$, where $s_1 = y^n$ $(n = 1, 2)$ is the ordinary pure strategy player 1 would use if $c_1 = a^1$, whereas y^t $(t = 1, 2)$ is the ordinary pure strategy he would use if $c_1 = a^2$. Likewise, player 2's normalized pure strategies will be of the form $s_2^* = z^{uv} = (z^u, z^v)$, where $s_2 = z^u$ $(u = 1, 2)$ is the ordinary pure strategy player 2 would use if $c_2 = b^1$, whereas $s_2 = z^v$ is the ordinary pure strategy he would use if $c_2 = b^2$.

It is easy to verify that the payoff matrix W for the normal form $\mathcal{N}(G)$ will be as follows.

TABLE 8

	z^{11}	z^{12}	z^{21}	z^{22}
y^{11}	7.6	8.8	6.2	7.4
y^{12}	7.0	9.1	1.0	3.1
y^{21}	8.8	13.6	14.6	19.4
y^{22}	8.2	13.9	9.4	15.1

The entries of this matrix are player 1's *total* (i.e., unconditional) payoff expectations corresponding to alternative pairs of normalized strategies (y^{nt}, z^{uv}). For instance, the entry for the strategy pair (y^{12}, z^{11}) is:

$$0.4 \times 2 + 0.1 \times (-24) + 0.2 \times 40 + 0.3 \times 2 = 7.0, \quad \text{etc.}$$

The only saddle point of this matrix W in pure strategies is at the point (y^{21}, z^{11}). (There is none in mixed strategies.) This means that in this game player 1's optimal strategy will be to use strategy y^2 if his attribute vector takes the value $c_1 = a^1$, and to use strategy y^1 if his attribute vector takes the value $c_1 = a^2$. On the other hand, player 2's optimal strategy will always be z^1, irrespective of the value of his attribute vector c_2.

It is easy to see that, in general, these optimal strategies are quite different from the optimal strategies associated with the four matrices stated in Tables 1 to 4. For instance, if the payoff matrix for case (a^1, b^1) is considered in isolation (see Table 1) then player 1's optimal strategy is y^1, though if we consider the game as a whole then his optimal normalized strategy y^{21} actually requires him to use strategy y^2 whenever his attribute vector has the value $c_1 = a^1$. Likewise, if the payoff matrix for case (a^2, b^1) is considered in isolation (see Table 3), then player 2's optimal strategy is z^2, though his optimal normalized strategy z^{11} for the game as a whole always requires him to use strategy z^1. These facts, of course, are not really surprising: they show only that the players will have to use different strategies if they do not know each other's attribute vectors, and different strategies if they do know them. For example, in case (a^1, b^1) player 1's optimal strategy would be y^1 only if he did know that his opponent's attribute vector is $c_2 = b^1$. But if all his information about player 2's attribute vector c_2 is what he can infer from the fact that his own attribute vector is $c_1 = a^1$, on the basis of the probability matrices of Tables 5 to 7, then his optimal strategy will be y^2 as prescribed by his optimal normalized strategy y^{21}, etc.

The strategy pair (y^{21}, z^{11}), being a saddle point, will also be an equilibrium point in Nash's sense for the normal form $\mathcal{N}(G)$ of the game. Thus, so long as player 2 is using strategy z^{11}, player 1 will be maximizing his *total* payoff expectation W_1 by choosing strategy y^{21}. Conversely, so long as player 1 is using strategy y^{21}, player 2 will be maximizing his total payoff expectation W_2 by choosing strategy z^{11}.

What is more important, by Theorem I of the previous section, the strategy pair (y^{21}, z^{11}) will also be a *Bayesian equilibrium point* in the standard form of game G (indeed it will be the only one since it is the only saddle point of the normal-form game). Hence, so long as player 2 uses strategy z^{11}, player 1 will maximize, not only his *total* payoff expectation W_1, but also his *conditional* payoff expectation $Z_1(\cdot|c_1)$, *given* the value of his own attribute vector c_1. To verify this we first list player 1's conditional payoff expectations when his attribute vector takes the value $c_1 = a^1$:

TABLE 9

	z^{11}	z^{12}	z^{21}	z^{22}
y^1	-3.2	-5.6	-0.8	-3.2
y^2	-0.8	4.0	16.0	20.8

We shall call this matrix $Z_1(\cdot|a^1)$. We do find that the highest entry of column z^{11} is in row y^2, corresponding to the *first* component of the normalized strategy $y^{21} = (y^2, y^1)$.

Next we list player 1's conditional payoff expectations when his attribute vector takes the value $c_1 = a^2$:

TABLE 10

	z^{11}	z^{12}	z^{21}	z^{22}
y^1	18.4	23.2	13.2	18.0
y^2	17.2	23.8	2.8	9.4

We shall call this matrix $Z_1(\cdot|a^2)$. Now we find that the highest entry of column z^{11} is in row y^1, corresponding to the *second* component of the normalized strategy $y^{21} = (y^2, y^1)$, as desired. Thus in both cases the ordinary strategy representing the appropriate component of the normalized strategy y^{21} maximizes player 1's conditional payoff expectation against the normalized strategy z^{11} of player 2, in accordance with Theorem I.

We leave it to the reader to verify by a similar computation that conversely the ordinary strategy z^1 [which happens to be both the first and the second component of the normalized strategy $z^{11} = (z^1, z^1)$] does maximize player 2's conditional payoff expectation against the normalized strategy y^{21} of player 1, both in case player 2's attribute vector takes the value $c_2 = b^1$ and in case it takes the value $c_2 = b^2$, as required by Theorem I.

In this particular example, each player's optimal strategy not only maximizes his security level in terms of his *total* payoff expectations W_i (which is true by definition), but also maximizes his security level in terms of his *conditional* payoff expectations $Z_i(\cdot|c_i)$. For instance, in

Table 9, strategy y^2 (which is the ordinary strategy prescribed for him by his optimal normalized strategy y^{21} in the case $c_1 = a^1$) is not only player 1's *best reply* to the normalized strategy z^{11} of his opponent, but is also his *maximin* strategy. Likewise, in Table 10, strategy y^1 is not only player 1's best reply to z^{11} but is also his maximin strategy.

However, we now propose to show by a counterexample that this relationship has no general validity for two-person zero-sum games with incomplete information. Consider a game G^0 where each player's attribute vector can again take two possible values, viz. $c_1 = a^1, a^2$ and $c_2 = b^1, b^2$. Suppose the payoff matrices in the resulting four possible cases are as follows:

TABLE 11

	z^1	z^2
y^1	8	-8
y^2	0	-4

In case (a^1, b^1):   Saddle point at (y^2, z^2).

TABLE 12

	z^1	z^2
y^1	-4	8
y^2	0	12

In case (a^1, b^2):   Saddle point at (y^2, z^1).

TABLE 13

	z^1	z^2
y^1	-8	12
y^2	-12	16

In case (a^2, b^1):   Saddle point at (y^1, z^1).

TABLE 14

	z^1	z^2
y^1	4	-4
y^2	0	-8

In case (a^2, b^2):   Saddle point at (y^1, z^2).

Suppose that basic probability matrix of this game G^0 is as follows:

TABLE 15

	$c_2 = b^1$	$c_2 = b^2$
$c_1 = a^1$	$r_{11} = 0.25$	$r_{12} = 0.25$
$c_1 = a^2$	$r_{21} = 0.25$	$r_{22} = 0.25$

The payoff matrix W for the normal form $\mathcal{N}(G^0)$ of G^0, listing player 1's *total* payoff expectations, will now be:

TABLE 16

	z^{11}	z^{12}	z^{21}	z^{22}
y^{11}	0	1	1	2
y^{12}	-2	-1	1	2
y^{21}	-1	0	3	4
y^{22}	-3	-2	3	4

The only saddle point of this matrix is at (y^{11}, z^{11}). Hence player 1's optimal normalized strategy is $y^{11} = (y^1, y^1)$ whereas player 2's is $z^{11} = (z^1, z^1)$. Player 1's *conditional* payoff expectations $Z_1(\cdot|c_1)$ in the case $c_1 = a^1$ will be as follows:

TABLE 17

	z^{11}	z^{12}	z^{21}	z^{22}
y^1	2	8	-6	0
y^2	0	6	-2	4

In accordance with Theorem I, as the table shows, player 1's best reply to the opponent's optimal normalized strategy z^{11} is strategy y^1, the ordinary strategy corresponding to his own optimal normalized strategy y^{11}. But player 1's security level would be maximized by y^2, and not by y^1.

This result has the following implication. Let G be a two-person zero-sum game with incomplete information. Then we may use the von Neumann-Morgenstern solution of the normal form $\mathcal{N}(G)$ of this game as the solution for the game G itself as originally defined in standard form. But, in general, the only real justification for this procedure will lie in the fact that every pair of optimal normalized strategies s_1^* and s_2^* will represent a Bayesian equilibrium point in G. However, we cannot justify this approach by the maximin and minimax properties of s_1^* and s_2^* because in general these strategies will not have such properties in terms of the players' *conditional* payoff expectations $Z_i(\cdot|c_i)$, which are the quantities the players will be really interested in and which they will treat as their true anticipated payoffs from the game.

10.

We now propose to illustrate on a third numerical example how our model enables us to deal with the problem of *exploiting the opponent's erroneous beliefs*. Let G again be a two-person zero-sum game where each player can belong to two different attribute classes, viz. to class $c_1 = a^1$ or to class $c_1 = a^2$, and to class $c_2 = b^1$ or to class $c_2 = b^2$, respectively. We shall assume that in the four possible combinations of attribute classes the payoffs of player 1 will be again the same as stated in Tables 1 to 4 of the previous section. However, the basic probability matrix $\{r_{km}\}$ of the game will now be assumed to be:

TABLE 18

	$c_2 = b^1$	$c_2 = b^2$
$c_1 = a^1$	$r_{11} = 0.01$	$r_{12} = 0.00$
$c_1 = a^2$	$r_{21} = 0.09$	$r_{22} = 0.90$

Accordingly, player 1 will now use the following conditional probabilities (subjective probabilities):

TABLE 19

	$c_2 = b^1$	$c_2 = b^2$
$\text{Prob}(c_2 = b^m \mid c_1 = a^1)$	$p_{11} = 1.00$	$p_{12} = 0.00$
$\text{Prob}(c_2 = b^m \mid c_1 = a^2)$	$p_{21} = 0.09$	$p_{22} = 0.91$

On the other hand, player 2 will now use the following conditional probabilities (subjective probabilities):

TABLE 20

	$\text{Prob}(c_1 = a^k \mid c_2 = b^1)$	$\text{Prob}(c_1 = a^k \mid c_2 = b^2)$
$c_1 = a^1$	$q_{11} = 0.10$	$q_{12} = 0.00$
$c_1 = a^2$	$q_{21} = 0.90$	$q_{22} = 1.00$

The payoff matrix W for the normal form of the game, stating players 1's *total* payoff expectations, will now be:

TABLE 21

	z^{11}	z^{12}	z^{21}	z^{22}
y^{11}	13.34	20.54	11.20	19.40
y^{12}	5.42	15.32	2.21	12.11
y^{21}	13.31	20.51	12.35	19.55
y^{22}	5.39	15.29	2.36	12.26

The only saddle point of this game is at (y^{21}, z^{21}). Thus player 1's optimal strategy is $y^{21} = (y^2, y^1)$ whereas player 2's optimal strategy is $z^{21} = (z^2, z^1)$.

Now suppose that player 1 actually belongs to attribute class $c_1 = a^1$. Then our result concerning the players' optimal strategies can be interpreted as follows. As $c_1 = a^1$, by Table 19 player 1 will be able to infer that player 2 must belong to attribute class $c_2 = b^1$. Hence, by Table 20 player 1 will be able to conclude that player 2 is assigning a near-unity probability (viz. $q_{21} = .90$) to the *mistaken hypothesis* that player 1 belongs to class $c_1 = a^2$, which would mean that case (a^2, b^1) would apply (since player 2's own attribute class is $c_2 = b^1$). Therefore, player 1 will expect player 2 to choose strategy z^2, which is the strategy the latter would use by Table 3 if he thought that case (a^2, b^1) represented the actual situation.

If case (a^2, b^1) did represent the actual situation, they by Table 3 player 1's best reply to strategy z^2 would be strategy y^1. But under our assumptions player 1 will know that the actual situation is (a^1, b^1), rather than (a^2, b^1). Hence, by Table 1 his actual best reply to z^2 will be y^2, and not y^1. Thus we can say that by choosing strategy y^2 player 1 will be able to *exploit* player 2's *mistaken belief* that player 1 probably belongs to class $c_1 = a^2$.

As Table 1 shows, if player 2 knew the true situation and acted accordingly then player 1 could obtain no more than the payoff $x_1 = 2$. If player 2 did make the mistaken assumption that player 1's attribute vector was $c_1 = a^2$, but if player 1 did not exploit this mistake (i.e., if player 1 used the same strategy y^1 he would use against a better informed opponent) then his payoff would be $x_1 = 5$. Yet if he does exploit player 2's mistake and counters the latter's strategy z^2 by his strategy y^2, then he will obtain the payoff $x_1 = 20$.

11.

The purpose of our last numerical example will be to show that the normal form $\mathcal{N}(G) = \mathcal{N}(G^*)$ is often an unsatisfactory representation of the corresponding I-game G (as originally defined in standard form), and is an unsatisfactory representation even for the corresponding Bayesian game G^*. More particularly, we shall argue that the solutions we would naturally associate with the normal-form game $\mathcal{N}(G) = \mathcal{N}(G^*)$ may often give highly counterintuitive results for G or G^*.

Let G be a two-person bargaining game in the sense of Nash [4] in which the two players have to divide \$100. If they cannot agree on their shares, then both of them will obtain zero payoffs. It will be assumed that both players have linear utility functions for money, so that their dollar payoffs can be taken to be the same as their utility payoffs.

Following Nash, we shall assume that the game will be played as follows. Each player i $(i = 1, 2)$ has to state his payoff demand y_i^* without knowing what the other player's payoff demand y_j^* will be. If $y_1^* + y_2^* \leq 100$, then each player will obtain a payoff $y_i = y_i^*$ equal to his own demand. If $y_1^* + y_2^* > 100$, then both players will obtain zero payoffs $y_1 = y_2 = 0$.

To introduce incomplete information in the game, we shall assume that either player i may have to pay half of his gross payoff y_i to a certain secret organization or he may not; and that he himself will always know in advance whether this is the case or not, but that neither player will know whether the other has to make this payment or not. We shall say that player i's attribute vector takes the value $c_i = a^1$ if he has to make this payment, and takes the value $c_i = a^2$ if this is not the case.

Thus, we can define the net payoffs x_i of both players i as:

$$x_i = \tfrac{1}{2}y_i \quad \text{if } c_i = a^1$$

$$x_i = y_i \quad \text{if } c_i = a^2 \qquad i = 1, 2. \tag{11.1}$$

Let X be the *payoff space* of the game, that is, the set of all possible net-payoff vectors $x = (x_1, x_2)$ that the players can achieve in the game.

If the players' attribute vectors are (a^1, a^1), then $X = X^{11}$ will be the triangular area satisfying the inequalities

$$x_i \geq 0 \qquad i = 1, 2 \tag{11.2}$$

and

$$x_1 + x_2 \leq 50. \tag{11.3}$$

If these vectors are (a^1, a^2), then $X = X^{12}$ will be the triangle satisfying (11.2) as well as

$$2x_1 + x_2 \leq 100. \tag{11.4}$$

If these vectors are (a^2, a^1), then $X = X^{21}$ will be the triangle satisfying (11.2) as well as

$$x_1 + 2x_2 \leq 100. \tag{11.5}$$

Finally, if these vectors are (a^2, a^2), then $X = X^{22}$ will be the triangle satisfying (11.2) as well as

$$x_1 + x_2 \leq 100 \tag{11.6}$$

(see Figures 1 to 4).

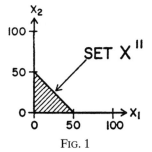

FIG. 1

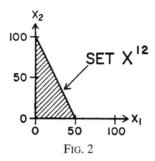

FIG. 2

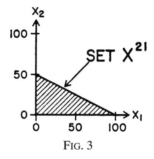

FIG. 3

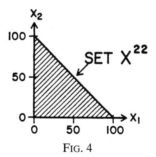

FIG. 4

Now suppose that the basic probability matrix $\{r_{km}\}$ of the game is as follows:

TABLE 22

	$c_2 = a^1$	$c_2 = a^2$
$c_1 = a^1$	$r_{11} = \frac{1}{2}\varepsilon$	$r_{12} = \frac{1}{2} - \frac{1}{2}\varepsilon$
$c_1 = a^2$	$r_{21} = \frac{1}{2} - \frac{1}{2}\varepsilon$	$r_{22} = \frac{1}{2}\varepsilon$

Here $r_{km} = \text{Prob}(c_1 = a^k$ and $c_2 = a^m)$ while ε is a very small positive number which for practical purposes can be taken to be zero.

Accordingly, both players i will use the following conditional probabilities (subjective probabilities):

TABLE 23

	$c_j = a^1$	$c_j = a^2$
$\text{Prob}(c_j = a^m \mid c_i = a^1)$	$p_{11} = q_{11} = \varepsilon$	$p_{12} = q_{21} = 1 - \varepsilon$
$\text{Prob}(c_j = a^m \mid c_i = a^2)$	$p_{21} = q_{12} = 1 - \varepsilon$	$p_{22} = q_{22} = \varepsilon$

That is, each player will assume a virtually complete negative correlation between his own attribute vector and the other player's: he will expect the other player to be free from the side-payment obligation if he himself is subject to it, and will expect the other player to be subject to this obligation if he himself is free from it.

In constructing the normal form of this game, we shall take $\varepsilon = 0$. Hence, the players will have $\frac{1}{2}$ probability of being able to select a payoff vector $x = x^*$ from the set X^{12}, and have $\frac{1}{2}$ probability of being able to select a payoff vector $x = x^{**}$ from the set X^{21}. Thus, their payoff vector from the normal-form game as a whole will be

$$x = \tfrac{1}{2}x^* + \tfrac{1}{2}x^{**} \quad \text{with} \quad x^* \in X^{12}, \quad x^{**} \in X^{21}. \tag{11.7}$$

The set $X^0 = \{x\}$ of all such vectors x is the area satisfying condition (11.2) as well as the two inequalities:

$$2x_1 + x_2 \leqq 75 \tag{11.8}$$

$$x_1 + 2x_2 \leqq 75. \tag{11.9}$$

(See Figure 5.)

Thus the normal form $\mathcal{N}(G)$ of this game will be a two-person bargaining game in which the players can choose any payoff vector $x = (x_1, x_2)$ from the set X^0, and obtain the payoffs $x_1 = 0$ and $x_2 = 0$ if they cannot agree on which particular vector x to choose.

Any solution concept satisfying the usual Efficiency and Symmetry Postulates will yield the solution $x_1 = x_2 = 50$. This, however, means that if case (a^1, a^2) arises, then the players will obtain the payoffs $x_1^* = 0$, $x_2^* = 100$, whereas if case (a^2, a^1) arises, then they will obtain the payoffs $x_1^{**} = 100$ and $x_2^{**} = 0$. This is so because, in view of

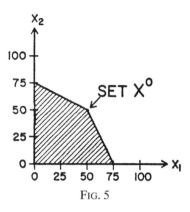

FIG. 5

condition (11.7), the only way we can obtain the vector $x = (50, 50)$ is by choosing $x^* = (0, 100)$ and $x^{**} = (100, 0)$.

If we interpret the normal form of the game in the usual way, and assume that the players will choose their strategies and will agree on the outcome *before* they know whether case (a^1, a^2) or case (a^2, a^1) will obtain, then as a solution of the normal form of the game this result makes very good sense. It means that in accordance with the Efficiency Postulate the players will try to *minimize* the expected value of the amount they would have to pay out to outsiders, and therefore will agree to assign a zero payoff to that player who would be under an obligation to hand over half of his payoff to the secret organization in question.

However, if we stick to the original definition of the game, under which the players can choose their strategies and can conclude agreements only *after* each player has observed the value of his own attribute vector c_i, then this result is highly counterintuitive. This is so because in this game there is near-perfect negative correlation between the two players' attribute vectors; so that each player will not only know the value of his own attribute vector but will also be able to infer with virtual certainty what the value of the other player's attribute vector is. Hence, even though formally the game is a game with *incomplete* information, it will always essentially reduce to a two-person bargaining game with *complete* information.

More particularly, if case (a^1, a^2) arises, then this game G will be essentially equivalent to a Nash-type bargaining game G^{12} where the payoff space is the set $X = X^{12}$, and where this fact is known to both players. Likewise, if case (a^2, a^1) arises, then game G will be essentially equivalent to a Nash-type bargaining game G^{21} where the payoff space is the set $X = X^{21}$, and where again this fact is known to both players. Yet, it would be quite counterintuitive to suggest that in game G^{12} player 1 would agree to a solution giving him a zero payoff $x_1^* = 0$, or that in game G^{21} player 2 would agree to a solution giving him a zero payoff $x_2^{**} = 0$.

In terms of the corresponding Bayesian game G^*, we can state our argument as follows. If the players can reach an agreement *before* a chance move decides whether case (a^1, a^2) or case (a^2, a^1) will occur, then each player may be quite willing to agree to an arrangement under which he would obtain \$100 in one of the two possible cases, and would obtain \$0 in the other case, both cases having the same probability $\frac{1}{2}$. But if the players can conclude an agreement only *after* the outcome of this chance event has come to be known to both of them, then neither player will voluntarily accept a payoff of \$0 in order that the other player can obtain all the \$100—even if such an arrangement would minimize the two players' total payment to the secret organization.

For definiteness let us assume that in actual fact in games G^{12} and G^{21} the players will use the Nash solution [4]. Then in G^{12} their payoffs will be $x_1^* = 25$ and $x_2^* = 50$, whereas in G^{21} their payoffs will be $x_1^{**} = 50$ and $x_2^{**} = 25$. Hence, by equation (11.7) the players' expected payoffs from game G^* as a whole will be $x_1 = \frac{1}{2}x_1^* + \frac{1}{2}x_1^{**} = 37.5$ and $x_2 = \frac{1}{2}x_2^* + \frac{1}{2}x_2^{**} = 37.5.$[2]

Of course, for game G^* as a whole, the vector $x = (37.5, 37.5)$ is an inefficient payoff vector because it is strongly dominated, e.g., by the payoff vector $x = (50, 50)$ previously considered. Yet in the case of game G^* the use of an inefficient solution is justified by the fact that G^* is not a cooperative game in the full sense of this term, precisely because the players are not allowed to conclude agreements at the very beginning of the game, before any chance move or personal move has occurred—but rather can conclude agreements only *after* the chance move deciding between cases (a^1, a^2) and (a^2, a^1) has been completed. Thus, by using the inefficient pay off vector $x = (37.5, 37.5)$ as the

---

[2] In terms of the terminology to be introduced in Part III, the bargaining game G under discussion can be *decomposed* into G^{12} and G^{21} as component games. Hence, according to Postulate 3 of Part III, whatever solution concept we may wish to use, the solution of game G must be equivalent to using the solution of game G^{12} or of game G^{21} depending on which of these two games represents the game really played by the two players.

solution for game G^*, we do not violate the principle that each fully cooperative game should have an efficient solution.[3, 4]

We can state our conclusion also as follows. Most games discussed in the game-theoretical literature are *games with immediate commitment*, in the sense that each player is free to commit himself to some particular strategy *before* any chance move or personal move has taken place. (In the special case where the game is a cooperative game, the players are also free to enter into fully binding and enforceable agreements at that point.)

In contrast, the Bayesian games G^* we are using in the analysis of games with incomplete information, must be interpreted as *games with delayed commitment*,[5] where the players can commit themselves to specific strategies (and can enter into binding agreements if the game they are playing is a cooperative game), only *after* certain moves have occurred in the game, viz., the chance moves determining the players' attribute vectors.[6]

By the very definition of the normal form of a game, von Neumann and Morgenstern's Normalization Principle [6] can apply only to games with immediate commitment. Hence, in general, we cannot expect that the normal form $\mathcal{N}(G^*) = \mathcal{N}(G)$ of a Bayesian game G^* will be a fully satisfactory representation of this game G^*—or of the incomplete-information game G Bayes-equivalent to G^*. This is the reason why in the analysis of these games we have found it necessary to replace von Neumann and Morgenstern's Normalization Principle by our weaker Semi-normalization Principle (Postulate 2 in Section 7 of Part I of this paper).

---

[3] To be sure, game G itself is a fully cooperative game, yet it is a cooperative game with incomplete information. But our point has been that, as we replace this I-game G by a Bayes-equivalent C-game G^*, the fully cooperative nature of the game is lost since the players' attribute vectors c_i come to be reinterpreted as chance moves *preceding* their choices of strategies.

[4] Our present discussion has been restricted to a very special case of two-person bargaining games with incomplete information, viz., to the case where almost perfect correlation exists between the two players' attribute vectors c_1 and c_2. The problem of defining a suitable solution concept in the general case will be discussed in [2].

[5] Quite apart from this connection with Bayesian games, the general class of games with delayed commitment seems to have many other applications, and would deserve further study.

[6] Note that the corresponding Selten games G^{**} do not give rise to this problem: they can always be interpreted as games with immediate commitment and therefore can be adequately represented by their normal forms $\mathcal{N}(G^{**})$.

REFERENCES

1. Gerard Debreu, "A Social Equilibrium Existence Theorem," *Proceedings of the National Academy of Sciences*, 38 (1952), 886–893.
2. John C. Harsanyi and Reinhard Selten, "A Generalized Nash Solution for Two-Person Bargaining Games with Incomplete Information," 1967 (unpublished).
3. John F. Nash, "Equilibrium Points in n-Person Games," *Proceedings of the National Academy of Sciences*, 36 (1950), 48–49.
4. ———, "The Bargaining Problem," *Econometrica*, 18 (1950), 155–162.
5. ———, "Non-Cooperative Games," *Annals of Mathematics*, 54 (1951), 286–295.
6. John von Neumann and Oskar Morgenstern, *Theory of Games and Economic Behavior*, Princeton: Princeton University Press, 1953.

Part III. The Basic Probability Distribution of the Game*[†][1]

JOHN C. HARSANYI

UNIVERSITY OF CALIFORNIA, BERKELEY

Parts I and II of this paper have described a new theory for the analysis of games with incomplete information. Two cases have been distinguished: consistent games in which there exists some basic probability distribution from which the players' subjective probability distributions can be derived as conditional probability distributions; and inconsistent games in which no such basic probability distribution exists. Part III will now show that in consistent games, where a basic probability distribution exists, it is essentially unique.

It will also be argued that, in the absence of special reasons to the contrary, one should try to analyze any given game situation with incomplete information in terms of a consistent-game model. However, it will be shown that our theory can be extended also to inconsistent games, in case the situation does require the use of an inconsistent-game model.

12.

We now propose to turn back to the questions raised in Section 5 of Part I of this paper. Given any arbitrarily chosen I-game G, is it always possible to construct some Bayesian game G^* Bayes-equivalent to G? And in cases where this is possible, is this Bayesian game G^* always unique? As we have seen in Section 5, these questions are equivalent to asking whether, for any arbitrarily chosen n-tuple of subjective probability distributions $R_1(c^1 \mid c_1), \ldots, R_n(c^n \mid c_n)$, there always *exists* some probability distribution $R^*(c_1, \ldots, c_n)$ satisfying the functional equation (5.3), and whether this distribution R^* is always *unique* in cases in which it does exist. The answers to these two questions are given by

---

* Received June 1965 and revised May 1966.

† Parts I and II of "Games with Incomplete Information Played by 'Bayesian' Players" have appeared in the preceding two issues of *Management Science: Theory*. A "Table of Contents" and a "Glossary" appeared with Part I.

[1] The author wishes again to express his indebtedness to the persons and institutions listed in Footnote 1 of Part I of this paper.

Theorem III below. But before stating the Theorem we shall have to introduce a few definitions.

In Section 3 of Part I we formally defined the range space $C_i = \{c_i\}$ of each attribute vector c_i $(i = 1, \ldots, n)$ as being the *whole* Euclidean space E^r of the required number of dimensions. For convenience we shall now relax this definition, and shall allow the range space C_i to be restricted to any proper *subset* C_i^* of the space E^r. (Obviously it makes little difference whether we say that in a given game G the vector c_i ranges only over some subset C_i^* of the Euclidean space E^r, or whether we say that formally the range of this vector c_i is the whole space E^r but that outside this set C_i^* the vector c_i has zero probability density everywhere.)

A given subset D of the range space $C = \{c\}$ for the composite vector $c = (c_1, \ldots, c_n)$ is called a *cylinder* if it is a Cartesian product of the form $D = D_1 \times \cdots \times D_n$, where $D_1, \ldots, D_n$ are subsets of the range spaces $C_1 = \{c_1\}, \ldots, C_n = \{c_n\}$ of the vectors $c_1, \ldots, c_n$. These sets $D_1, \ldots, D_n$ are called the *factor sets* of cylinder D. Two cylinders D and D' are said to be *separated* if all their corresponding factor sets D_i and D_i' $(i = 1, \ldots, n)$ are disjoint.

Suppose that in a given I-game G (as assessed by player j) each of the range spaces C_i $(i = 1, \ldots, n)$ can be partitioned into two or more nonempty subsets $D_i^1, D_i^2, \ldots$, such that each player i can always infer with probability one that every other player's information vector c_k has taken a value $c_k = \gamma_k$ from the set D_k^m $(m = 1, 2, \ldots)$ whenever his own information vector c_i has taken a value $c_i = \gamma_i$ from set D_i^m. In symbols this means that

$$R_i(c_k \in D_k^m \mid c_i) = 1 \quad \text{for all } c_i \in D_i^m, \tag{12.1}$$

and for all pairs of players i and k. In this case we shall say that game G can be *decomposed* into the component games $G^1, G^2, \ldots$, in which G^m $(m = 1, 2, \ldots)$ is defined as the game resulting from G if the range space C_i of each information vector c_i $(i = 1, \ldots, n)$ is restricted to the set $C_i^m = D_i^m$. Accordingly, the range space C of the composite vector $c = (c_1, \ldots, c_n)$ in each component game G^m is restricted to the cylinder $C^m = D^m = D_1^m \times \cdots \times D_n^m$. We shall call D^m the *defining cylinder* of game G^m and shall write $G^m = G(D^m)$.

If G can be decomposed into two or more component games $G^1, G^2, \ldots$ then it is said to be *decomposable*, while in the opposite case it is said to be *indecomposable*.

This terminology is motivated by the fact that whenever a given decomposable game G is being played, the players in actual fact will always play one of its component games $G^1, G^2, \ldots$. Moreover, they will always know, *before* making any move in the game, and before choosing their strategies s_i, *which* of these component games has to be played on that particular occasion, since each player can always observe whether his own information vector c_i lies in set D_i^1, or D_i^2, etc.

We shall use the same terminology also in the case of C-games G^* given in standard form, except that condition (12.1) will then have to be replaced by its analogue:

$$R^*(c_k \in D_k^m \mid c_i) = 1 \quad \text{for all } c_i \in D_i^m. \tag{12.2}$$

The concept of decomposition suggests the following postulate.

Postulate 3. Decomposition of games. Suppose that a given I-game G (as assessed by player j)—or a given C-game G^*—can be decomposed into two or more component games $G^1 = G(D^1), G^2 = G(D^2), \ldots$. Let $\sigma^1, \sigma^2, \ldots$ be the solutions (in terms of any particular solution concept) of $G^1, G^2, \ldots$ if they are regarded as independent games. Then the solution σ of the composite game G—or G^*—itself will be equivalent to using solution σ^1 whenever the players' information vectors c_i take values from the sets D_i^1, to using solution σ^2 whenever these vectors c_i take values from the sets D_i^2, etc.

In other words, playing game G—or G^*—in practice always means playing one of its component games G^m. Hence in each case the players will use strategies appropriate for that particular component game G^m, and their strategy choices will be unaffected by the fact that, had their information vectors c_i taken different values, they might now have to play a different component game $G^{m'} \neq G^m$. (To repeat, the justification of this postulate is wholly dependent on the fact that, under our definition of games G and G^*, at the time of their strategy choices the players will always be assumed to know which particular component game G^m they are actually playing. If the players could choose their strategies *before* they were informed about the values of their own information vectors c_i—and so before they knew which component game G^m they would be playing—then Postulate 3 would be inappropriate.)

Let $R_1 = R_1(c^1 \mid c_1), \ldots, R_n = R_n(c^n \mid c_n)$ be the subjective probability distributions that player j uses in analyzing a given I-game G. We shall say that these distributions $R_1, \ldots, R_n$ are *mutually consistent* if there exists some probability distribution $R^* = R^*(c)$ [not necessarily unique] satisfying equation (5.3) with respect to $R_1, \ldots, R_n$. In this case, we shall say that game G (as assessed by player j) is itself *consistent*. Moreover, each probability distribution R^* satisfying (5.3) will be called *admissible*.

On the basis of these definitions we can now state the following theorem.

THEOREM III. *Let G be an I-game (as assessed by player j) in which the player's subjective probability distributions are discrete or absolutely continuous or mixed (but have no singular components). Then three cases are possible*:

1. *Game G may be inconsistent, in which case no admissible probability distribution R^* will exist.*
2. *Game G may be consistent and indecomposable, in which case exactly one admissible probability distribution R^* will exist.*
3. *Game G may be consistent and decomposable. In this case the following statements hold.*
    (a) *Game G can be decomposed into a finite or infinite number of component games $G^1 = G(D^1)$, $G^2 = G(D^2), \ldots$, themselves indecomposable.*
    (b) *There will be an infinite set of different admissible probability distributions R^*.*
    (c) *But for each indecomposable component game $G^m = G(D^m)$, every admissible distribution R^* will generate the same uniquely determined conditional probability distribution*

    $$R^m = R^m(c) = R^*(c \mid c \in D^m), \qquad (12.3)$$

    whenever this conditional probability distribution R^m generated by this admissible distribution R^* is well defined.
    (d) *Every admissible probability distribution R^* will be a convex combination of the conditional probability distributions $R^1, R^2, \ldots$, and different admissible distributions R^* will differ only in the probability weights $R^*(D^m)$ $= R^*(c \in D^m)$ assigned to each component R^m. Conversely, every convex combination of the distributions $R^1, R^2, \ldots$ will be admissible distribution R^*.*

13.

We shall prove the theorem only for games where the distributions $R_1, \ldots, R_n$ are discrete (so that all admissible distributions R^*, if there are any, will also be discrete.)[2] For games with absolutely continuous probability distributions the proof is essentially the same, except that the probability mass functions used in the proof have to be replaced by probability density functions. Once the theorem has been proven for both discrete and continuous distributions, the proof for mixed distributions is quite straightforward. Before proving the theorem itself, we shall prove a few lemmas.

Let G be an I-game (as assessed by player j) where all n players i have discrete subjective probability distributions $R_i = R_i(c^i \mid c_i)$. For each probability distribution R_i we can define the corresponding probability mass function r_i as

$$r_i = r_i(\gamma^i \mid \gamma_i) = R_i(c^i = \gamma^i \mid c_i = \gamma_i) \qquad i = 1, \ldots, n, \quad (13.1)$$

where $\gamma^i = (\gamma_1, \ldots, \gamma_{i-1}, \gamma_{i+1}, \ldots, \gamma_n)$ and γ_i stand for specific values of the vectors c^i and c_i, respectively. Likewise, for each admissible probability distribution R^* (if there is one) we can define the corresponding probability mass function r^* as

$$r^* = r^*(\gamma) = R^*(c = \gamma) \qquad (13.2)$$

where $\gamma = (\gamma_1, \ldots, \gamma_n)$ stands for specific values of vector c. If R^* is admissible then the corresponding probability mass function r^* will also be called admissible. The set of all admissible functions r^* will be called $\mathscr{R}$.

Equation (5.3) now can be written in the form

$$r^*(\gamma) = r_i(\gamma^i \mid \gamma_i) \cdot r^*(\gamma_i) \quad \text{for all } r^* \in \mathscr{R} \qquad (13.3)$$

and for all values $c = \gamma$, $c^i = \gamma^i$, and $c_i = \gamma_i$. Here $r^*(\gamma_i)$ denotes the marginal probability

$$r^*(\gamma_i) = \sum_{(\gamma^i \in c^i)} r^*(\gamma_i, \gamma^i). \qquad (13.4)$$

We shall call any point $c = \gamma = (\gamma_i, \gamma^i)$ of the range space C a *null point* if one or more of the expressions $r_1(\gamma^1 \mid \gamma_1), \ldots, r_n(\gamma^n \mid \gamma_n)$ are

---

[2] Section 13 contains only the proof of Theorem III and may be omitted without loss of continuity.

zero. We shall call a point $c = \gamma$ a *nonnull point* if all these expressions are positive.

In a consistent game G a given value $c_i = \gamma_i$ of vector c_i will be called an *impossible value* if every admissible function r^* assigns to this value $c_i = \gamma_i$ the marginal probability $r^*(\gamma_i) = 0$. We shall adopt the convention that in any consistent game G each range space $C_i = \{c_i\}$, $i = 1, \ldots, n$, will be defined with the exclusion of all impossible values $c_i = \gamma_i$ of vector c_i, from C_i. Therefore, we shall assume that

$$\gamma_i \in C_i \quad \text{implies} \quad r^*(\gamma_i) > 0 \quad \text{for some } r^* \in \mathcal{R}. \qquad (13.5)$$

Consequently at a null point $c = \gamma$ all n expressions $r_1(\gamma^1 \mid \gamma_1)$, $\ldots, r_n(\gamma^n \mid \gamma_n)$ will be zero because, in view of (13.3) and (13.5), $r_i(\gamma^i \mid \gamma_i) = 0$ implies $r_k(\gamma^k \mid \gamma_k) = 0$ for $k = 1, \ldots, n$.

LEMMA 1. *Suppose that in a consistent game G at a given point $c = \gamma$ all admissible functions r^* identically take the value*

$$r^*(\gamma) = 0. \qquad (13.6)$$

This will be the case if and only if $c = \gamma$ is a null point.

The lemma again follows from (13.3) and (13.5).

Let $c = \alpha = (\alpha_1, \ldots, \alpha_n)$, $c = \beta = (\beta_1, \ldots, \beta_n)$, $c = \gamma = (\gamma_1, \ldots, \gamma_n)$, and $c = \delta = (\delta_1, \ldots, \delta_n)$ be four nonnull points, such that $\alpha_i = \beta_i$ and $\gamma_i = \delta_i$, whereas $\alpha_k = \gamma_k$ and $\beta_k = \delta_k$. Let r^* be an admissible function taking a positive value at the point $c = \alpha$. Then, by equation (13.3), the quantities $r^*(\alpha_i) = r^*(\beta_i)$ and $r^*(\alpha_k) = r^*(\gamma_k)$ will also be positive, and, therefore, the same will be true of the quantities $r^*(\beta)$ and $r^*(\gamma)$. This in turn implies that the quantities $r^*(\beta_k) = r^*(\delta_k)$ and $r^*(\gamma_i) = r^*(\delta_i)$, as well as the quantity $r^*(\delta)$, will also be positive. Consequently in view of equation (13.3) we have:

$$\frac{r^*(\alpha)}{r^*(\delta)} = \frac{r^*(\alpha)}{r^*(\beta)} \cdot \frac{r^*(\beta)}{r^*(\delta)} = \frac{r_i(\alpha^i \mid \alpha_i)}{r_i(\beta^i \mid \beta_i)} \cdot \frac{r_k(\beta^k \mid \beta_k)}{r_k(\delta^k \mid \delta_k)}, \qquad (13.7)$$

but also

$$\frac{r^*(\alpha)}{r^*(\delta)} = \frac{r^*(\gamma)}{r^*(\delta)} \cdot \frac{r^*(\alpha)}{r^*(\gamma)} = \frac{r_i(\gamma^i \mid \gamma_i)}{r_i(\delta^i \mid \delta_i)} \cdot \frac{r_k(\alpha^k \mid \alpha_k)}{r_k(\gamma^k \mid \gamma_k)}. \qquad (13.8)$$

If the probability mass functions r_i and r_k—or, equivalently, the probability distributions R_i and R_k—are chosen arbitrarily, then equations (13.7) and (13.8) may become inconsistent, so that no admissible probability mass function r^* taking a positive value $r^*(\gamma) > 0$ at the nonnull point $c = \gamma$ will exist. But then, by Lemma 1, the game will be inconsistent. Thus, we can state:

LEMMA 2. *If the subjective probability distributions* $R_1, \ldots, R_n$ *of a given game G are chosen arbitrarily, then they may turn out to be mutually inconsistent, so that no admissible probability distribution* R^* *will exist.*

Now suppose again that G is a consistent game with discrete probability distributions. Any pair of nonnull points $c = \gamma$ and $c = \delta$ will be called *similar* if all admissible probability mass functions r^* yield the same unique ratio $r^*(\gamma)/r^*(\delta)$ whenever this ratio is well defined. (That is, this must be true for all admissible functions r^* not taking zero values $r^*(\gamma) = r^*(\delta) = 0$ at *both* of these two points.) Clearly the relation of similarity partitions the set C^* of all nonnull points into equivalence classes, which we shall call *similarity classes*. All points $c = \gamma$ belonging to the same similarity class E will be similar, and all points $c = \gamma$ and $c = \delta$ belonging to two different similarity classes E and E' will be dissimilar.

LEMMA 3. *Let* $c = \gamma$ *and* $c = \delta$ *be two nonnull points of a consistent game G, agreeing in their ith component* $c_i = \gamma_i = \delta_i$. *Then* γ *and* δ *will belong to the same similarity class E.*

PROOF. By Lemma 1, we can find some admissible function $r^* = r^0$ such that

$$r^0(\gamma) > 0. \tag{13.9}$$

In view of (13.3) this implies that

$$r^0(\gamma_i) = r^0(\delta_i) > 0. \tag{13.10}$$

As both γ and δ are nonnull points we also have

$$r_i(\gamma^i \mid \gamma_i) > 0 \text{ and } r_i(\delta^i \mid \delta_i) > 0. \tag{13.11}$$

Now let $r^* = r^{00}$ be any other admissible function for which the ratio $r^{00}(\delta)/r^{00}(\gamma)$ is well defined. In view of (13.3) and (13.10), we can write

$$r^0(\delta)/r^0(\gamma) = r_i(\delta^i \mid \delta_i)/r_i(\gamma^i \mid \gamma_i) = r^{00}(\delta)/r^{00}(\gamma). \tag{13.12}$$

By (13.9) and (13.11), the first two ratios in the equation will also be well defined. As this equation will hold for *any* admissible function $r^* = r^{00}$ for which the last ratio is well defined, we can infer that the points δ and γ will belong to the same similarity class E, as desired.

For a given set $E \subseteq C$, let D_i be the set of all vectors $c_i = \gamma_i$ occurring as the ith component of some vector(s) $c = \gamma = (\gamma_1, \ldots, \gamma_i, \ldots, \gamma_n)$ in set E. Then D_i is called the *projection* of set E on vector space C_i. Moreover, if $D_1, \ldots, D_n$ are the projections of set E on the vector spaces $C_1, \ldots, C_n$ then the cylinder $D = D_1 \times \cdots \times D_n$ is called the cylinder *spanned* by set E. Obviously, $E \subseteq D$.

LEMMA 4. *Let* $D^1, D^2, \ldots$ *be the cylinders spanned by the similarity classes* $E^1, E^2, \ldots$ *of a given consistent game G. Then these cylinders* $D^1, D^2, \ldots$ *will be mutually separated.*

PROOF. All we have to prove is that for any given value of i $(i = 1, \ldots, n)$ the projections D_i^m and $D_i^{m'}$ of two different similarity classes E^m and $E^{m'}$ are always disjoint. Now suppose this would not be the case. Then D_i^m and $D_i^{m'}$ would have some common point $c_i = \gamma_i$. This in turn would imply that γ_i would be the ith component of some point $c = \gamma = (\gamma_i, \gamma^i)$ in set E^m and also of some point $c = \delta = (\delta_i, \delta^i) = (\gamma_i, \delta^i)$ in set $E^{m'}$. But then, by Lemma 3, γ and δ would belong to the same similarity class, and could not belong to two different similarity classes E^m and $E^{m'}$, which is inconsistent with the assumptions made. Hence, D^m and $D^{m'}$ cannot have any common point $c_i = \gamma_i$.

LEMMA 5. *If the nonnull points of a given consistent game G can be partitioned into two or more similarity classes* $E^1, E^2, \ldots$, *then game G itself can be decomposed into corresponding component games* $G^1 = G(D^1)$, $G^2 = G(D^2), \ldots$, *where the defining cylinder* D^m $(m = 1, 2, \ldots)$ *of component game* $G^m = G(D^m)$ *is the cylinder spanned by the similarity class* E^m.

PROOF. Let $c_i = \alpha_i$ and $c_k = \beta_k$ be specific values of vectors c_i and c_k, such that $\alpha_i \in D_i^m$ but $\beta_k \notin D_k^m$. Let $c = \gamma = (\gamma_1, \ldots, \alpha_i, \ldots, \beta_k, \ldots, \gamma_n)$ be any point having $\gamma_i = \alpha_i$ as its ith component and having $\gamma_k = \beta_k$ as its kth component. Then, by Lemma 4, γ will be a null point so that

$$r_i(\gamma^i \mid \alpha_i) = r_i(\beta_k, \gamma^{ik} \mid \alpha_i) = 0, \qquad (13.13)$$

where γ^{ik} is the vector which remains if the ith and kth components are both omitted from $\gamma = (\gamma_1, \ldots, \gamma_n)$. Therefore,

$$r_i(\beta_k \mid \alpha_i) = R_i(c_k = \beta_k \mid c_i = \alpha_i) = 0 \tag{13.14}$$

whenever $\alpha_i \in D_i^m$ but $\beta_k \notin D_k^m$. Consequently,

$$R_i(c_k \notin D_k^m \mid c_i) = 0 \quad \text{if } c_i \in D_i^m \tag{13.15}$$

and so

$$R_i(c_k \in D_k^m \mid c_i) = 1 \quad \text{if } c_i \in D_i^m. \tag{13.16}$$

Hence, for each cylinder D^m $(m = 1, 2, \ldots)$, every pair of factor sets D_i^m and D_k^m $(i, k = 1, \ldots, n)$ will satisfy condition (12.1), and so game G can be decomposed into the component games $G^m = G(D^m)$, $m = 1, 2, \ldots$.

LEMMA 6. *A given consistent game G is decomposable if and only if it has more than one admissible probability distribution R^*.*

Proof of the "if" part of the lemma. If all nonnull points of game G belong to the same similarity class E, then, for all pairs of nonnull points γ and δ, the ratios $r^*(\gamma)/r^*(\delta)$ are uniquely determined. Hence, the game can have only one admissible probability mass function r^* and only one admissible probability distribution R^*.

Thus, if game G does have more than one admissible probability distribution R^*, then it must contain two or more similarity classes $E^1, E^2, \ldots$; and, therefore, by Lemma 5 it will be decomposable.

Proof of the "only if" part of the lemma. We have to show that G will have more than one admissible probability distribution R^* (or more than one admissible probability mass function r^*) if it is a decomposable game.

Suppose to the contrary that a given game G can be decomposed into the component games $G^1 = G(D^1)$, $G^2 = G(D^2), \ldots$ but has only one admissible probability mass function $r^* = r^0$. Then, by Lemma 1, this function r^0 must take a positive value $r^0(\gamma) > 0$ at each nonnull point $c = \gamma$ of G. Let r^m $(m = 1, 2, \ldots)$ be the conditional probability mass function

$$r^m(\gamma) = r^0(\gamma \mid \gamma \in D^m)$$

$$= r^0(\gamma) \Big/ \sum_{(c \in D^m)} r^0(c) \quad \text{for all } \gamma \in D^m, \qquad (13.17)$$

$$r^m(\gamma) = 0 \qquad\qquad\qquad \text{for all } \gamma \notin D^m.$$

Since $r^0(c) > 0$ at all nonnull points c, this function r^m will be everywhere well defined. In view of (13.3) and (13.17) we can write

$$r^m(\gamma) = r_i(\gamma^i \mid \gamma_i) \cdot r^m(\gamma_i), \qquad (13.18)$$

which means that the function $r^* = r^m$ $(m = 1, 2, \dots)$ satisfies condition (13.3). Hence, the functions $r^1, r^2, \dots$ are just as much admissible probability mass functions as r^0 itself is, which contradicts the assumption that $r^* = r^0$ is the only admissible probability mass function of the game. This completes the proof of Lemma 6.

LEMMA 7. *A consistent game G is indecomposable if and only if all its nonnull points belong to the same similarity class E.*

Proof of the "if" part. If all nonnull points are similar then the ratios $r^*(\gamma)/r^*(\delta)$ are uniquely determined for all nonnull points γ and δ. Hence, the game can have only one admissible probability mass function r^*, and so, by Lemma 6, it is indecomposable.

Proof of the "only if" part. If game G is indecomposable, then, by Lemma 6, it will have only one admissible function r^*. Hence, the ratios $r^*(\gamma)/r^*(\delta)$ will be uniquely determined for all pairs of nonnull points γ and δ, and so all of these points will be similar.

LEMMA 8. *Let G be a decomposable consistent game, and let $G^1 = G(D^1)$, $G^2 = G(D^2), \dots$ be its component games as defined by Lemma 5. Then*

1. *Each component game G^m will be itself an indecomposable game.*
2. *If $r^* = r^0$ and $r^* = r^{00}$ are two admissible probability mass functions for G then, for each component game G^m $(m = 1, 2, \dots)$, both $r^* = r^0$ and $r^* = r^{00}$ will generate the same conditional probability mass function*

$$r^m(\gamma) = r^0(\gamma \mid \gamma \in D^m) = r^{00}(\gamma \mid \gamma \in D^m) \qquad (13.19)$$

whenever these conditional probability mass functions generated by r^0 and r^{00} are well defined—that is, whenever the two functions r^0 and r^{00} assign a positive probability mass to the event $E = \{\gamma \in D^m\}$.

3. *Every admissible probability mass function r^* for the game will be a convex combination of these functions $r^1, r^2, \ldots$ corresponding to the component games $G^1, G^2, \ldots$; and any convex combination of these functions $r^1, r^2, \ldots$ will be an admissible probability mass function r^*.*

PROOF. Parts 1 and 2 follow from Lemma 7 in view of the fact that all nonnull points γ lying in any given cylinder D^m belong to the same similarity class E^m.

Part 3 follows from equations (13.17) and (13.3).

Lemmas 2, 5, 6, 7, and 8 together imply Theorem I as restricted to games with discrete probability distributions. We have already indicated (in the first paragraph of this section) how the proof can be extended to the general case.

14.

By Theorem III, if a given player j participating in some I-game G does not take special care to base his analysis of the game on mutually consistent subjective probability distributions $R_1, \ldots, R_j, \ldots, R_n$, then, in general, no admissible basic probability distribution R^* will exist; and, therefore, there will be no Bayesian game G^* Bayes-equivalent to G. In view of this fact we now propose the following postulate.

Postulate 4. Mutual consistency. Unless he has special reasons to believe that the subjective probability distributions $R_1, \ldots, R_j, \ldots, R_n$ used by the n players are, in fact, mutually inconsistent, every player j of an I-game G will always analyze the game in terms of a set of some mutually consistent probability distributions $R_1, \ldots, R_j, \ldots, R_n$.

Postulate 4 can be supported by the following considerations:

(i) Player j can greatly simplify his analysis of the game situation by using n mutually consistent distributions $R_1, \ldots, R_n$, because this will enable him to reduce the problem of analyzing the I-game G to the much easier problem of analyzing an equivalent C-game G^*. Therefore, player j will be well advised to assume n mutually consistent distributions $R_1, \ldots, R_n$, unless he feels that the information he has about the game situation is incompatible with this assumption.

(ii) Admittedly, if player j does not restrict his choice to mutually consistent distributions R_i, then he will have a freer hand in postulating suitable differences between the different players' subjective probability distributions R_i, and so will have a freer hand in providing mathematical representation for any differences he may wish to assume between different players i as to their beliefs and expectations about the game situation. However, even if he does restrict his choice to mutually consistent R_i's, this need not prevent him from finding suitable representation for such interplayer differences in his mathematical model of the game; for such differences can often be given fully adequate representation by assuming corresponding differences between the relevant players' attribute vectors c_i, and by choosing a basic probability distribution $R^* = R^*$ $(c_1, \ldots, c_n)$ which is a sufficiently asymmetric function of the different players' attribute vectors c_i; so that the conditional probability distributions $R_i = R^*(c^i \mid c_i)$ for different players i will have appropriately different mathmematical forms. Mutually inconsistent R_i's will have to be assumed only in cases in which player j comes to the definite conclusion that the interplayer differences in question could not be represented by mutually consistent R_i's.

(iii) According to the prior-lottery model (see Section 6 of Part I of this paper), every I-game G can be considered to be a result of a random social process which has selected the n actual players of the game from some hypothetical populations $\Pi_1, \ldots, \Pi_n$ of possible players, where the probability that this random process will select n players with any given combination of specific attribute vectors $c_1^0, \ldots, c_n^0$ is governed by some (objective) probability distribution $R^* = R^*(c_1, \ldots, c_n)$. However, in general, the true distribution R^* governing this random process will be unknown to the players.

Let $R_i^* = R^*(c^i \mid c_i)$, for $i = 1, \ldots, n$, be the conditional probability distributions generated by this unknown distribution R^*. Clearly the n distributions $R_1^*, \ldots, R_n^*$ will always be mutually consistent, since by definition they are conditional distributions generated by the same basic distribution R^*. The n players' subjective probability distributions $R_1, \ldots, R_n$ can be regarded as their *estimates* of these mutually consistent, but unknown, conditional probability distributions $R_1^*, \ldots, R_n^*$.

Accordingly, it is natural to argue (at least in cases where player j has no special reasons for using inconsistent distributions R_i) that, instead of trying to estimate each player's subjective probability distribution R_i separately, player j should rather try to estimate directly the objective probability distribution R^* governing the random social process in question; and then he should choose the n distributions R_i he will use in the analysis of the game situation, by setting $R_i = R^*(c^i \mid c_i)$, where R^* now refers to his *estimate* of the true distribution R^*. Obviously, if player j uses this estimation procedure, he will always come up with n mutually consistent distributions $R_1, \ldots, R_n$.

We have seen in Section 6 (Part I) that the prior-lottery model pictures the game situation in the way in which it would be seen by an intelligent and "properly informed" outside observer.[3] Consequently, when player j is trying to estimate the probability distribution R^*, he will have to ask himself the question, "What probability distribution R^* would be chosen by a randomly selected intelligent and properly informed outside observer, as his estimate of the true probability distribution governing the random social process underlying this game situation G?" (If player j feels that he cannot give a unique answer to this question because different intelligent and properly informed observers are likely to choose different probability distributions R^*, then he can construct his own probability distribution R^* by some averaging process over the various distributions R^* he thinks different observers might choose.)

To put it differently, in trying to estimate the probability distribution R^*, player j should try to use only the information *common* to all n-players. Of course, in his analysis at some point or another he should make use of *all* information he has about the game situation, including any special information which may be available to him but may not be available to the other players. But under our model all such special information must be represented by incorporating it into player j's own information vector c_j. In contrast, the basic probability distribution R^* itself is meant to represent only information common to all n players. (The same is true for the n subjective probability distributions $R_1, \ldots, R_n$ and the n payoff functions $V_1, \ldots, V_n$: by definition they are meant to contain only such information that, in player j's opinion, is common property of the n players—see Section 3 in Part I.)

To be sure, if player j follows the suggested estimation procedure this will guarantee only that his own subjective probability distribution R_j and the subjective probability distributions $R_1, \ldots, R_{j-1}, R_{j+1}, \ldots, R_n$ he will *ascribe* to the other $(n - 1)$ players will satisfy equation (5.3).

---

[3] We have defined a "properly informed" observer as a person possessing all information *shared* by all n players but having no extra information unavailable to the players, or available to some particular player(s) yet unavailable to the other player(s).

But, of course, it will not guarantee that his own subjective probability distribution R_j and the subjective probability distributions $R_1, \ldots,$ $R_{j-1}, R_{j+1}, \ldots, R_n$ the other $(n-1)$ players will in *actual fact* use in the game will satisfy equation (5.3). In other words, the suggested estimation procedure will guarantee *internal consistency* among the n probability distributions $R_1, \ldots, R_n$ that player j himself will use in analyzing the game, but will not and cannot guarantee *external consistency* among the probability distributions used by the different players. Indeed, so long as each player has to choose his subjective probabilities (probability estimates) independently of the other players, no conceivable estimation procedure can ensure consistency among the different players' subjective probabilities.

More particularly, there is no way of ensuring that the different players will choose the *same* subjective probability distribution as their estimate of the objective probability distribution R^* underlying the game situation. For, by the very nature of subjective probabilities, even if two individuals have exactly the same information and are at exactly the same high level of intelligence, they may very well assign different subjective probabilities to the very same events.

At the same time, while the suggested estimation procedure does not in any way guarantee external consistency among the different players' probability estimates, we feel it does go as far as an estimation procedure can go in promoting such external consistency. By asking each player to take the point of view of an outside observer in estimating the objective probability distribution R^*, it asks him to choose an estimate as close as possible to the estimates other intelligent people might be expected to choose, and to make his estimate as independent as possible of his own personal prejudices and idiosyncrasies.

15.

So far our analysis has been restricted to the case in which player j has no special information suggesting mutual inconsistencies among the probability distributions $R_1, \ldots, R_j, \ldots, R_n$ of the players. Now what happens if player j does have such information? For instance, what happens if in an international conflict situation country j obtains quite convincing reports from its diplomatic representatives, intelligence agencies, newspaper correspondents, etc., to the effect that in analyzing the international situation country i is using a very different political and social philosophy, and a very different model of the world, from country j's own; and that as a result the subjective probability distribution R_i used by country i is clearly inconsistent with the subjective probability distribution R_j used by country j?

In a situation such as this, one obvious possibility is to take this information at face value and to conclude that the players' subjective probability distributions *are* in fact inconsistent, which means that every player j simply has to resign himself to the necessity of analyzing the game situation in terms of n mutually inconsistent subjective probability distributions $R_1, \ldots, R_j, \ldots, R_n$.

As Reinhard Selten has pointed out,[4] his posterior-lottery model (see Section 6, Part I of this paper) can be extended to this case of mutually inconsistent subjective probability distributions. All we have to do is to assume that when all K players have chosen their strategies there will be a *separate* lottery $L(i_m)$ for every player i_m, instead of there being merely *one* grand lottery L^* for all K players. For each player i_m in role class i his lottery $L(i_m)$ will choose $(n - 1)$ players [one player from each of the $(n - 1)$ role classes $1, \ldots, i - 1, i + 1, \ldots, n$] as his "partners" in the game. If player i_m himself has the attribute vector $c_i = c_i^0$, then the probability of his $(n - 1)$ partners' having any specific combination of attribute vectors $c_1 = c_1^0; \ldots; c_{i-1} = c_{i-1}^0; c_{i+1} = c_{i+1}^0; \ldots; c_n = c_n^0$ will be governed by the subjective probability distribution $R_i = R_i(c^i \mid c_i = c_i^0)$ which player i himself entertains. Player i_m will receive the payoff:

$$x_i = V_i(s_1^0, \ldots, s_i^0, \ldots, s_n^0; c_1^0, \ldots, c_i^0, \ldots, c_n^0) \qquad (15.1)$$

where $s_1^0, \ldots, s_n^0$ are the strategies chosen by player i_m and by his partners, whereas $c_1^0, \ldots, c_n^0$ are the attribute vectors of these same n players.

Note that, under this model, "partnership" is not necessarily a symmetric relationship. If lottery $L(i_m)$ of player i_m chooses a given player k_r as a partner for player i_m, then it does not follow that lottery $L(k_r)$ of player k_r will likewise choose player i_m as a partner[5] for player

---

[4] In private communication (see Footnote 1 in Part I).

[5] This fact does not give rise to any logical difficulty. It only means that, owing to the outcome of lottery $L(i_m)$, player i_m's payoff may come to depend on the strategy s_{k^0} and the attribute vector c_{k^0} of some player k_r—even though, owing to the outcome of lottery $L(k_r)$, player k_r's own payoff will not be dependent on the strategy s_i^0 and the attribute vector c_i^0 of player i_m, but rather will be dependent on the strategy s_i^{00} and the attribute vector c_i^{00} of another player $i_t \neq i_m$ in role class i. (Of course, when the players choose their strategies they will not know who will be whose partner and whether any such relation of partnership will be reciprocal or not.)

k_r. As partnership in general, will not be a symmetric relationship, and as the lotteries associated with different players will be quite independent, this model does not presuppose any mutual consistency in the sense of Postulate 4 among the n probability distributions $R_1 = R_1(c^1 \mid c_1), \ldots, R_n = R_n(c^n \mid c_n)$ governing the lotteries conducted for the players in the n role classes $1, \ldots, n$.

Thus, a Selten game G^{**} will exist even for an inconsistent I-game G, for which no Bayesian game G^* exists.

16.

It is, however, always questionable whether any given information suggesting inconsistencies among the different players' subjective probability distributions $R_1, \ldots, R_n$ should really be taken at face value.

First of all, information about other players' assessment of probabilities (and indeed about their internal beliefs and attitudes in general) tends to be very unreliable, because the players will often have a vested interest in misleading the other players about their real ways of thinking. But even if the empirical facts themselves about the other players' probability judgments are quite correct, they will be usually open to alternative interpretations.

More particularly, any inconsistency among the various players' subjective probability distributions R_i is always a result of discrepancies among the basic probability distributions $R^* = R^{(i)}$ used by different players i (see Theorem IV below). On the other hand, these discrepancies among the probability distributions $R^{(i)}$ will themselves often admit of explanation in terms of the *differences in the information* available to different players i—in which case, as we shall see, the game can be *reinterpreted* as a game involving mutually *consistent* subjective probability distributions $R_1, \ldots, R_n$ on the part of the n players.

We now propose to state the following simple theorem.

THEOREM IV. *Let*

$$R^*(c_1, \ldots, c_i, \ldots, c_n) = R^{(i)}(c_1, \ldots, c_i, \ldots, c_n) \qquad (16.1)$$

be player i's basic probability distribution ($i = 1, \ldots, n$), defined as his estimate of the probability distribution governing the random social process which gave rise to the current game situation. Suppose that

$$R^{(1)} = \cdots = R^{(n)} = R^0. \qquad (16.2)$$

Then the n players' subjective probability distributions $R_1, \ldots, R_n$ will be mutually consistent.

PROOF. Suppose player i thinks that the social process in question is governed by the probability distribution $R^* = R^{(i)}$, and knows that his own attribute vector has the value $c_i = c_i^0$. Then he must use the conditional probability distribution $R^*(c^i \mid c_i) = R^{(i)}(c^i \mid c_i)$, generated by this distribution $R^* = R^{(i)}$, for evaluating the probability that the other $(n-1)$ players will have any specific combination of attribute vectors $c_1 = c_1^0; \ldots; c_{i-1} = c_{i-1}^0; c_{i+1} = c_{i+1}^0; \ldots; c_n = c_n^0$. Hence, player i's subjective probability distribution R_i will be defined by this conditional distribution[6]

$$R_i = R_i(c^i \mid c_i) = R^*(c^i \mid c_i) = R^{(i)}(c^i \mid c_i). \qquad (16.3)$$

This will be true for all players $i = 1, \ldots, n$. Consequently, in view of equation (16.2) we can write

$$R_i = R^0(c^i \mid c_i) \qquad i = 1, \ldots, n. \qquad (16.4)$$

Hence, all n functions $R_1, \ldots, R_n$ will have the nature of conditional probability distributions generated by the same basic probability distribution R^0, which means that they will be mutually consistent.

By Theorem IV, if the functions $R_1, \ldots, R_n$ are in fact mutually inconsistent, then the basic probability distributions $R^{(1)}, \ldots, R^{(n)}$ used by the n players cannot be all identical; and it will be the discrepancies existing among these basic probability distributions $R^{(1)}, \ldots, R^{(n)}$ that account for the inconsistency among the subjective probability distributions $R_1, \ldots, R_n$. Now, our contention is that the discrepancies existing among the different players' basic probability distributions $R^{(i)}$ will often admit of explanation in terms of differences in the information that different players have about the nature of the social process underlying the current game situation.

For instance, in our previous example, if the subjective probability distributions R_i and R_j used by countries i and j are mutually inconsistent, this will indicate that they assess probabilities in terms of two quite different basic probability distributions $R^{(i)}$ and $R^{(j)}$. More particularly,

---

[6] In the case $k \neq i$, the relationship $R_i = R^{(k)}(c^i \mid c_i)$ will hold in general only if $R^{(k)} = R^{(i)}$ as required by equation (16.2). However, in the case $k = i$, the relationship will always hold, by the very definitions of the subjective probability distributions R_i and of the basic probability distributions $R^* = R^{(i)}$.

e.g., country i may assign much higher probabilities than country j does to the success of violent social revolutions in various parts of the world. Now it will often be a very natural hypothesis that these differences in the two countries' assessment of probabilities are due to differences in the information they have acquired about the world as a result of their divergent historical experiences—more particularly as a result of their divergent experience about the prospects of violent social revolutions on the one hand, and about the chances of achieving social reforms by peaceful nonrevolutionary methods on the other hand.

Of course, if player j makes the assumption that the discrepancies among the probability distributions $R^{(i)}$ used by different players i are due to differences in the information available to them, this will mean under our terminology that these distributions $R^{(i)}$ are not really the *basic* probability distribution of these players, because in this case, obviously, these distributions $R^{(i)}$ cannot be assumed any longer to represent exclusively the information *common* to all n players. Instead, each distribution $R^{(i)}$ will now have to be interpreted as a *conditional* probability distribution, conditioned by some special information available to player i himself but not available to all n players.

On the other hand, if player j is willing to assume that *all* discrepancies among the distributions $R^{(i)}$ used by different players i are due to these differences in the information available to them, then he can consider all these distributions $R^{(i)}$ to be conditional distributions derived from *one* and the *same* unconditional distribution $R^{*\prime}$, which means that this distribution $R^{*\prime}$ will be the true basic distribution of the game.

In symbols, let d_i be a vector summarizing all special information about the relevant variables which makes any given player i adopt the probability distribution $R^{(i)} = R^{(i)}(c_1, \ldots, c_i, \ldots, c_n)$—instead of adopting some probability distribution $R^{(k)} = R^{(k)}(c_1, \ldots, c_i, \ldots, c_n)$ used by another player k. Then player i's true information vector (or attribute vector) will not be the vector c_i but rather will be the larger vector $c_i' = (c_i, d_i)$ summarizing the information contained in *both* vectors c_i and d_i.

Consequently, the true basic probability distribution $R^{*\prime}$ of the game will be the joint probability distribution of these larger vectors c_i', and so will have the form

$$R^{*\prime} = R^{*\prime}(c_1', \ldots, c_n') = R^{*\prime}(c_1, d_1; \ldots; c_n, d_n). \qquad (16.5)$$

In contrast, each distribution $R^* = R^{(i)}$ will have the nature of a conditional probability distribution (or more exactly of a conditional marginal probability distribution) of the form

$$R^{(i)} = R^{(i)}(c_1, \ldots, c_i, \ldots, c_n) = R^{*\prime}(c_1, \ldots, c_i, \ldots, c_n \mid d_i). \quad (16.6)$$

In other words, player j will find that, as soon as he redefines the players' attribute vectors as the larger vectors $c_i' = (c_i, d_i)$, then the information he has about the other players' subjective probabilities will become compatible with the hypothesis that all n players are actually using the *same* basic probability distribution $R^{*\prime} = R^{*\prime}(c_1', \ldots, c_j', \ldots, c_n')$. Consequently, by Theorem IV, this information will be compatible also with the hypothesis that the n players are using *mutually consistent* subjective probability distributions $R_1', \ldots, R_j', \ldots, R_n'$ of the form

$$R_i' = R_i'\big((c^i)' \mid c_i'\big) = R^{*\prime}\big((c^i)' \mid c_i'\big) \qquad i = 1, \ldots, j, \ldots, n, \quad (16.7)$$

where

$$(c^i)' = (c_1', \ldots, c_{i-1}', c_{i+1}', \ldots, c_n').$$

To sum up, if the information player j has about the other players' subjective probabilities seems to suggest that the n players are using mutually *inconsistent* subjective probability distributions $R_1, \ldots, R_j, \ldots, R_n$, then more careful reanalysis of the game situation may very well lead him to the conclusion that, in actual fact, the players' subjective probability distributions can be better represented by another set of probability distributions $R_1', \ldots, R_j', \ldots, R_n'$, mutually *consistent* with one another.

Note that, in reanalyzing the game situation in this way, player j must use not only the information he has obtained independently about the game, but also any information he can infer from what he knows about the other players' assessment of probabilities. For instance, suppose that player j himself would be inclined to assign the probability $P_j(e)$ to a given event e, but then obtains the information that another player i assigns a very different probability $P_i(e) \neq P_j(e)$ to this event. Then player j cannot rest the matter with the conclusion that he and player i have apparently been assessing the probability of this event on the basis of two rather different sets of information. Instead, he also has to reach a decision on whether his own assessment of this probability is likely to be based on more correct and more complete information than player i's assessment is, or whether the opposite is the case; and, in particular, he has to decide whether player i's quite different assessment of this probability is or is not a good enough reason for him himself also to change his own assessment of this probability from $P_j(e)$ to some number closer to the probability $P_i(e)$ which player i is assigning to event e.

Of course, this method of reconciling the different players' assessments of probabilities in terms of the same basic probability distribution $R^{*\prime}$ will work only if player j feels that the empirical facts which he knows about the other players' probability assessments are favorable to, or at least are compatible with, the hypothesis that the discrepancies among the various players' probability judgments can be reasonably explained in terms of differences in their information. In our own view, in most cases the empirical facts will be at least compatible with this hypothesis. But if in any given I-game G player j reaches the opposite conclusion, then he will have to fall back upon Selten's model (see Section 15), which permits mutually inconsistent subjective probability distributions on the part of the players.

17.

According to Theorem III (Section 12), even if player j uses mutually consistent subjective probability distributions $R_1, \ldots, R_n$ in his analysis of a given I-game G, he may still have the problem of there being an infinite set of equally admissible probability distributions R^* satisfying equation (5.3) [Part I of this paper]; which means that there will be an infinite number of Bayesian games G^* Bayes-equivalent to this I-game G.

However, by Postulate 3, this fact will not give rise to any difficulties because, if player j uses the solution σ of *any* of these games G^* as the solution of game G, he will always obtain the *same* solution σ. This is so because, by Theorem III, even if there are many Bayesian games G^* Bayes-equivalent to G, there will always be only *one* Bayesian game $(G^m)^*$ Bayes-equivalent to any given indecomposable *component game* G^m of G (since G^m will have only one unique admissible probability distribution $R^*(G^m) = R^m$). Hence, no ambiguity will result if he uses the solution σ^m of this Bayesian game $(G^m)^*$ as the solution of this component game G^m. On the other hand, by Postulate 3, the solutions

σ^m of these component games G^m will completely determine the solution σ of the whole game G.

Yet, even though the possible multiplicity of Bayesian games G^* Bayes-equivalent to a given I-game G causes no real problem, for some purposes it will be convenient if we can make the assumption that there is always only *one* Bayesian game G^* Bayes-equivalent to each I-game G. We can achieve this by using the fact that, under the procedure proposed in Section 14, each player j will start his analysis of a given I-game G by first choosing a basic probability distribution $R^* = R^*(c_1, \ldots, c_n)$, and then will define the n probability distributions $R_1 = R_1(c^1 \mid c_1), \ldots, R_n = R_n(c^n \mid c_n)$ as the conditional distributions $R_1 = R^*(c^1 \mid c_1); \ldots; R_n = R^*(c^n \mid c_n)$ derived from this distribution R^*. This make it natural to define the I-game G itself (as assessed by player j) in terms of this one distribution R^* chosen by player j at the beginning of his analysis, rather than in terms of the n distributions $R_1, \ldots, R_n$ derived from this distribution R^*.

Formally, this will mean that not only a C-game G^* in standard form (i.e., a Bayesian game G^*), but also any consistent I-game G in standard form, will now be defined as the ordered set

$$G = G^* = \{S_1, \ldots, S_n; C_1, \ldots, C_n; V_1, \ldots, V_n; R^*\}, \qquad (17.1)$$

instead of being defined as the ordered set

$$G = \{S_1, \ldots, S_n; C_1, \ldots, C_n; V_1, \ldots, V_n; R_1, \ldots, R_n\} \qquad (17.2)$$

in accordance with equation (3.18), which was our original suggestion. (Of course, in the case of an inconsistent I-game G we have to go on using equation (17.2) as our formal definition of the game.)

This now completes our general discussion of incomplete-information games G and of their Bayesian-game analogues G^*, which represent games with complete but imperfect information involving certain chance moves. We have discussed the actual solution concepts our approach selects for these games only in the case of a few simple illustrative examples (see Sections 9–11, Part II). More systematic discussion of these solution concepts for a wider range of games with incomplete information must be left for other forthcoming publications.

THE BIG MATCH[1]

DAVID BLACKWELL AND T. S. FERGUSON

UNIVERSITY OF CALIFORNIA, BERKELEY AND LOS ANGELES

1. INTRODUCTION

Extending Shapley's work on stochastic games [2], Gillette [1] has studied the following situation. We are given three nonempty finite sets S, I, J, a real-valued function a defined for all triples (s, i, j), $s \in S$, $i \in I$, $j \in J$, and a function p which associates with each triple (s, i, j) a probability distribution $p(\cdot|s, i, j)$ on S. These five quantities S, I, J, a, p define a two-person zero-sum game, played as follows. We start with some initial state $s \in S$, known to both players. Player 1 choose $i \in I$ and, simultaneously, player 2 chooses $j \in J$. Player 1 is then awarded $a(s, i, j)$ points, and the game moves to state s' selected according to $p(\cdot|s, i, j)$. The new state s' is announced to both players, who then choose i', j', giving player 1 $a(s', i', j')$ points, and causing the game to move to the state s'' selected according to $p(\cdot|s', i', j')$, etc. The payoff to player 1 from the infinite sequence of choices is

$$\limsup_{n \to \infty} (a_1 + \cdots + a_n)/n,$$

where a_m is the point score 1 obtained on the mth round. Whether such games, the finite stochastic games, always have a value is not known. In this paper we consider one interesting example of Gillette, the big match, and show that it does have a value.

The big match is played as follows. Every day player 2 chooses a number 0 or 1, and player 1 tries to predict 2's choice, winning a point if he is correct. This continues as long as player 1 predicts 0. But if he ever predicts 1, all future choices for both players are required to be the same as that day's choices: if player 1 is correct on that day, he wins a point every day thereafter; if he is wrong on that day, he wins zero every

Received 13 June 1967.

[1] Supported in part by the Office of Naval Research under Contract Nonr-222 (53), and in part by National Science Foundation Grant GP-5224.

day thereafter. The payoff to 1 is

$$\limsup_{n \to \infty} (a_1 + \cdots + a_n)/n,$$

where a_m is the number of points he wins on the mth day. The big match is the finite stochastic game with

$$S = \{0, 1, 2\}, \qquad I = J = \{0, 1\},$$

$$a(2, i, j) = \delta_{ij}, \qquad a(s, i, j) = s \quad \text{for } s = 0, 1,$$

$$p(2, 0, j) = \delta(2), \qquad p(2, 1, j) = \delta(j),$$

$$p(s, i, j) = \delta(2) \quad \text{for } s = 0, 1, \quad \text{and initial position } s = 2.$$

2.   SOLUTION OF THE BIG MATCH

THEOREM 1. *The value of the big match is $\frac{1}{2}$. An optimal strategy for player 2 is to toss a fair coin every day. Player 1 has no optimal strategy, but for any nonnegative integer N he can get*

$$V(N) = N/2(N + 1)$$

by using strategy N, defined as follows: having observed player 2's first n choices $x_1, \ldots, x_n$, $n \geq 0$, calculate the excess k_n of 0's over 1's among $x_1, \ldots, x_n$, and predict 1 with probability $p(k_n + N)$, where $p(m) = 1/(m + 1)^2$.

PROOF. Clearly if player 2 tosses a fair coin every day, player 1's expected payoff is $\frac{1}{2}$, no matter what he does: if he ever predicts 1 his payoff is equally likely to be 0 or 1, and if he predicts 0 forever, the strong law of large numbers gives him payoff $\frac{1}{2}$ with probability 1.

Next, notice that strategy N predicts 1 with certainty whenever the excess is $-N$, i.e., whenever N more 1's than 0's have occurred. Suppose we have verified that strategy N produces expected payoff at least $V(N)$ against every sequence of 0's and 1's which eventually achieves an excess of $-N$, and consider any sequence $\omega = (x_1, x_2, \ldots)$ which never achieves an excess as low as $-N$. We show that strategy N yields at least $V(N)$ against ω. Denote the number of observations after which player 1 first predicts 1 by t (so $t = \infty$ if he never predicts 1).

Define

$$\lambda(m) = P\{t < m \text{ and } x_{t+1} = 0\},$$

$$\mu(m) = P\{t < m \text{ and } x_{t+1} = 1\},$$

$$\lambda = \lim_{m \to \infty} \lambda(m), \qquad \mu = \lim_{m \to \infty} \mu(m).$$

Player 1's expected income is at least

$$\mu + (1 - \lambda - \mu)(\tfrac{1}{2}),$$

since the sequence ω, never having excess as low as $-N$, satisfies

$$(x_1 + \cdots + x_n)/n < \tfrac{1}{2} + N/2n, \quad \text{for all } n \tag{1}$$

so that 1's income when $t = \infty$, which is

$$\limsup_{n \to \infty} ((1 - x_1) + \cdots + (1 - x_n))/n,$$

is at least $\tfrac{1}{2}$ (this would be true even with lim sup replaced by lim inf).
   To show that $\mu + (1 - \lambda - \mu)(\tfrac{1}{2}) \geq V(N)$, consider player 1's income from strategy N against the following strategy for 2: choose $x_1, \ldots, x_m$, and toss a fair coin thereafter. With probability 1, the resulting sequence will eventually reach excess $-N$ so, by assumption, player 1's expected income is at least $V(N)$. But his expected income is exactly

$$\mu(m) + \tfrac{1}{2}(1 - \lambda(m) - \mu(m)),$$

which therefore is at least $V(N)$. Letting $m \to \infty$ completes the proof that it suffices to study sequences ω which eventually achieve an excess of $-N$.
   Take such an $\omega = (x_1, x_2, \ldots)$, let player 1 use strategy N against it, and denote by t the number of observations after which he first predicts 1. Define

$$E(m) = \{t \geq m, \quad \text{or} \quad t < m \quad \text{and} \quad x_{t+1} = 1\}, \qquad m = 1, 2, \ldots.$$

We show, inductively on m, that

$$P_N(E(m)) \geq V(N), \quad \text{for all } N. \tag{1}$$

(a) $m = 1$. If $x_1 = 1$, $P_N(E(1)) = 1 > V(N) = N/2(N + 1)$. If $x_1 = 0$, $P_N(E(1)) = P_N\{t \geq 1\} = 1 - p(N) = N(N + 2)/(N + 1)^2 \geq V(N)$.

(b) Suppose $P_N(E(m)) \leq V(N)$, for all N. If $x_1 = 1$, $P_N(E(m + 1)) = p(N) + [1 - p(N)]P_{N-1}(E(m)) \geq p(N) + [1 - p(N)]V(N - 1) = V(N)$, where $P_{N-1}(E(m)) \geq V(N - 1)$ by induction since using strategy N against $\omega = (1, x_2, x_3, \dots)$ is equivalent to predicting 1 initially with probability $p(N)$ and, with probability $1 - p(N)$ predicting 0 initially and thereafter using strategy $N - 1$ against $\omega' = (x_2, x_3, \dots)$. Similarly, if $x_1 = 0$,

$$P_N(E(m + 1)) = [1 - p(N)]P_{N+1}(E(m))$$

$$\geq [1 - p(N)]V(N + 1) = V(N).$$

So (1) is proved. Since $t < \infty$ with probability 1, letting $m \to \infty$ in (1) yields

$$P_N\{x_{t+1} = 1\} \geq V(N),$$

strategy N yields at least $V(N)$ against every sequence, and the value of the game is $\frac{1}{2}$.

To show that 1 has no optimal strategy, consider any strategy σ for him. If σ never predicts 1 with positive probability against $\omega^* = (1, 1, 1, \dots)$, it wins 0 against ω^* so is not optimal. If not, say $m \geq 0$ is the smallest initial number of 1's after which σ predicts 1 with positive probability, say ϵ. Player 2 can counter σ by choosing m initial 1's, then 0, and tossing a fair coin thereafter, giving 1 an expected income of $(1 - \epsilon)\frac{1}{2}$.

The above argument that 1 has no optimal strategy is due to Lester Dubins. We are indebted to him, David Freedman, and Volker Strassen for stimulating conversations about the big match.

3. OTHER NEAR OPTIMAL STRATEGIES FOR PLAYER 1

Let $0 < \epsilon < 1$, and let $\{\alpha_n\}$, $n = 0, 1, 2, \dots$, be a sequence of numbers satisfying the conditions

(a) $\alpha_n \geq \alpha_{n+1}$ for $n = 0, 1, 2, \dots$,
(b) $(1 - \epsilon)\alpha_n \leq \alpha_{n+1}$ for $n = 0, 1, 2, \dots$,
(c) $\sum_0^\infty \alpha_n = 1$.

(Automatically, $\alpha_n > 0$ for all n.) For a given $0 < \epsilon < 1$ and $\{\alpha_n\}$ satisfying (a), (b), and (c), we define a strategy ψ for player 1 by defining the distribution of t, the time at which player 1 first predicts 1, via the functions

$$\psi_n(x_1, \ldots, x_n) = P\{t = n + 1 | x_1, \ldots, x_n\}, \qquad n = 0, 1, 2, \ldots .$$

Note that ψ_n represents the (unconditional) probability that $t = n + 1$, whereas in Theorem 1 we referred to the conditional probabilities, $P\{t = n + 1 | t > n, x_1, \ldots, x_n\}$.

The functions ψ_n are defined inductively by letting $\psi_0 = \epsilon \alpha_0$ and

$$\psi_n(x_1, \ldots, x_n) = \epsilon \alpha_{k_n}, \qquad\qquad \text{if } \sum_0^n \epsilon \alpha_{k_j} \leq 1,$$

$$= 1 - \sum_0^{n-1} \psi_j(x_1, \ldots, x_j), \quad \text{otherwise,}$$

where k_n, as before, represents the excess of 0's over 1's among $x_1, \ldots, x_n$ ($k_n = \sum_1^n (1 - 2x_i)$), and where for $j < 0$, α_j is defined equal to α_0. We denote by $\mathscr{C}_\epsilon$ the class of strategies ψ formed in this manner for some sequence $\{\alpha_n\}$ satisfying (a), (b), and (c).

The main property of the class $\mathscr{C}_\epsilon$ is as follows:

THEOREM 2. *If $\psi \in \mathscr{C}_\epsilon$, $\sum_0^\infty (1 - 2x_{n+1}) \psi_n(x_1, \ldots, x_n) \leq 3\epsilon$.*

PROOF. If $x_1, x_2, \ldots$ is such that $\sum_0^\infty \epsilon \alpha_{k_n} \leq 1$ (so that $\psi_n = \epsilon \alpha_{k_n}$), there is an integer M such that $\epsilon \sum_0^{M-1} \alpha_{k_n} \geq \epsilon \sum_0^\infty \alpha_{k_n} - \epsilon$. If not, let M denote the smallest integer for which $\sum_0^M \epsilon \alpha_{k_n} > 1$ (so that $\psi_n = \epsilon \alpha_{k_n}$ for $n < M$). In either case,

$$\sum_0^\infty (1 - 2x_{n+1}) \psi_n(x_1, \ldots, x_n) \leq \epsilon \sum_0^{M-1} (1 - 2x_{n+1}) \alpha_{k_n} + \epsilon.$$

Let $J = k_M$. We assume $J \geq 0$. (For $J < 0$, a similar argument obviously works.) Let

$$I_1 = \left\{ n: k_{n+1} > \max_{0 \leq j \leq m} k_j, \quad k_n < J, \quad 0 \leq n \leq M - 1 \right\},$$

$$I_2 = \{ n: x_{n+1} = 0, \quad n \notin I_1, \quad 0 \leq n \leq M - 1 \},$$

$$I_3 = \{ n: x_{n+1} = 1, \quad 0 \leq n \leq M - 1 \}.$$

There is a one-to-one correspondence between I_2 and I_3: If $k_i \leq J$, $i \in I_3$ is paired with the smallest $j \in I_2$ such that $j > i$ and $k_j = k_i - 1$. If $k_i > J$, $i \in I_3$ is paired with the largest $j \in I_2$ such that $j < i$ and $k_j = k_i - 1$. In this pairing, $\alpha_{k_i} \geq (1 - \epsilon)\alpha_{k_j}$. Hence,

$$\sum_0^{M-1} (1 - 2x_{n+1})\alpha_{k_n} = \sum_{I_1} \alpha_{k_n} + \sum_{I_2} \alpha_{k_j} - \sum_{I_3} \alpha_{k_i} \leq 1 + \epsilon \sum_{I_2} \alpha_{k_j} \leq 2,$$

completing the proof.

In the big match, the expected payoff to player 1 given $x_1, x_2, \ldots$ is, in terms of the functions $\{\psi_n\}$,

$$\sum_0^\infty x_{n+1}\psi_n(x_1, \ldots, x_n) + \left(1 - \sum_0^\infty \psi_n(x_1, \ldots, x_n)\right) \limsup(1 - \bar{x}_n),$$

where $\bar{x}_n$ represents $(x_1 + \cdots + x_n)/n$. Theorem 2 implies that for $\psi \in \mathscr{C}_\epsilon$ this expected payoff is at least

$$\tfrac{1}{2}\sum_0^\infty \psi_n(x_1, \ldots, x_n) - \tfrac{3}{2}\epsilon + \left(1 - \sum_0^\infty \psi_n(x_1, \ldots, x_n)\right) \limsup(1 - \bar{x}_n).$$

Hence, if $\limsup(1 - \bar{x}_n) \geq \tfrac{1}{2}$, the expected payoff is at least $\tfrac{1}{2} - (\tfrac{3}{2})\epsilon$. On the other hand, if $\limsup(1 - \bar{x}_n) < \tfrac{1}{2}$, then k_n is negative infinitely often so that $\sum_0^\infty \psi_n(x_1, \ldots, x_n) = 1$ for any $\psi \in \mathscr{C}_\epsilon$. Thus, in any case, the expected payoff is at least $\tfrac{1}{2} - (\tfrac{3}{2})\epsilon$. In other words, every $\psi \in \mathscr{C}_\epsilon$ is $(\tfrac{3}{2})\epsilon$-optimal for the big match.

In closing, we mention a related problem, in which further restrictions are placed on the strategies available to the player. We require of player 2 that he choose a sequence $\omega = (x_1, x_2, \ldots)$ such that $\lim \bar{x}_n = \tfrac{1}{2}$, and we require of player 1 that he choose a distribution of t such that $P\{t < \infty | x_1, x_2, \ldots\} = 1$ for all ω available to player 2. The value of this modified game is still $\tfrac{1}{2}$, as may be seen from the following considerations. Player 2's optimal strategy for the big match is still available to him in the modified game, so the upper value is at most $\tfrac{1}{2}$. To see that the lower value is at least $\tfrac{1}{2}$, let player 1 play as follows. He chooses a small $\delta > 0$ and produces a private random sequence, $y_1, y_2, \ldots$ of independent variables with $P\{y_i = 1\} = \delta$ and $P\{y_i = 0\} = 1 - \delta$. He defines $z_n = \max(x_n, y_n)$ and uses any strategy in $\mathscr{C}_\epsilon$ (or the strategy in Theorem 1), pretending that 2's sequence is $z_1, z_2, \ldots$. Since

$\bar{x}_n \to \frac{1}{2}$, $\bar{z}_n \to (1 + \delta)/2$, so that $t < \infty$ with probability one. Then, $P\{z_{t+1} = 1\} > \frac{1}{2} - (\frac{3}{2})\epsilon$. But since $P\{y_{t+1} = 1\} = \delta$, we have $P\{x_{t+1} = 1\} > \frac{1}{2} - (\frac{3}{2})\epsilon - \delta$, showing the lower value is at least $\frac{1}{2}$.

REFERENCES

[1] Gillette, Dean (1957). Stochastic games with zero stop probabilities. *Contributions to the Theory of Games* **3**. Princeton University Press.
[2] Shapley, L. S. (1953). Stochastic games. *Proc. Nat. Acad. Sci.* **39** 1095–1100.

17

ON MARKET GAMES[1]

LLOYD S. SHAPLEY

The RAND Corporation, Santa Monica, California 90406

AND

MARTIN SHUBIK

Department of Administrative Sciences, Yale University,
New Haven, Connecticut 06520

The "market games"—games that derive from an exchange economy in which the traders have continuous concave monetary utility functions, are shown to be the same as the "totally balanced games"—games which with all their subgames possess cores. (The core of a game is the set of outcomes that no coalition can profitably block.) The coincidence of these two classes of games is established with the aid of explicit transformations that generate a game from a market and vice versa. It is further shown that any game with a core has the same *solutions*, in the von Neumann-Morgenstern sense, as some totally balanced game. Thus, a market may be found that reproduces the solution behavior of any game that has a core. In particular, using a recent result of Lucas, a ten-trader ten-commodity market is described that has no solution.

1. INTRODUCTION

Recent discovery of *n*-person games in the classical theory which either possess no solutions [8, 9], or have unusually restricted classes of solutions [6, 7, 10, 18], has raised the question of whether these games are mere mathematical curiosities or whether they could actually arise in application. Since the most notable applications of *n*-person game theory to date have been to economic models of exchange, or exchange

---

[1] This research is supported by the United States Air Force under Project RAND—Contract No. F44620-67-C-0045—monitored by the Directorate of Operational Requirements and Development Plans, Deputy Chief of Staff, Research and Development, HQ USAF. Views and conclusions expressed herein should not be interpreted as representing the official opinion or policy of the United States Air Force or of The RAND Corporation.

and production [3, 13, 15, 16, 19–23], the question may be put in a more concrete form: Are there markets, or other basic economic systems, that when interpreted as n-person games give rise to the newly-discovered counterexamples? If so, can they be distinguished from the ordinary run of market models, on some economic, heuristic, or even formal grounds —i.e., do they give any advance warning of their peculiar solution properties?

These questions stimulated the present investigation. The answers are "bad news": Yes, the games can arise in economics; No, there are no outwardly distinguishing features. In reaching these conclusions, how-ever, we were led to a positive result: a surprisingly simple mathematical criterion that tells precisely which games can arise from economic models of exchange (with money). In fact, this criterion identifies a very fundamental class of games, called "totally balanced," whose further study seems merited quite apart from any consideration of solution abnormalities. Of technical interest, our derivation of the basic proper-ties of these games and their solutions makes a substantial application of the recently developed theory of *balanced sets* [1, 2, 11, 17], as well as of the older work of Gillies on *domination-equivalence* [4, 5].

In the present note we confine our attention to the classical theory "with side payments" [23]; this corresponds in the economic interpreta-tion to the assumption that an ideal money, free from income effects or transfer costs, is available.[2] We further restrict ourselves to exchange economies, without explicit production or consumption processes, in which the commodities are finite in number and perfectly divisible and transferable, and in which the traders, also finite in number, are motivated only by their own final holdings of goods and money, their utility functions being continuous and concave and additive in the money term.

For our immediate purpose, these strictures do not matter, since the anomalous games are already attainable within the limited class of ideal markets considered. But for our larger purpose—that of initiating a systematic study of "market games" as distinct from games in general—some relaxation may be desirable, particularly with regard to money. The prospects for significant generalizations in this direction appear good, and we intend to pursue them in subsequent work.

---

[2] For a discussion of this assumption, see [20], pp. 807–808.

1.1 Outline of the Contents

The notions of game, core, and balanced set are reviewed. A game is called "balanced" if it has a core, and "totally balanced" if all of the subgames obtained by restricting the set of players have cores as well (Section 2).

A "market" is defined as an exchange economy with money, in which the traders have utility functions that are continuous and concave. The method of passing from a market to its "market game" is described. The market games with n traders form a closed convex cone in the space of all n-person games with side payments. Every market game not only has a core, but is totally balanced (Section 3).

A canonical market form—the "direct market"—is introduced, in which the commodities are in effect the traders themselves, made infinitely divisible, and the utility functions are all the same and are homogeneous of degree one. The method of passing from any game to its direct market is described, the utilities being based upon the optimal assignment of "fractional players" to the various coalitional activities. The "cover" of a game is defined as the market game of its direct market; the cover is at least as profitable to all coalitions as the original game. Every totally balanced game is its own cover, and hence is a market game. This shows that *the class of market games and the class of totally balanced games are the same*. Moreover, every market is game-theoretically equivalent to a direct market (Section 4).

The notions of imputation, domination, and solution are reviewed. Games are "d-equivalent" if they have identical domination relations on identical imputation spaces. They therefore have identical solutions (or lack of solutions), and their cores, if any, are the same. It is shown that *every balanced game is d-equivalent to a totally balance game*. Hence, for every game with a core, there is a market that has precisely the same set of solutions (Section 5).

Using Lucas's solutionless game [8, 9], a direct market is constructed that has ten traders and ten commodities (plus money), and that has no solution. Another version in the form of a production economy is also presented. Several other examples of market games with unusual solution properties are mentioned, and in one case, where the solutions contain arbitrary components, the utility function is worked out explicitly (Section 6).

2. GAMES AND CORES

For the purpose of this note, a *game* is an ordered pair $(N; v)$, where N is a finite set [the players] and v is a function from the subsets of N [coalitions] to the reals satisfying $v(O) = 0$, called the *characteristic function*. A *payoff vector* for $(N; v)$ is a point α in the $|N|$-dimensional vector space E^N whose coordinates α^i are indexed by the elements of N. If $\alpha \in E^N$ and $S \subseteq N$, we shall write $\alpha(S)$ as an abbreviation for $\Sigma_{i \in S} \alpha^i$.

The *core* of $(N; v)$ is the set of all payoff vectors α, if any, such that

$$\alpha(S) \geq v(S), \quad \text{all } S \subseteq N, \tag{2-1}$$

and

$$\alpha(N) = v(N). \tag{2-2}$$

If no such α exists, we shall say that $(N; v)$ *has no core*. (Thus, in this usage, the core may be nonexistent, but is never empty.)

2.1 Balance Sets of Coalitions

A *balanced set* $\mathscr{B}$ is defined to be a collection of subsets S of N with the property that there exist positive numbers γ_S, $S \in \mathscr{B}$, called "weights," such that for each $i \in N$ we have

$$\sum_{\substack{S \in \mathscr{B} \\ S \ni i}} \gamma_S = 1. \tag{2-3}$$

If all $\gamma_S = 1$, we have a partition of N; thus, balanced sets may be regarded as generalized partitions.

For example, if $N = \overline{1234}$, then $\{\overline{12}, \overline{13}, \overline{14}, \overline{234}\}$ is a balanced set, by virtue of the weights $1/3, 1/3, 1/3, 2/3$.

A game $(N; v)$ is called *balanced* if

$$\sum_{S \in \mathscr{B}} \gamma_S v(S) \leq v(N) \tag{2-4}$$

holds for every balanced set $\mathscr{B}$ with weights $\{\gamma_S\}$.[3]

---

[3] These conditions are heavily redundant; it suffices to assert (2-4) for the *minimal* balanced set $\mathscr{B}$ (which moreover have unique weights). In the case of a superadditive game, only the minimal balanced sets that contain no disjoint elements are needed. (See [17].)

THEOREM 1. *A game has a core if and only if it is balanced.*

This is proved in [17]. In Scarf's generalization to games without transferable utility [11], all balanced games have cores, but some games with cores are not balanced. If our present results can be generalized in this direction, we conjecture that it will be the balance property, rather than the core property, that plays the central role.

2.2 Totally Balanced Games

By a *subgame* of $(N; v)$ we shall mean a game $(R; v)$ with $O \subset R \subseteq N$. Here v is the same function, but implicitly restricted to the domain consisting of the subsets of R. A game will be said to be *totally balanced* if all of its subgames are balanced. In other words, all subgames of a totally balanced game have cores.

Not all balanced games are totally balanced. For example, let $N = \overline{1234}$ and define $v(S) = 0, 0, 1, 2$ for $|S| = 0, 1, 3, 4$ respectively, and, for $|S| = 2$:

$$v(\overline{12}) = v(\overline{13}) = v(\overline{23}) = 1$$

$$v(\overline{14}) = v(\overline{24}) = v(\overline{34}) = 0.$$

This game has a core, including the vector $(\frac{1}{2}, \frac{1}{2}, \frac{1}{2}, \frac{1}{2})$ among others. But it is not totally balanced, since the subgame $(\overline{123}; v)$ has no core.

3. MARKETS AND MARKET GAMES

For the purpose of this note, a *market* is a special mathematical model, denoted by the symbol (T, G, A, U). Here T is a finite set [the traders]; G is the nonnegative orthant of a finite-dimensional vector space [the commodity space]; $A = \{a^i : i \in T\}$ is an indexed collection of points in G [the initial endowments]; and $U = \{u^i : i \in T\}$ is an indexed collection of continuous, concave functions from G to the reals [the utility functions]. When we wish to indicate that $u^i \equiv u$, all $i \in T$ [the special case of "equal tastes"], we shall sometimes denote the market by the more specific symbol $(T, G, A, \{u\})$.

If S is any subset of T, an indexed collection $X^S = \{x^i : i \in S\} \subset G$ such that $\sum_S x^i = \sum_S a^i$ will be called a *feasible S-allocation* of the market (T, G, A, U).

A market (T, G, A, U) can be used to "generate" a game $(N; v)$ in a natural way. We set $N = T$, and define v by

$$v(S) = \max_{X^S} \sum_{i \in S} u^i(x^i), \quad \text{all } S \subseteq N, \tag{3-1}$$

where the maximum runs over all feasible S-allocations. Any game that can be generated in this way from some market is called a *market game*.[4]

In the special case of identical utility functions $u^i \equiv u$, we have

$$v(S) = |S| u \left(\sum_S a^i / |S| \right), \quad \text{all } S \subseteq N; \tag{3-2}$$

this is a simple consequence of concavity. In the still more special case where u is homogeneous of degree 1, we have simply

$$v(S) = u \left(\sum_S a^i \right), \quad \text{all } S \subseteq N. \tag{3-3}$$

3.1 Some Elementary Properties

The following two theorems are of a routine nature; they show that the property of being a market game is invariant under "strategic equivalence," and that the set of all market games on N forms a convex cone in the $(2^{|N|} - 1)$-dimensional space of all games on N.

THEOREM 2. *If $(N; v)$ is a market game, if $\lambda \geq 0$, and if c is an additive set function on N, then $(N; \lambda v + c)$ is a market game.*

PROOF. We need merely take any market that generates $(N; v)$ and replace each utility function $u^i(x)$ by $\lambda u^i(x) + c(\{i\})$.   Q.E.D.

THEOREM 3. *If $(N; v')$ and $(N; v'')$ are market games, then $(N, v' + v'')$ is a market game.*

PROOF. Let (N, G', A', U') and (N, G'', A'', U'') be markets that generate $(N; v')$ and $(N; v'')$ respectively. We shall superimpose these two markets, keeping the two sets of commodities distinct. Specifically, let G be the set of all ordered pairs (x', x'') of points from G' and G''

---

[4] For examples, see [13, 15, 16, 20, 21, 22]. The abstract definition of "market game" proposed in [13] is not equivalent to the present one, however.

respectively; let A be the set of pairs (a'^i, a''^i) of correspondingly indexed elements of A' and A''; and let U be the set of sums:

$$u^i((x', x'')) = u'^i(x') + u''^i(x'')$$

of correspondingly indexed elements of U' and U''. One can then verify without difficulty that the elements of U are continuous and concave on the domain G (which is a nonnegative orthant in its own right), so that (N, G, A, U) is a market. Finally, one can verify without difficulty that (N, G, A, U) generates the game $(N; v' + v'')$.   Q.E.D.

3.2 The Core Theorem

THEOREM 4. *Every market game has a core.*

This theorem is well known, and has been generalized well beyond the limited class of markets we are now considering. Nevertheless we shall give two proofs, both short, for the sake of the insights they provide. In the first, we in effect determine a competitive equilibrium for the generating market (a simple matter when there is transferable utility), and then show that the competitive payoff vector lies in the core. In the second proof, we show directly that the game is balanced, and then apply Theorem 1.

PROOF 1. Let $(N; v)$ be a market game and let (N, G, A, U) be a market that generates it. Let $B = \{b^i : i \in N\}$ be a feasible N-allocation that achieves the value $v(N)$ in (3-1) for $S = N$. The maximization in (3-1) ensures the existence of a vector p [competitive prices—but possibly negative!] such that for each $i \in N$, the expression

$$u^i(x^i) - p \cdot (x^i - a^i), \qquad x^i \in G, \qquad (3\text{-}4)$$

is maximized at $x^i = b^i$. Define the payoff vector β by

$$\beta^i = u^i(b^i) - p \cdot (b^i - a^i);$$

we assert that β is in the core. Indeed, let S be any nonempty subset of N, and let Y^S be a feasible S-allocation that achieves the maximum in (3-1), so that $v(S) = \Sigma_S u^i(y^i)$. Since b^i maximizes (3-4), we have

$$\beta^i \geq u^i(y^i) - p \cdot (y^i - a^i).$$

Summing over $i \in S$, we obtain

$$\beta(S) \geq \sum_S u^i(y^i) - p \cdot 0 = v(S),$$

as required by (2-1). Moreover, if $S = N$ we may take $Y^S = B$ and obtain $\beta(N) = v(N)$, as required by (2-2).

PROOF 2. Let (N, G, A, U) be a generating market for $(N; v)$, and, for each $S \subseteq N$, let $Y^S = \{y_S^i : i \in S\}$ a maximizing S-allocation in (3-1). Let $\mathscr{B}$ be balanced, with weights $\{\gamma_S^i : S \in \mathscr{B}\}$. Then we have

$$\sum_{S \in \mathscr{B}} \gamma_S v(S) = \sum_{S \in \mathscr{B}} \sum_{i \in S} \gamma_S u^i(y_S^i) = \sum_{i \in N} \sum_{\substack{S \in \mathscr{B} \\ S \ni i}} \gamma_S u^i(y_S^i).$$

Now define

$$z^i = \sum_{\substack{S \in \mathscr{B} \\ S \ni i}} \gamma_S y_S^i \in G, \quad \text{all } i \in N.$$

Note that z^i is a center of gravity of the points y_S, by virtue of (2-3). Hence, by concavity,

$$\sum_{S \in \mathscr{B}} \gamma_S v(S) \leq \sum_{i \in N} u^i(z^i). \tag{3-5}$$

But $Z = \{z^i : i \in N\}$ is a feasible N-allocation, since

$$\sum_{i \in N} z^i = \sum_{S \in \mathscr{B}} \sum_{i \in S} \gamma_S y_S^i = \sum_{S \in \mathscr{B}} \gamma_S \sum_{i \in S} a^i = \sum_{i \in N} a^i.$$

Hence the right side of (3-5) is $\leq v(N)$, and we conclude from (2-4) that the game is balanced and from Theorem 1 that it has a core.
Q.E.D.

COROLLARY. *Every market game is totally balanced.*

PROOF. If $(N; v)$ is generated by the market (N, G, A, U), and if $O \subset R \subseteq N$, then we may define a market (R, G, A', U'), where A' and U' come from A and U by simply omitting all a^i and u^i for i not in R. This market clearly generates the game $(R; v)$. Hence $(R; v)$ is balanced.
Q.E.D.

Our next objective will be to prove the converse of this corollary—i.e., that every totally balanced game is a market game.

4. DIRECT MARKETS

A special class of markets, called *direct markets*, will play an important role in the sequel. They have the form

$$(T, E_+^T, I^T, \{u\}),$$

where u is homogeneous of degree 1 as well as concave and continuous. Here E_+^T denotes the nonnegative orthant of the vector space E^T with coordinates indexed by the members of T, and I^T denotes the collection of unit vectors of E^T—in effect, the identity matrix on T.

Thus, in a direct market, each trader starts with one unit of a personal commodity [e.g., his time, his labor, his participation, "himself"]. When it is brought together with other personal commodities, we may imagine that some desirable state of affairs is created, having a total value to the traders that is independent (because of homogeneity and equal tastes) of how they distribute the benefits.

Let e^S denote the vector in E^N in which $e_i^S = 1$ or 0 according as $i \in S$ or $i \notin S$; geometrically, these vectors represent the vertices of the unit cube in E_+^N. Then the characteristic function of the market game generated by a direct market can be put into a very simple form:

$$v(S) = u(e^S), \quad \text{all } S \subseteq N \tag{4-1}$$

(compare (3-3)). Note that only finitely many commodity bundles are involved in this expression.

4.1 The Direct Market Generated by a Game

Thus far we used markets to generate games. We now go the reverse route, associating with any game (not necessarily a market game) a certain "market of coalitions." Specifically, we shall say that the game $(N; v)$ "generates" the direct market $(N, E_+^N, I^N, \{u\})$, with u given by

$$u(x) = \max_{\{\gamma_S\}} \sum_{S \subseteq N} \gamma_S v(S), \quad \text{all } x \in E_+^N, \tag{4-2}$$

maximized over all sets of nonnegative γ_S satisfying

$$\sum_{S \ni i} \gamma_S = x_i, \quad \text{all } i \in N. \tag{4-3}$$

To explain this market,[5] we may imagine that each coalition S has an activity $\mathscr{A}_S$ that can earn $v(S)$ dollars if all the members of S participate fully. More generally, it earns $\gamma_S v(S)$ dollars if each member of S devotes the fraction γ_S of "himself" to $\mathscr{A}_S$. The maximization in (4-2) is then nothing but an optimal assignment of activity levels γ_S to the various $\mathscr{A}_S$'s, subject to the condition (4-3) that each player, i, distribute exactly the amount x_i of "himself" among his activities, including of course the "solo" activity $\mathscr{A}_{(i)}$.

The utility function defined by (4-2) is obviously homogeneous of degree 1, as required for a direct market. But before we can claim to have defined a market, let along a direct market, we must also establish that (4-2) is continuous and concave. Continuity gives no trouble. To show concavity, it suffices (with homogeneity) to prove that

$$u(x) + u(y) \leq u(x + y), \quad \text{all } x, y \in E_+^N.$$

This is not difficult. By definition, there exist sets of nonnegative coefficients $\{\gamma_S\}$ and $\{\delta_S\}$ such that

$$u(x) = \sum_{S \subseteq N} \gamma_S v(S), \qquad u(y) = \sum_{S \subseteq N} \delta_S v(S);$$

and

$$\sum_{S \ni i} \gamma_S = x_i, \qquad \sum_{S \ni i} \delta_S = y_i, \quad \text{all } i \in N.$$

Hence $\{\gamma_S + \delta_S\}$ is admissible for $x + y$, and (4-2) yields

$$u(x + y) \geq \sum (\gamma_S + \delta_S) v(S) = u(x) + u(y),$$

as required.

[5] The essence of this model was suggested by D. Cantor and M. Maschler (private correspondence, 1962).

4.2 The Cover of a Game

We shall now use the direct market generated by a game $(N; v)$ to generate in turn a new game $(N; \bar{v})$—schematically:

arbitrary game → direct market → market game.

We shall call $(N; \bar{v})$ the *cover* of $(N; v)$.

Combining (4-1) with (4-2) and (4-3), we obtain the following relation between v and $\bar{v}$:

$$\bar{v}(R) = \max_{\{\gamma_S\}} \sum_{S \subseteq R} \gamma_S v(S), \quad \text{all } R \subseteq N \tag{4-4}$$

maximized over $\gamma_S \geq 0$ such that

$$\sum_{\substack{S \subseteq R \\ S \ni i}} \gamma_S = 1, \quad \text{all } i \in R. \tag{4-5}$$

Note that we could have taken (4-4), (4-5) as the definition of "cover," bypassing the intermediate market. Indeed, the cover of a game proves to be a useful mathematical concept quite apart from the present economic application.

We see immediately that

$$\bar{v}(R) \geq v(R), \quad \text{all } R \subseteq N, \tag{4-6}$$

since one of the admissible choices for $\{\gamma_S\}$ in (4-4) is to take $\gamma_R = 1$ and all other $\gamma_S = 0$. Moreover, the equality cannot always hold in (4-6); indeed, $\bar{v}$ comes from a market game while v was arbitrary. Thus, the mapping $v \to \bar{v}$ takes an arbitrary characteristic function and, by perhaps increasing some values, turns it into the characteristic function of a market game.

LEMMA 1. *If $(N; v)$ has a core, then $\bar{v}(N) = v(N)$, and conversely.*

PROOF. Let α be in the core of $(N; v)$. Then

$$\bar{v}(N) = \max_{\{\gamma_S\}} \sum_{S \subseteq N} \gamma_S v(S)$$

$$\leq \max_{\{\gamma_S\}} \sum_{S \subseteq N} \gamma_S \alpha(S) = \max_{\{\gamma_S\}} \sum_{i \in N} \alpha^i \sum_{S \ni i} \gamma_S$$

$$= \max_{\{\gamma_S\}} \sum_{i \in N} \alpha^i = \alpha(N)$$

$$= v(N),$$

the successive lines being justified by (4-4), (2-1), (4-5), and (2-2). In view of (4-6) we therefore have $\bar{v}(N) = v(N)$.

Conversely, if $(N; v)$ has no core, then (2-4) fails for some balanced set $\mathcal{B}$ with weights $\{\gamma_S\}$. Defining $\gamma_S = 0$ for $S \notin \mathcal{B}$, we see that (4-5) holds (for $R = N$). Then (4-4) and the denial of (2-4) give us

$$\bar{v}(N) \geqq \sum_{S \subseteq N} \gamma_S v(S) = \sum_{S \in \mathcal{B}} \gamma_S v(S) > v(N).$$

Hence $\bar{v}(N) \neq v(N)$.   Q.E.D.

LEMMA 2. *A totally balanced game is equal to its cover.*

PROOF. Let $(N; \bar{v})$ be the cover of $(N; v)$, and let $O \subset R \subseteq N$. Then it is clear from the definitions that the cover of $(R; v)$ is $(R; \bar{v})$. But if $(N; v)$ is totally balanced, then $(R; v)$ has a core and $\bar{v}(R) = v(R)$ by Lemma 1. Hence $\bar{v} = v$.   Q.E.D.

THEOREM 5. *A game is a market game if and only if it is totally balanced.*

PROOF. We proved earlier (corollary to Theorem 4) that market games are totally balanced. We have just now shown that totally balanced games are equal to their covers, which are market games.

Q.E.D.

4.3 Equivalence of Markets

There is one more result of some heuristic interest that we can extract from the present discussion, before entering the realm of solution theory. This time, we follow the scheme:

arbitrary market → market game → direct market.

Let us call two markets *game-theoretically equivalent* if they generate the same market game. Then the two markets in the above scheme are equivalent in this way, since the cover of the game in the middle is just the market game of the market on the right, and these two games are equal by Lemma 2. This proves

THEOREM 6. *Every market is game-theoretically equivalent to a direct market.*

5. SOLUTIONS

An *imputation* for a game $(N; v)$ is a payoff vector α that satisfies

$$\alpha(N) = v(N) \tag{5-1}$$

and

$$\alpha^i \geqq v(\{i\}), \quad \text{all } i \in N. \tag{5-2}$$

A comparison with (2-1) and (2-2) shows that the imputation set is certainly not empty if the game has a core.[6]

Classical solution theory [23] rests upon a relation of "domination" between imputations. If α and β are imputations for $(N; v)$, then α is said to *dominate* β (written $\alpha \dashv \beta$) if there is some nonempty subset S of N such that

$$\alpha^i > \beta^i, \quad \text{all } i \in S, \tag{5-3}$$

and

$$\alpha(S) \leqq v(S). \tag{5-4}$$

A *solution* of $(N; v)$ is defined to be any set of imputations, mutually undominating, that collectively dominate all other imputations. Our only concern with this definition, technically, is to observe that it depends only on the concepts of imputation and domination; any further information conveyed by the characteristic function is disregarded.

The core is also closely dependent on these concepts. In fact, the core, when it exists, is precisely the set of undominated imputations. The converse is not universally true—there are some games that have undominated imputations but no core.[7] We can rule this out, however,

---

[6] Some approaches to solution theory omit (5-2), relying on the solution concept itself to impose whatever "individual rationality" the situation may demand [5, 12]. This modification in the definition of solution would make little difference to our present discussion, except for eliminating the fussy condition (5-5). In particular, Theorem 7 and all of Section 6 would remain correct as written.

[7] We are indebted to Mr. E. Kohlberg for this observation.

by imposing the very weak condition:

$$v(S) + \sum_{N-S} v(\{i\}) \leqq v(N), \quad \text{all } S \subseteq N, \tag{5-5}$$

which is satisfied by all games likely to be met in practice.[8]

5.1 Domination-Equivalence

Two games will be called *d-equivalent* (domination-equivalent) if they have the same imputation sets and the same domination relations on them. It follows that *d*-equivalent games have precisely the same solutions, or lack of solutions. Also, if they have cores, they have the same cores; moreover, within the class of games satisfying (5-5) the property of being balanced is preserved under *d*-equivalence. However, the property of being totally balanced is not so preserved, as the following lemma reveals.

LEMMA 3. *Every balanced game is d-equivalent to its cover.*

PROOF. Let $(N; v)$ be balanced. By Lemma 1, $\bar{v}(N) = v(N)$ and by (4-4), $\bar{v}(\{i\}) = v(\{i\})$; hence the two games have the same imputations. Denote the respective domination relations by $\dashv$ and $\dashv'$. By (4-6) and (5-4) we see at once that the latter is, if anything, stronger than the former—i.e., $\alpha \dashv \beta$ implies $\alpha \dashv' \beta$. It remains to prove the converse.

Assume, *per contra*, that α and β are imputations satisfying $\alpha \dashv' \beta$ but not $\alpha \dashv \beta$. Then for some nonempty subset R of N we have

$$\alpha^i > \beta^i, \quad \text{all } i \in R,$$

and

$$\alpha(R) \leq \bar{v}(R). \tag{5-6}$$

To avoid $\alpha \dashv \beta$ we must have

$$\alpha(S) > v(S) \tag{5-7}$$

for *all* S, $O \subset S \subseteq R$. Referring to the definition of $\bar{v}$, we see that there

---

[8] Thus, (5-5) is implied by either superadditivity or balancedness, but is weaker than both. For a game in normalized form, i.e., with $v(\{i\}) \equiv 0$, it merely states that no coalition is worth more than N.

are nonnegative weights γ_S, $S \subseteq R$, such that

$$\bar{v}(R) = \sum_{S \subseteq R} \gamma_S v(S)$$

and

$$\sum_{\substack{S \subset R \\ S \ni i}} \gamma_S = i, \quad \text{all } i \in R.$$

Hence, using (5-7)

$$\bar{v}(R) < \sum_{S \subseteq R} \gamma_S \alpha(S) = \alpha(R).$$

The strict inequality here contradicts (5-6).   Q.E.D.

By a "solution" of a market, we shall mean a solution of the associated market game.

THEOREM 7. *If $(N; v)$ is any balanced game whatever, then there is a market that has precisely the same solutions as $(N; v)$.*

PROOF. The main work has been done in Lemma 3. Indeed, let $(N, E_+^T, I^T, \{u\})$ be the direct market generated by $(N; v)$. Then the solutions of this market are the solutions of $(N; \bar{v})$, which by the lemma is d-equivalent to $(N; v)$ and hence has the same solutions.   Q.E.D.

5.2 A Technical Remark

The notion of d-equivalence is essentially due to Gillies [4, 5], though he works with a broader definition of imputation, not tied to the characteristic function by (5-1). He defines a *vital* coalition as one that achieves some domination that no other coalition can achieve, and shows that two games are d-equivalent (in the present sense) if and only if they have (i) the same imputation sets, (ii) the same vital coalitions, and (iii) the same v-values on their vital coalitions.

A necessary (but not sufficient) condition for a coalition to be vital is that it cannot be partitioned into proper subsets, the sum of whose v-values equals or exceeds its own v-value. Sufficiency would require the generalized partitioning provided by balanced sets.

Given a game $(N; v)$, we can define its "least superadditive majorant" $(N; \tilde{v})$ by

$$\tilde{v}(S) = \max_h \sum_h v(S_h), \qquad (5\text{-}8)$$

the maximization running over all partitions $\{S_h\}$ of S. (Compare (4-4), (4-5).) It can be shown that $\tilde{v}(N) = v(N)$ if and only if $(N; v)$ has a core (cf. Lemma 1 above), in which case the two games are d-equivalent. Thus, every game with a core is d-equivalent to a superadditive game.

However, as Gillies observes, d-equivalence can also hold nontrivially among superadditive games. That is, it may be possible to push the v-value of some nonvital coalition *higher* than the value demanded by superadditivity, without making the coalition vital.[9] We are using the full power of this observation, since the cover $\bar{v}$ can be thought of as the "greatest d-equivalent majorant" of v. Thus, $v \le \tilde{v} \le \bar{v}$, and all three may be different.

6. EXAMPLES

Lucas's 10-person game [8, 9] with no solution has players $N = \overline{1234567890}$ and the following characteristic function:

$$
\begin{aligned}
&v(\overline{12}) = v(\overline{34}) = v(\overline{56}) = v(\overline{78}) = v(\overline{90}) = 1 \\
&v(\overline{137}) = v(\overline{139}) = v(\overline{157}) = v(\overline{159}) = v(\overline{357}) = v(\overline{359}) = 2 \\
&v(\overline{1479}) = v(\overline{2579}) = v(\overline{3679}) = 2 \\
&v(\overline{1379}) = v(\overline{1579}) = v(\overline{3579}) = 3 \\
&v(\overline{13579}) = 4 \\
&\quad v(N) = 5, \quad \text{and} \\
&\quad v(S) = 0, \quad \text{all other } S \subseteq N.
\end{aligned}
\right\}
$$

$$(6\text{-}1)$$

The game has a core, containing among others the imputation that gives each "odd" player 1.[10] It is not superadditive (for example,

_____

[9] For example, at the end of Sec. 2, $v(\overline{123})$ may be increased from 1 to $3/2$ without making 123 vital.

[10] The full core is a five-dimensional polyhedron, having vertices e^S: $S = \overline{13579}, \overline{23579}, \overline{14579}, \overline{13679}, \overline{13589},$ and $\overline{13570}$.

$v(\overline{12}) + v(\overline{34}) > v(\overline{1234})$); however it is d-equivalent to its least super-additive majorant $(N, \tilde{v})$, which can be calculated without difficulty, using (5-8). Moreover, one can verify that the latter is totally balanced, i.e., that $\tilde{v} = \bar{v}$ in this case. Thus, $(N, \tilde{v})$, defined by (6-1) and (5-8), is a market game with no solution.

The corresponding *market* with no solution, provided by Theorem 7, has ten traders and ten commodities, plus money. The traders have identical continuous concave homogeneous utilities $u(x)$, which may be calculated by applying (4-2), (4-3) to (6-1). Note that positive weights γ_S need be considered only for the eighteen vital coalitions and the ten singletons.[11] Of course, this is not the only utility function that works, since only a finite set of its values are actually used (cf. (4-1)).

6.1 A Production Model

Perhaps the most straightforward economic realization of Lucas's game is in the form of a production economy. (Compare the "activity" description in Section 4.1.) The production possibilities are generated by 18 specific processes, which produce the same consumer good (at constant returns to scale) out of various combinations of the raw materials (see Table I). Each entrepreneur starts with one unit of the correspondingly indexed raw material. The utility is simply the consumer good: $u(x) \equiv x_{11}$; hence it is not necessary to postulate a separate money.

This type of construction is perfectly general: a production model can be set up in a similar fashion for any other game in characteristic function form, one activity being required for each vital coalition. The market game generated by such a model will be the cover of the original game, and will have the same core and solutions provided that the original game was balanced.

6.2 Other Examples

Lucas [6, 7, 10] gives several examples of games in which the solution is unique but does not coincide with the core. In [10] he also describes a symmetric 8-person game, very similar to the above 10-person game, that has an infinity of solutions but none that treats the symmetric players symmetrically. Shapley [18] describes a 20-person game, of the

[11] The singleton weights are needed as slack variables, because we need " = " in (4-3) instead of " ≤ ."

TABLE I

			Inputs							Output
x_1	x_2	x_3	x_4	x_5	x_6	x_7	x_8	x_9	x_{10}	x_{11}
1	1									1
		1	1							1
				1	1					1
						1	1			1
								1	1	1
1		1				1				2
1		1						1		2
1				1		1				2
1				1				1		2
		1		1		1				2
		1		1				1		2
1			1			1		1		2
	1			1		1		1		2
		1			1	1		1		2
1		1				1		1		3
1				1		1		1		3
		1		1		1		1		3
1		1		1		1		1		4

same general type, every one of whose many solutions consists of the core, which is a straight line, plus an infinity of mutually disjoint closed sets that intersect the core in a dense point-set of the first category. A common feature of all these "pathological" examples is the existence of a core; hence, by Theorem 7, they are d-equivalent to market games that have the same solution behavior.

We close with another "pathological" example, of an older vintage [14], which because of its simple form leads to a direct market with utilities that we can write down explicitly. The game has players $N = \overline{123\ldots n}$, with $n \geq 4$, and its characteristic function is given by

$$\left\{ \begin{array}{c} v(N - \{1\}) = v(N - \{2\}) = v(N - \{3\}) = v(N) = 1 \\ v(S) = 0, \quad \text{all other coalitions } S. \end{array} \right\} \quad (6\text{-}2)$$

Thus, to win anything requires the participation of a majority of $\overline{123}$, plus all of the "veto" players $4, \ldots, n$. The core is the set of all imputations α that satisfy $\alpha_1 = \alpha_2 = \alpha_3 = 0$. It is easily verified that the game is totally balanced: $v = \bar{v}$. There are many solutions; but the remarkable feature of the game is a certain subclass of solutions, as follows:

Let B_e denote the set of imputations α that satisfy $\alpha_1 = 0$, $\alpha_2 = \alpha_3 \geq e > 0$. Thus, B_e is a $(n - 3)$-dimensional closed convex subset of the imputation space. In [14] it was shown that one may start with any closed subset of B_e whatever, and extend it to a solution of the game by adding only imputations that are at least $e/2$ distant from B_e.[12] The arbitrary starting set remains a distinct, isolated portion of the full solution. For example, if $n = 4$ (the simplest case), an arbitrary closed set of points on a certain line can be used.

To determine the direct market of this game, we apply (4-2), (4-3) to (6-2) and obtain the utility function

$$u(x) = \max_{(\gamma^i)} (\gamma^1 + \gamma^2 + \gamma^3),$$

maximized subject to

$$\left. \begin{array}{ccc} \gamma^1 \geq 0, & \gamma^2 \geq 0, & \gamma^3 \geq 0; \\ \gamma^2 + \gamma^3 \leq x_1, & \gamma^1 + \gamma^3 \leq x_2, & \gamma^1 + \gamma^2 \leq x_3; \\ \gamma^1 + \gamma^2 + \gamma^3 \leq x_i, & i = 4, \ldots, n; \end{array} \right\}$$

where γ^i abbreviates $\gamma_{N-\{i\}}$. This reduces to the closed form:

$$u(x) = \min \left[x_1 + x_2, x_1 + x_3, x_2 + x_3, \frac{x_1 + x_2 + x_3}{2}, x_4, \ldots, x_n \right].$$

$$(6\text{-}3)$$

We see that u is the envelope-from-below of $n + 1$ very simple linear functions.

Thus, an n-trader n-commodity market having the solutions containing arbitrary components, as described above, is obtained by giving the ith trader one unit of the ith commodity, $i = 1, \ldots, n$, and assigning them all the utility function (6-3).

---

[12] The metric used here is $\rho(\alpha, \beta) = \max_i |\alpha_i - \beta_i|$. Our present claim entails a slight change in the construction given in [14], which merely keeps the rest of the solution away from the arbitrary subset of B_e, rather than from B_e itself.

REFERENCES

1. Bondareva, O. N., Nekotorye primeneniia metodov linejnogo program-mirovaniia k teorii kooperativnykh igr (Some applications of linear pro-gramming methods to the theory of cooperative games). *Problemy Kiber-netiki* **10** (1963), 119–139.

2. Charnes, A. and Kortanek, K. O., On balanced sets, cores, and linear programming. *Cahiers du Centre d'Etudes de Recherche Opérationelle* **9** (1967), 32–43.

3. Debreu, G. and Scarf, H., A limit theorem on the core of an economy, *International Economic Review* **4** (1963), 235–246.

4. Gillies, D. B., Some theorems on *n*-person games, Ph.D. Thesis, Princeton University, June 1953.

5. Gillies, D. B., Solutions to general non-zero-sum games. *Ann. Math. Study* **40** (1959), 47–85.

6. Lucas, W. F., A counterexample in game theory, *Management Science* **13** (1967), 766–767.

7. Lucas, W. F., *Games with Unique Solutions which are Nonconvex*. The RAND Corporation, RM-5363-PR, May 1967.

8. Lucas, W. F., A game with no solution. *Bull. Am. Math. Soc.* **74** (1968), 237–239; also The RAND Corporation, RM-5518-PR, November 1967.

9. Lucas, W. F., *The Proof that a Game may not have a Solution*. The RAND Corporation, RM-5543-PR, January 1968 (to appear in *Trans. Am. Math. Soc.*).

10. Lucas, W. F., *On Solutions for n-Person Games*. The RAND Corporation, RM-5567-PR, January 1968.

11. Scarf, H., The core of an *n*-person game. *Econometrica* **35** (1967), 50–69.

12. Shapley, L. S., *Notes on the n-Person Game*—III: *Some Variants of the von Neumann-Morgenstern Definition of Solution*. The RAND Corporation, RM-817, April 1952.

13. Shapley, L. S., *Markets as Cooperative Games*. The RAND Corporation, P-629, March 1955.

14. Shapley, L. S., A solution containing an arbitrary closed component. *Ann. Math. Study* **40** (1959), 87–93; also The RAND Corporation, RM-1005, December 1952, and P-888, July 1956.

15. Shapley, L. S., The solutions of a symmetric market game. *Ann. Math. Study* **40** (1959), 145–162; also The RAND Corporation, P-1392, June 1958.

16. Shapley, L. S., *Values of Large Games*—VII: *A General Exchange Economy with Money*. The RAND Corporation, RM-4248, December 1964.

17. Shapley, L. S., On balanced sets and cores. *Naval Research Logistics Quarterly* **14** (1967), 453–560; also The RAND Corporation, RM-4601-PR, June 1965.

18. Shapley, L. S., *Notes on n-Person Games*—VIII: *A Game with Infinitely "Flaky" Solutions*. The RAND Corporation, RM-5481-PR (to appear).

19. Shapley, L. S. and Shubik, M., "Concepts and theories of pure competition." *Essays in Mathematical Economics: In Honor of Oskar Morgenstern*, Martin Shubik (ed.). Princeton University Press, Princeton, New Jersey, 1967, 63–79; also The RAND Corporation, RM-3553-PR, May 1963.

20. Shapley, L. S. and Shubik, M., Quasi-cores in a monetary economy with nonconvex preferences. *Econometrica* **34** (1966), 805–827; also The RAND Corporation, RM-3518-1, October 1965.
21. Shapley, L. S. and Shubik, M., *Pure Competition, Coalition Power, and Fair Division*. The RAND Corporation, RM-4917-1-PR, March 1967 (to appear in *International Economic Review*).
22. Shubik, M., Edgeworth market games. *Ann. Math. Study* **40** (1959), 267–278.
23. von Neumann, J. and Morgenstern, O., *Theory of Games and Economic Behavior*. Princeton University Press, Princeton, New Jersey, 1944, 1947, 1953.

Received: October 8, 1968.

18

REEXAMINATION OF THE PERFECTNESS CONCEPT FOR EQUILIBRIUM POINTS IN EXTENSIVE GAMES

R. SELTEN, Bielefeld[1])

1. INTRODUCTION

The concept of a perfect equilibrium point has been introduced in order to exclude the possibility that disequilibrium behavior is prescribed on unreached subgames [Selten 1965 and 1973]. Unfortunately this definition of perfectness does not remove all difficulties which may arise with respect to unreached parts of the game. It is necessary to reexamine the problem of defining a satisfactory noncooperative equilibrium concept for games in extensive form. Therefore a new concept of a perfect equilibrium point will be introduced in this paper[2]).

In retrospect the earlier use of the word "perfect" was premature. Therefore a perfect equilibrium point in the old sense will be called "subgame perfect." The new definition of perfectness has the property that a perfect equilibrium point is always subgame perfect but a subgame perfect equilibrium point may not be perfect.

It will be shown that every finite extensive game with perfect recall has at least one perfect equilibrium point.

Since subgame perfectness cannot be detected in the normal form, it is clear that for the purpose of the investigation of the problem of perfectness, the normal form is an inadequate representation of the extensive form. It will be convenient to introduce an "agent normal form" as a more adequate representation of games with perfect recall.

[1]) Professor R. Selten, Institute of Mathematical Economics, University of Bielefeld, Schloß Rheda, 484 Rheda, Germany.

[2]) The idea to base the definition of a perfect equilibrium point on a model of slight mistakes as described in Section 7 is due to John C. Harsanyi. The author's earlier unpublished attempts at a formalization of this concept were less satisfactory. I am very grateful to John C. Harsanyi who strongly influenced the content of this paper.

2. EXTENSIVE GAMES WITH PERFECT RECALL

In this paper the words *extensive game* will always refer to a finite game in extensive form. A game of this kind can be described as a sextuple:

$$\Gamma = (K, P, U, C, p, h) \qquad (1)$$

where the *constituents* K, P, U, C, p, and h of Γ are as follow[3]):

The Game Tree

The game tree K is a finite tree with a distinguished vertex o, the *origin* of K. The sequence of vertices and edges which connects o with a vertex x is called the *path* to x. We say that x comes before y or that y comes after x if x is different from y and the path to y contains the path to x. An *endpoint* is a vertex z with the property that no vertex comes after z. The set of all endpoints is denoted by Z. A path to an endpoint is called a play. The edges are also called *alternatives*. An alternative at x is an edge which connects x with a vertex after x. The set of all vertices of K which are not endpoints, is denoted by X.

The Player Partition

The player partition $P = (P_0, \ldots, P_n)$ partitions X into *player sets*. P_i is called player i's player set (Player 0 is the "random" player who represents the random mechanisms responsible for the random decisions in the game.) A player set may be empty. The player sets P_i with $i = 1, \ldots, n$ are called *personal player* sets.

The Information Partition

For $i = 1, \ldots, n$ a subset u of P_i is called *eligible* (as an information set) if n is not empty, if every play intersects u at most once and if the number of alternatives at x is the same for every $x \in u$. A subset $u \in P_0$ is called *eligible* if it contains exactly one vertex. The *information partition U* is a refinement of the player partition P into eligible subsets u of the player sets. These sets u are called *information sets*. The information sets u with $u \subseteq P_i$ are called information sets of player i. The set of all information sets of player i is denoted by U_i. The information sets of player $1, \ldots, n$ are called *personal* information sets.

---

[3]) The notation is different from that used by Kuhn [1953].

The Choice Partition

For $u \in U$ let A_u be the set of all alternatives at vertices $x \in u$. We say that a subset c of A_u is *eligible* (as a choice) if it contains exactly one alternative at x for every vertex $x \in u$. The *choice partition* C partitions the set of all edges K into eligible subsets c of the A_u with $u \in U$. These sets c are called *choices*. The choices c which are subsets of A_u are called choices at u. The set of all choices at u is denoted by C_u. A choice at a personal information set is called a personal choice. A choice which is not personal is a *random choice*. We say that the vertex x comes after the choice c if one of the edges in c is on the path to x. In this case we also say that c is *on* the path to x.

The Probability Assignment

A probability distribution p_u over C_u is called *completely mixed* if it assigns a positive probability $p_u(c)$ to every $c \in C_u$. The *probability assignment* p is a function which assigns a completely mixed probability distribution p_u over C_u to very $u \in U_0$ (p specifies the probabilities of the random choices.)

The Payoff Function

The *payoff function* h assigns a vector $h(z) = (h_1(z), \ldots, h_n(z))$ with real numbers as components to every endpoint z of K. The vector $h(z)$ is called the *payoff vector* at z. The component $h_i(z)$ is player i's *payoff* at z.

Perfect Recall

An extensive game $\Gamma = (K, P, U, C, p, h)$ is called an extensive game with *perfect recall* if the following condition is satisfied for every player $i = 1, \ldots, n$ and any two information sets u and v of the same player i: if one vertex $y \in v$ comes after a choice c at u then every vertex $x \in v$ comes after this choice c^4).

Interpretation

In a game with perfect recall a player i who has to make a decision at one of his information sets v knows which of his other information sets have been reached by the previous course of the play and which choices

---

^{4}) The concept of perfect recall has been introduced by H. W. Kuhn [1953].

have been taken there. Obviously a player always must have this knowledge if he is a person with the ability to remember what he did in the past. Since game theory is concerned with the behavior of absolutely rational decision makes whose capabilities of reasoning and memorizing are unlimited, a game, where the players are individuals rather than teams, must have perfect recall.

Is there any need to consider games where the players are teams rather than individuals? In the following we shall try to argue that at least as far as strictly noncooperative game theory is concerned the answer to this question is no. In principle it is always possible to model any given interpersonal conflict situation in such a way that every person involved is a single player. Several persons who form a team in the sense that all of them pursue the same goals can be regarded as separate players with identical payoff functions. Against this view one might object that a team may be united by more than accidentally identical payoffs. The team may be a preestablished coalition with special cooperative possibilities not open to an arbitrary collection of persons involved in the situation. This is not a valid objection. Games with preestablished coalitions of this kind are outside the framework of strictly noncooperative game theory. In a strictly noncooperative game the players do not have any means of cooperation or coordination which are not explicitly modelled as parts of the extensive form. If there is something like a preestablished coalition, then the members must appear as separate players and the special possibilities of the team must be a part of the structure of the extensive game.

In view of what has been said no room is left for strictly noncooperative extensive games without perfect recall. In the framework of strictly noncooperative game theory such games can be rejected as misspecified models of interpersonal conflict situations.

3. Strategies, Expected Payoff and Normal Form

In this section several definitions are introduced which refer to an extensive game $\Gamma = (K, P, U, C, p, h)$.

Local Strategies

A local strategy b_{iu} at the information set $u \in U_i$ is a probability distribution over the set C_u of the choices at u; a probability $b_{iu}(c)$ is assigned to every choice c at u. A local strategy b_{iu} is called *pure* if it assigns 1 to one choice c at u and 0 to the other choices. Wherever this can be done without danger of confusion no distinction will be made between the choice c and the pure local strategy which assigns the probability 1 to c.

Behavior Strategies

A behavior strategy b_i of a personal player i is a function which assigns a local strategy b_{iu} to every $u \in U_i$. The set of all behavior strategies of player i is denoted by B_i.

Pure Strategies

A *pure strategy* π_i of player i is a function which assigns a choice c at u (a pure local strategy) to every $u \in U_i$. Obviously a pure strategy is a special behavior strategy. The set of all pure strategies of player i is denoted by Π_i.

Mixed Strategies

A mixed strategy q_i of player i is a probability distribution over Π_i; a probability $q_i(\pi_i)$ is assigned to every $\pi_i \in \Pi_i$. The set of all mixed strategies q_i of player i is denoted by Q_i. Wherever this can be done without danger of confusion no distinction will be made between the pure strategy π_i and the mixed strategy q_i which assigns 1 to π_i. Pure strategies are regarded as special cases of mixed strategies.

Behavior Strategy Mixtures

A behavior strategy mixture s_i for player i is a probability distribution over B_i which assigns positive probabilities $s_i(b_i)$ to a finite number of elements of B_i and zero probabilities to the other elements of B_i. No distinction will be made between the behavior strategy b_i and the behavior strategy mixture which assigns 1 to b_i. The set of all behavior strategy mixtures of player i is denoted by S_i. Obviously pure strategies, mixed strategies, and behavior strategies can all be regarded as special behavior strategy mixtures.

Combinations

A combination $s = (s_1, \ldots, s_n)$ *of behavior strategy mixtures* is an n-tuple of behavior strategy mixtures $s_i \in S_i$, one for each personal player. *Pure strategy combinations* $\pi = (\pi_1, \ldots, \pi_n)$, *mixed strategy combinations*, and *behavior strategy combinations* are defined analogously.

Realization Probabilities

A player i who plays a behavior strategy mixture s_i behaves as follows: He first employs a random mechanism which selects one of the behavior strategies b_i with the probabilities $s_i(b_i)$. He then in the course of the play at every $u \in U_i$ which is reached by the play selects one of the choices c at u with the probabilities $b_{iu}(c)$. Let $s = (s_1, \ldots, s_n)$ be a combination of behavior strategy mixtures. On the assumption that the s_i are played by the players we can compute a *realization probability* $\rho(x, s)$ of x under s for every vertex $x \in K$. This probability $\rho(x, s)$ is the probability that x is reached by the play, if s is played. Since these remarks, make it sufficiently clear, how $\rho(x, s)$ is defined, a more precise definition of $\rho(x, s)$ will not be given here.

Expected Payoffs

With the help of the realization probabilities an *expected payoff vector* $H(s) = (H_1(s), \ldots, H_n(s))$ can be computed as follows:

$$H(s) = \sum_{z \in Z} \rho(z, s) h(z). \tag{2}$$

Since pure strategies, mixed strategies, and behavior strategies are special cases of behavior strategy mixtures, the expected payoff definition (2) is applicable here, too.

Normal Form

A normal form $G = (\Pi_1, \ldots, \Pi_n; H)$ consists of n finite nonempty and pairwise nonintersecting *pure strategy sets* Π_i and an *expected payoff function* H defined on $\Pi = \Pi_1 \times \cdots \times \Pi_n$. The expected payoff function H assigns a payoff vector $H(\pi) = (H_1(\pi), \ldots, H_n(\pi))$ with real numbers as components to every $\pi \in \Pi$. For every extensive game Γ the pure strategy sets and the expected payoff function defined above generate the *normal form of* Γ.

In order to compute the expected payoff vector for a mixed strategy combination, it is sufficient to know the normal form of Γ. The same is not true for combinations of behavior strategies. As we shall see, in the transition from the extensive form to the normal form some important information is lost.

4. Kuhn's Theorem

H. W. Kuhn [1953, p. 213] has proved an important theorem on games with perfect recall. In this section Kuhn's theorem will be restated in a slightly changed form. For this purpose some further definitions must be introduced. As before, these definitions refer to an extensive game $\Gamma = (K, P, U, C, p, h)$.

Notational Convention

Let $s = (s_1, \ldots, s_n)$ be a combination of behavior strategy mixtures and let t_i be a behavior strategy mixture for player i.

The combination $(s_1, \ldots, s_{i-1}, t_I, s_{i+1}, \ldots, s_n)$ which results from s, if s_i is replaced by t_i and the other components of s remain unchanged, is denoted by s/t_i. The same notational convention is also applied to other types of strategy combinations.

Realization Equivalence

Let s_i' and s_i'' be two behavior strategy mixtures for player i. We say that s_i' and s_i'' are *realization equivalent* if for every combination s of behavior strategy mixtures we have:

$$\rho(x, s/s_i') = \rho(x, s/s_i'') \quad \text{for every } x \in K. \tag{3}$$

Payoff Equivalence

Let s_i' and s_i'' be two behavior strategy mixtures for player i. We say that s_i' and s_i'' are *payoff equivalent* if for every combination s of behavior strategy mixtures we have

$$H(s/s_i') = H(s/s_i''). \tag{4}$$

Obviously s_i' and s_i'' are payoff equivalent if they are realization equivalent, since (3) holds for the endpoints z.

Theorem 1 (Kuhn's theorem): *In every extensive game with perfect recall a realization equivalent behavior strategy b_i can be found for every behavior strategy mixture s_i of a personal player i.*

In order to prove this theorem we introduce some further definitions.

Conditional Choice Probabilities

Let $s = (s_1, \ldots, s_n)$ be a combination of behavior strategy mixtures and let x be a vertex in an information set u of a personal player i, such that $\rho(x, s) > 0$. For every choice c at u we define a *conditional choice probability* $\mu(c, x, s)$. The choice c contains an edge e at x; this edge e connects x with another vertex y. The probability $\mu(c, x, s)$ is computed as follows:

$$\mu(c, x, s) = \frac{\rho(y, s)}{\rho(x, s)}. \tag{5}$$

The probability $\mu(c, x, s)$ is the conditional probability that the choice c will be taken if s is player and x has been reached.

LEMMA 1: *In every extensive game* Γ *(with or without perfect recall) on the region of those triplets* (c, x, s) *where the conditional choice probability* $\mu(c, x, s)$ *is defined the conditional choice probabilities* $\mu(c, x, s)$ *with* $x \in u \in U_i$ *do not depend on the components* s_j *of s with* $i \neq j$.

PROOF: Let $b_i^1, \ldots, b_i^k$ be the behavior strategies which are selected by s_i with positive probabilities $s_i(b_i^j)$. For $\rho(x, s) > 0$ an outside observer, who knows that c has been reached by the play but does not know which of the b_i^j has been selected before the beginning of the game, can use this knowledge in order to compute posterior probabilities $t_i(b_i^j)$ from the prior probabilities $s_i(b_i^j)$. The posterior probability $t_i(b_i^j)$ is proportional to $s_i(b_i^j)$ multiplied by the product of all probabilities assigned by b_i^j to choices of player i on the path to x. Obviously the $t_i(b_i^j)$ depend on s_i but not on the other components of s. The conditional choice probability $\mu(c, x, s)$ can be written as follows:

$$\mu(c, x, s) = \sum_{j=1}^{k} t_i(b_i^j) b_{iu}^j(c). \tag{6}$$

This shows that $\mu(c, x, s)$ does not depend on the s_j with $i \neq j$.

LEMMA 2: *In every extensive game* Γ *with perfect recall, on the region of those triples* (c, x, s) *where the conditional choice probability* $\mu(c, x, s)$ *is defined, we have*

$$\mu(c, x, s) = \mu(c, y, s) \quad \text{for } x \in u \text{ and } y \in u. \tag{7}$$

PROOF: In a game with perfect recall for $x \in u$, $y \in u$, and $u \in U_i$ player i's choices on the path to x are the same choices as his choices on the path to y. (This is not true for games without perfect recall). Therefore at x and y the posterior probabilities for the behavior strategies b_i^j occurring in player i's behavior strategy mixture s_i are the same at both vertices. Consequently (7) follows from (6).

PROOF OF KUHN'S THEOREM: In view of the Lemma 1 and Lemma 2 the conditional choice probabilities at the vertices x in the player set P_i of a personal player can be described by a function $\mu_i(c, u, s_i)$ which depends on his behavior strategy mixture s_i and the information set u with $x \in u$.

With the help of $\mu_i(c, u, s_i)$ we construct the behavior strategy b_i whose existence is asserted by the theorem. If for at least one $s = (s_1, \ldots, s_n)$ with s_i as component we have $\mu(x, s) > 0$ for some $x \in u$, we define

$$b_{iu}(c) = \mu_i(c, u, s_i). \tag{8}$$

The construction of b_i is completed by assigning arbitrary local strategies b_{iu} to those $u \in U_i$ where no such s can be found.

It is clear that this behavior strategy b_i and the behavior strategy mixture s_i are realization equivalent.

The Significance of Kuhn's Theorem

The theorem shows that in the context of extensive games with perfect recall one can restrict one's attention to behavior strategies. Whatever a player can achieve by a mixed strategy or a more general behavior strategy mixture can be achieved by the realization equivalent and therefore also payoff equivalent behavior strategy whose existence is secured by the theorem.

5. SUBGAME PERFECT EQUILIBRIUM POINTS

In this section we shall introduce some further definitions which refer to an extensive game $\Gamma = (K, P, U, C, p, h)$ with perfect recall. In view of Kuhn's theorem only behavior strategies are important for such games. Therefore the concepts of a best reply and an equilibrium point are formally introduced for behavior strategies only.

Best Reply

Let $b = (b_1, \ldots, b_n)$ be a combination of behavior strategies for Γ. A behavior strategy $\tilde{b}_i$ of player i as a *best reply* to b if we have

$$H_i\left(b/\tilde{b}_i\right) = \max_{b_i' \in B_i} H_i(b/b_i'). \tag{9}$$

A combination of behavior strategies $\tilde{b} = (\tilde{b}_1, \ldots, \tilde{b}_n)$ is called a *best reply* to b if for $i = 1, \ldots, n$ the behavior strategy $\tilde{b}_i$ is a best reply to b.

Equilibrium Point

A behavior strategy combination $b^* = (b_1^*, \ldots, b_n^*)$ is called an *equilibrium point* if b^* is a best reply to itself.

Remark: The concepts of a best reply and an equilibrium point can be defined analogously for behavior strategy mixtures. In view of Kuhn's theorem it is clear that for games with perfect recall an equilibrium point in behavior strategies is a special case of an equilibrium point in behavior strategy mixtures. The existence of an equilibrium point in behavior strategies for every extensive game with perfect recall is an immediate consequence of Kuhn's theorem together with Nash's well-known theorem on the existence of an equilibrium point in mixed strategies for every finite game [Nash, 1951].

Subgame

Let $\Gamma = (K, P, U, C, p, h)$ be an extensive game with or without perfect recall. A subtree K' of K consists of a vertex x of K together with all vertices after x and all edges of K connecting vertices of K'. A subtree K' is called *regular* in Γ, if every information set in Γ, which contains at least one vertex of K', does not contain any vertices outside of K'. For every regular subtree K' a *subgame* $\Gamma' = (K', P', U', C', p', h')$ is defined as follows: P', U', C', p', and h' are the restrictions of the partitions P, U, C and the functions p and h to K'.

Induced Strategies

Let Γ' be a subgame of Γ and let $b = (b_1, \ldots, b_n)$ be a behavior strategy combination for Γ. The restriction of b_i to the information sets of player i in Γ' is a strategy b_i' of player i for Γ'. This strategy b_i' is called *induced* by b_i on Γ' and the behavior strategy combination $b' = (b'_1, \ldots, b'_n)$ defined in this way is called *induced* by b on Γ'.

Subgame Perfectness

A *subgame perfect* equilibrium point $b^* = (b_i^*, \ldots, b_n^*)$ of an extensive game Γ is an equilibrium point (in behavior strategies) which induces an equilibrium point on every subgame of Γ.

6.   A NUMERICAL EXAMPLE

The definition of a subgame perfect equilibrium point excludes some cases of intuitively unreasonable equilibrium points for extensive games. In this section we shall present a numerical example which shows that not every intuitively unreasonable equilibrium point is excluded by this definition. The discussion of the example will exhibit the nature of the difficulty.

The numerical example is the game of Figure 1. Obviously this game has no subgames. Every player has exactly one information set. The game is a game with perfect recall.

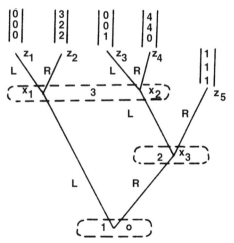

FIG. 1. A numerical example. Information sets are represented by dashed lines. Choices are indicated by the letters L and R (standing for "left" and "right"). Payoff vectors are indicated by column vectors above the corresponding endpoints.

Since every player has two choices, L and R, a behavior strategy of player i can be characterized by the probability with which he selects R. The symbol p_i will be used for this probability. A combination of behavior strategies is represented by a triple (p_1, p_2, p_3).

As the reader can verify for himself without much difficulty the game of Figure 1 has the following two types of equilibrium points:

$$\text{Type 1:} \quad p_1 = 1, p_2 = 1, 0 \le p_3 \le \tfrac{1}{4}.$$

$$\text{Type 2:} \quad p_1 = 0, \tfrac{1}{3} \le p_2 \le 1, p_3 = 1.$$

Consider the equilibrium points of type 2. Player 2's information set is not reached, if an equilibrium point of this kind is played. Therefore his expected payoff does not depend on his strategy. This is the reason why his equilibrium strategy is best reply to the equilibrium strategies of the other players.

Now suppose that the players believe that a specific type 2 equilibrium point, say $(0, 1, 1)$ is the rational way to play the game. Is it really reasonable to believe that player 2 will choose R if he is reached? If he believes that player 3 will choose R as prescribed by the equilibrium point, then it is better for him to select L where he will get 4 instead of R where he will get 1. The same reasoning applies to the other type 2 equilibrium points, too.

Clearly, the type 2 equilibrium points cannot be regarded as reasonable. Player 2's choices should not be guided by his payoff expectations in the whole game but by his conditional payoff expectations at x_3. The payoff expectation in the whole game is computed on the assumption that player 1's choice is L. At x_3 this assumption has been shown to be wrong. Player 2 has to assume that player 1's choice was R.

For every strategy combination (p_1, p_2, p_3) it is possible to compute player 2's conditional payoff expectations for his choices L and R on the assumption that his information set has been reached. The same cannot be done for player 3. Player 3's information set can be reached in two ways. Consider an equilibrium point of type 1, e.g, the equilibrium point $(1, 1, 0)$. Suppose that $(1, 1, 0)$ is believed to be the rational way to play the game and assume that contrary to the expectations generated by this belief, player 3's information set is reached. In this case player 3 must conclude that either player 1 or player 2 must have deviated from the rational way of playing the game but he does not know which one. He has no obvious way of computing a conditional probability distribution over the vertices in his information set, which tells him, with which probabilities he is at x_1 and at x_2 if he has to make his choice.

In the next section a model will be introduced which is based on the idea that with some very small probability a player will make a mistake. These mistake probabilities do not directly generate a conditional probability distribution over the vertice of player 3's information set. As we shall see in Section 8 the introduction of slight mistakes may lead to a strategic situation where the rational strategies add some small voluntary deviations to the mistakes.

7. A Model of Slight Mistakes

There cannot be any mistakes if the players are absolutely rational. Nevertheless, a satisfactory interpretation of equilibrium points in extensive games seems to require that the possibility of mistakes is not completely excluded. This can be achieved by a point of view which looks at complete rationality as a limiting case of incomplete rationality.

Suppose that the personal players in an extensive game Γ with perfect recall are subject to a slight imperfection of rationality of the following kind. At every information set u there is a small positive probability ε_u for the breakdown of rationality. Whenever rationality breaks down, every choice c at u will be selected with some positive probability q_c which may be thought of as determined by some unspecified psychological mechanism. Each of the probabilities ε_u and q_c is assumed to be independent of all the other ones.

Suppose that the rational choice at u is a local strategy which selects c with probability p_c. Then the total probability of the choice c will be

$$\hat{p}_c = (1 - \varepsilon_u)p_c + \varepsilon_u q_c. \qquad (4)$$

The introduction of the probabilities ε_u and q_c transforms the original game into a changed game $\hat{\Gamma}$ where the players do not completely control their choices. A game of this kind will be called a *perturbed game* of Γ.

Obviously, it is not important whether the p_c or the $\hat{p}_c$ are considered to be the strategic variables of the perturbed game $\hat{\Gamma}$. In the following we shall take the latter point of view. This means that in $\hat{\Gamma}$ every player *i* selects a behavior strategy which assigns probability distributions over the choices *c* at *u* to the information sets *u* of player *i* in such a way that the probability $\hat{p}_c$ assigned to a choice *c* at *u* always satisfies the following condition:

$$\hat{p}_c \geq \varepsilon_u q_c. \tag{10}$$

The probability $\hat{p}_c$ is also restricted by the upper bound $1 - \varepsilon_u(1 - q_c)$; it is not necessary to introduce this upper bound explicitly since it is implied by the lower bounds on the probabilities of the other choices at the same information set. With the help of the notation

$$\eta_c = \varepsilon_u q_c \tag{11}$$

condition (10) can be rewritten as follows:

$$\hat{p}_c \geq \eta_c \quad \text{for every personal choice } c. \tag{12}$$

Consider a system of positive constants η_c for the personal choices *c* in Γ such that

$$\sum_{c \text{ at } u} \eta_c < 1. \tag{13}$$

Obviously for every system of this kind we can determine positive probabilities ε_u and q_c which generate a perturbed game $\hat{\Gamma}$ whose conditions (10) coincide with the conditions (12). Therefore we may use the following definition of a perturbed game.

DEFINITION: A *perturbed game* $\hat{\Gamma}$ is a pair (Γ, η) where Γ is an extensive game with perfect recall and η is a function which assigns a positive probability η_c to every personal choice *c* in Γ such that (13) is satisfied.

The probabilities η_c are called minimum probabilities. For every choice *c* at a personal information set *u* define

$$\mu_c = 1 + \eta_c - \sum_{c' \text{ at } u} \eta_{c'}. \tag{14}$$

Obviously μ_c is the upper bound of $\hat{p}_c$ implied by the conditions (7). This probability μ_c is called the *maximum probability* of *c*.

Strategies

A *local strategy* for the perturbed game $\hat{\Gamma} = (\Gamma, \eta)$ is a local strategy for Γ which satisfies the conditions (12). A *behavior strategy* of player i in $\hat{\Gamma}$ is a behavior strategy of player i in Γ which assigns local strategies for $\hat{\Gamma}$ to the information sets of player i. The set of all behavior strategies of player i for $\hat{\Gamma}$ is denoted by $\hat{B}_i$. A *behavior strategy combination* for $\hat{\Gamma}$ is a behavior strategy combination $\hat{b} = (\hat{b}_1, \ldots, \hat{b}_n)$ for Γ whose components are behavior strategies for $\hat{\Gamma}$. The set of all behavior strategy combinations for $\hat{\Gamma}$ is denoted by $\hat{B}$.

Best Replies

Let $b = (b_1, \ldots, b_n)$ be a behavior strategy combination for $\hat{\Gamma}$. A behavior strategy $\tilde{b}_i$ of player i for $\hat{\Gamma}$ is called a *best reply to b in* $\hat{\Gamma}$ if we have

$$H_i(b/\tilde{b}_i) = \max_{b_i' \in \hat{B}_i} H_i(b/b_i'). \tag{15}$$

A behavior strategy combination $\tilde{b} = (\tilde{b}_1, \ldots, \tilde{b}_n)$ for $\hat{\Gamma}$ is called a *best reply to b in* $\hat{\Gamma}$ if every component $\tilde{b}_i$ of $\tilde{b}$ is a best reply to b in $\hat{\Gamma}$.

Equilibrium Point

An equilibrium point of $\hat{\Gamma}$ is a behavior strategy combination for $\hat{\Gamma}$ which is a best reply to itself in $\hat{\Gamma}$.

Remark: Note that there is a difference between a best reply in Γ and a best reply in $\hat{\Gamma}$. The strategy sets $\hat{B}_i$ are subsets of the strategy sets B_i. Pure strategies are not available in $\hat{\Gamma}$.

8. Perfect Equilibrium Points

The difficulties which should be avoided by a satisfactory definition of a perfect equilibrium point are connected to unreached information sets. There cannot be any unreached information sets in the perturbed game. If b is a behavior strategy combination for the perturbed game then the realization probability $\rho(x, b)$ is positive for every vertex x of K. This makes it advantageous to look at a game Γ as a limiting case of perturbed games $\hat{\Gamma} = (\Gamma, \eta)$. In the following a perfect equilibrium point will be defined as a limit of equilibrium points for perturbed games.

Sequences of Perturbed Games

Let Γ be an extensive game with perfect recall. A sequence $\hat{\Gamma}^1, \hat{\Gamma}^2, \ldots$ where for $k = 1, 2, \ldots$ the game $\hat{\Gamma}^k = (\Gamma, \eta^k)$ is a perturbed game of Γ, is called a test sequence for Γ, if for every choice c of the personal players in Γ the sequence of the minimum probabilities η_c^k assigned to c by η^k converges to 0 for $k \to \infty$.

Let $\hat{\Gamma}^1, \hat{\Gamma}^2, \ldots$ be a test sequence for Γ. A behavior strategy combination b^* for Γ is called a *limit equilibrium point* of this test sequence if for $k = 1, 2, \ldots$ an equilibrium point $\hat{b}^k$ of $\hat{\Gamma}^k$ can be found such that for $k \to \infty$ the sequence of the $\hat{b}^k$ converges to b^*.

LEMMA 3: *A limit equilibrium point b^* of a test sequence $\hat{\Gamma}^1, \hat{\Gamma}^2, \ldots$ for an extensive game Γ with perfect recall is an equilibrium point of Γ.*

PROOF: The fact that the b^k are equilibrium points of the $\hat{\Gamma}^k$ can be expressed by the following inequalities

$$H_i(\hat{b}^k) \geq H_i(\hat{b}^k/b_i) \quad \text{for every } b_i \in \hat{B}_i^k \quad \text{and} \quad \text{for } i = 1, \ldots, n. \quad (16)$$

Let B_i^m be the intersection of all $\hat{B}_i^k$ with $k \geq m$. For $k \geq m$ we have

$$H_i(\hat{b}^k) \geq H_i(\hat{b}^k/b_i) \quad \text{for every } b_i \in B_i^m. \quad (17)$$

Since the expected payoff depends continuously on the behavior strategy combination this inequality remains valid if on both sides we take the limits for $k \to \infty$. This yields:

$$H_i(b^*) \geq H_i(b^*/b_i) \quad \text{for every } b_i \in B_i^m. \quad (18)$$

Inequality (18) holds for every m. The closure of the union of all B_i^m is B_i. This together with the continuity of H_i yields:

$$H_i(b^*) \geq H_i(b^*/b_i) \quad \text{for every } b_i \in B_i. \quad (19)$$

Inequality (19) shows that b^* is an equilibrium point of Γ.

Perfect Equilibrium Point

Let Γ be an extensive game with perfect recall. A *perfect equilibrium point* of Γ is a behavior strategy combination $b^* = (b_1^*, \ldots, b_n^*)$ for Γ with the property that for at least one test sequence $\hat{\Gamma}^1, \hat{\Gamma}^2, \ldots$ the combination b^* is a limit equilibrium point of $\hat{\Gamma}^1, \hat{\Gamma}^2, \ldots$.

Interpretation

A limit equilibrium point b^* of a test sequence has the property that it is possible to find equilibrium points of perturbed games as close to b^* as desired. The definition of a perfect equilibrium point is a precise statement of the intuitive idea that a reasonable equilibrium point should have an interpretation in terms of arbitrarily small imperfections of rationality. A test sequence which has b^* as limit equilibrium point provides an interpretation of this kind. If b^* fails to be the limit equilibrium point of at least one test sequence b^* must be regarded as instable against very small deviations from perfect rationality.

Up to now it has not been shown that perfectness implies subgame perfectness. In order to do this we need a lemma on the subgame perfectness of equilibrium points for perturbed games.

Subgames of Perturbed Games

Let $\hat{\Gamma} = (\Gamma, \eta)$ be a perturbed game of Γ. A *subgame* $\hat{\Gamma}' = (\Gamma', \eta')$ of $\hat{\Gamma}$ consists of a subgame Γ' of Γ and the restriction η' of η to the personal choices of Γ'. We say that $\hat{\Gamma}'$ is generated by Γ'. An equilibrium point $\hat{b}$ of $\hat{\Gamma}$ is called *subgame perfect* if an equilibrium point $\hat{b}'$ is induced on every subgame $\hat{\Gamma}'$ of $\hat{\Gamma}$.

LEMMA 4: *Let Γ be an extensive game with perfect recall and let $\hat{\Gamma} = (\Gamma, \eta)$ be a perturbed game of Γ. Every equilibrium point of $\hat{\Gamma}$ (in behavior strategies) is subgame perfect.*

PROOF: Let $\hat{b}'$ be the behavior strategy combination induced by an equilibrium point $\hat{b}$ of $\hat{\Gamma}$ on a subgame Γ' of Γ. Obviously $\hat{b}'$ is a behavior strategy combination for the subgame $\hat{\Gamma}' = (\Gamma', \eta')$ generated by Γ'. Suppose that $\hat{b}'$ fails to be an equilibrium point of $\hat{\Gamma}'$. It follows that for some personal player j a behavior strategy b_j' for $\hat{\Gamma}'$ exists, such that player j's expected payoff for $\hat{b}'/b_j'$ in $\hat{\Gamma}'$ is greater than his expected payoff for $\hat{b}'$ in $\hat{\Gamma}'$. Consider the behavior strategy b_j for $\hat{\Gamma}$

which agrees with b_j' on Γ' and with player j's strategy $\hat{b}_j$ in $\hat{b}$ everywhere else. Since the realization probabilities in $\hat{\Gamma}$ are always positive, player j's expected payoff for $\hat{b}/b_j$ must be greater than his expected payoff for $\hat{b}$. Since a behavior strategy b_j with this property does not exist, $\hat{b}'$ is an equilibrium point of $\hat{\Gamma}'$.

THEOREM 2: *Let Γ be an extensive game with perfect recall and let $\tilde{b}$ be a perfect equilibrium point of Γ. On every subgame Γ' of Γ a perfect equilibrium point $\tilde{b}'$ is induced by $\tilde{b}$ on Γ'.*

COROLLARY: *Every perfect equilibrium point of an extensive game Γ with perfect recall is a subgame perfect equilibrium point of Γ.*

PROOF: Let $\hat{\Gamma}^1, \hat{\Gamma}^2, \ldots$ be a test sequence for Γ which has $\hat{b}$ as limit equilibrium point. Let $\hat{b}^1, \hat{b}^2, \ldots$ be a sequence of equilibrium points $\hat{b}^k$ of $\hat{\Gamma}^k$. It follows from the subgame perfectness of the $\hat{b}^k$ that the subgames of $\hat{\Gamma}^k$ generated by Γ' form a test sequence for Γ' with $\hat{b}'$ as a limit equilibrium point. Therefore $\hat{b}'$ is a perfect equilibrium point of Γ'.

The corollary is an immediate consequence of the fact that a perfect equilibrium point is an equilibrium point. (See Lemma 4.)

9.   A SECOND LOOK AT THE NUMERICAL EXAMPLE

In this section we shall first look at a special test sequence of the numerical example of Figure 1 in order to compute its limit equilibrium point. The way in which this limit equilibrium point is approached exhibits an interesting phenomenon which is important for the interpretation of perfect equilibrium points. Later we shall show that every equilibrium point of type 1 is perfect.

Let $\varepsilon_1, \varepsilon_2, \ldots$ be a monotonically decreasing sequence of positive probabilities with $\varepsilon_1 < \frac{1}{4}$ and $\varepsilon_k \to 0$ for $k \to \infty$. Let Γ be the game of Figure 1. Consider the following test sequence $\hat{\Gamma}^1, \hat{\Gamma}^2, \ldots$ for Γ. For $k = 1, 2, \ldots$ the perturbed game $\hat{\Gamma}^k = (\Gamma, \eta^k)$ is defined by $\eta_c^k = \varepsilon_k$ for every choice c of Γ.

As in Section 7 let p_i be the probability of player i's choice R. A behavior strategy combination can be represented by a triple $p = (p_1, p_2, p_3)$. The behavior strategy combinations for $\hat{\Gamma}^k$ are restricted by the condition

$$1 - \varepsilon_k \geq p_i \geq \varepsilon_k \quad \text{for } i = 1, 2, 3. \tag{20}$$

As we shall see, the perturbed game $\hat{\Gamma}^k$ has only one equilibrium point $p^k = (p_1^k, p_2^k, p_3^k)$ whose components p_i^k are as follows:

$$p_1^k = 1 - \varepsilon_k \tag{21}$$

$$p_2^k = 1 - \frac{2\varepsilon_k}{1 - \varepsilon_k} \tag{22}$$

$$p_3^k = \tfrac{1}{4}. \tag{23}$$

Equilibrium Property of p^k

In the following it will be shown that p^k is an equilibrium point of $\hat{\Gamma}^k$. Let us first look at the situation of player 3. For any $p = (p_1, p_2, p_3)$ the realization probabilities $\rho(x_1, p)$ and $\rho(x_2, p)$ of the vertices x_1 and x_2 in the information set of player 3 are given by (24) and (25):

$$\rho(x_1, p) = 1 - p_1 \tag{24}$$

$$\rho(x_2, p) = p_1(1 - p_2). \tag{25}$$

Player 3's expected payoff under the condition that his information set is reached is $2\rho(x_p p)$ if he takes his choice R and $\rho(x_2, p)$ if he takes his choice L. Therefore p_3 is a best reply to p in $\hat{\Gamma}^k$ if and only if the following is true:

$$p_3 = \varepsilon_k \qquad \text{for } 2(1 - p_1) < p_1(1 - p_2) \tag{26}$$

$$\varepsilon_k \le p_3 \le 1 - \varepsilon_k \quad \text{for } 2(1 - p_1) = p_1(1 - p_2) \tag{27}$$

$$p_3 = 1 - \varepsilon_k \qquad \text{for } 2(1 - p_1) > p_1(1 - p_2). \tag{28}$$

In the case of p^k we have

$$\rho(x_1, p^k) = \varepsilon_k \tag{29}$$

$$\rho(x_2, p^k) = 2\varepsilon_k. \tag{30}$$

Therefore it follows by (27) that p_3^k is a best reply to p^k. Let us now look at the situation of player 2. Here we can see that p_2 is a best reply to p in $\hat{\Gamma}^k$ if and only if the following is true:

$$p_2 = \varepsilon_k \qquad \text{for } p_3 > \tfrac{1}{4} \qquad (31)$$

$$\varepsilon_k \le p_2 \le 1 - \varepsilon_k \quad \text{for } p_3 = \tfrac{1}{4} \qquad (32)$$

$$p_2 = 1 - \varepsilon_k \qquad \text{for } p_3 < \tfrac{1}{4}. \qquad (33)$$

p_2^k is a best reply to p^k in view of (32).

p_1 is a best reply to p in $\hat{\Gamma}^k$ if and only if the following is true:

$$p_1 = \varepsilon_k \qquad \text{for } 3p_3 > 4(1 - p_2)p_3 + p_2 \qquad (34)$$

$$\varepsilon_k \le p_1 \le 1 - \varepsilon_k \quad \text{for } 3p_3 = 4(1 - p_2)p_3 + p_2 \qquad (35)$$

$$p_1 = 1 - \varepsilon_k \qquad \text{for } 3p_3 < 4(1 - p_2)p_3 + p_2. \qquad (36)$$

p_i^k is a best reply to p^k in view of (36).

Uniqueness of the Equilibrium Point

In the following it will be shown that p^k is the only equilibrium point of $\hat{\Gamma}^k$. We first exclude the possibility $p_3 \ne 1/4$. Suppose that p is an equilibrium point with $p_3 < 1/4$. It follows by (33) that we have $p_2 = 1 - \varepsilon_k$. Consequently $3p_3$ is smaller than p_2 and (36) yields $p_1 = 1 - \varepsilon_k$. Therefore (28) applies to p_3. We have $p_3 = 1 - \varepsilon_k$ contrary to the assumption $p_3 < 1/4$.

Now we suppose that p is an equilibrium point with $p_3 > 1/4$. Condition (31) yields $p_2 = \varepsilon_k$. In view of $1 - p_2 > 3/4$ condition (36) applies to p_1. It follows that (26) applies to p_3 contrary to the assumption $p_3 > 1/4$.

We know now that an equilibrium point p of $\hat{\Gamma}^k$ must have the property $p_3 = \tfrac{1}{4}$. Obviously (36) applies to an equilibrium point p. We must have $p_1 = 1 - \varepsilon_k$. Moreover neither (26) nor (28) are satisfied by p_3. Therefore in view of (27) an equilibrium point p has the following property:

$$2(1 - p_1) = p_1(1 - p_2). \qquad (37)$$

This together with $p_1 = 1 - \varepsilon_k$ yields

$$p_2 = \frac{2\varepsilon_k}{1 - \varepsilon_k}. \tag{38}$$

Voluntary Deviations from the Limit Equilibrium Point

For $k \to \infty$ the sequence p^k converges to $p^* = (1, 1, 1/4)$. This is the only limit equilibrium point of the test sequence $\hat{\Gamma}^1, \hat{\Gamma}^2, \ldots$.

Note that p_1^k is as near as possible to $p_1^* = 1$ since p_1^k is the maximum probability $1 - \varepsilon_k$. Contrary to this p_2^k is not as near as possible to p_2^*. The probability p_2^* is smaller than $1 - \varepsilon_k$ by $\varepsilon_k(1 + \varepsilon_k)/(1 - \varepsilon_k)$. The rules of the perturbed game force player 2 to take his choice L with a probability of at least ε_k but to this minimum probability he adds the "voluntary" probability $\varepsilon_k(1 + \varepsilon_k)/(1 - \varepsilon_k)$. In this sense we can speak of a voluntary deviation from the limit equilibrium point.

The voluntary deviation influences the realization probabilities $\rho(x_1, p^k)$ and $\rho(x_2, p^k)$. The conditional probabilities for x_1 and x_2, if the information set of player 3 is reached by p^k, are $1/3$ and $2/3$ for every k. It is natural to think of these conditional probabilities as conditional probabilities for the limit equilibrium point p^*, too. The assumptions on the probabilities of slight mistakes which are embodied in the test sequence $\hat{\Gamma}^1, \hat{\Gamma}^2, \ldots$ do not directly determine these conditional probabilities but indirectly via the equilibrium points p^k.

Perfectness of the Equilibrium Points of Type 1

In the following it will be shown that every equilibrium point of type 1 is perfect. Let $p^* = (1, 1, p_3^*)$ be one of these equilibrium points. We construct a test sequence $\hat{\Gamma}^1, \hat{\Gamma}^2, \ldots$ with the property that p^* is a limit equilibrium point of $\hat{\Gamma}^1, \hat{\Gamma}^2, \ldots$. Let $\varepsilon_1, \varepsilon_2, \ldots$ be a decreasing sequence of positive numbers with $\varepsilon_1 < p_3^*/2$ and $\varepsilon_k \to 0$ for $k \to \infty$. The minimum probabilities η_c^k for the perturbed game $\hat{\Gamma}^k = (\Gamma, \eta^k)$ are defined as follows:

$$\eta_c^k = \begin{cases} \varepsilon_k & \text{if } c \text{ is a choice of player 1 or player 3} \\ \dfrac{2\varepsilon_k}{1 - \varepsilon_k} & \text{if } c \text{ is a choice of player 2.} \end{cases} \tag{39}$$

With the help of arguments similar to those which have been used in the subsection "equilibrium property of p^k," it can be shown that for $k = 1, 2, \ldots$ the following behavior strategy combination $\hat{p}^k = (\hat{p}_1^k, \hat{p}_2^k, \hat{p}_3^k)$ is an equilibrium point of $\hat{\Gamma}^k$:

$$\hat{p}_1 = 1 - \varepsilon_k \tag{40}$$

$$\hat{p}_2 = 1 - \frac{2\varepsilon_k}{1 - \varepsilon_k} \tag{41}$$

$$\hat{p}_3 = p_3^*. \tag{42}$$

The sequence $\hat{p}^1, \hat{p}^2, \ldots$ converges to p^*. Therefore p^* is a perfect equilibrium point.

Imperfectness of the Equilibrium Points of Type 2

In the following it will be shown that the equilibrium points of type 2 fail to be perfect. Let $p^* = (0, p_2^*, 1)$ be an equilibrium point of type 2 and let $\hat{\Gamma}^1, \hat{\Gamma}^2, \ldots$ be a test sequence which has p^* as limit equilibrium point. Let $p^1, p^2, \ldots$ be a sequence of equilibrium points p^* of $\hat{\Gamma}^k$ which for $k \to \infty$ converges to p^*. For every $\varepsilon > 0$ we can find a number $m(\varepsilon)$ such that for $k > m(\varepsilon)$ the following two conditions (a) and (b) are satisfied: (a) Every minimum probability η_c^k in $\hat{\Gamma}^k = (\Gamma, \eta^k)$ is smaller than ε. (b) For $i = 1, 2, 3$ we have $|p_i^* - p_i^k| < \varepsilon$. For sufficiently small ε it follows from (a) and (b) that p_2^k is not a best reply to p^k; we must have $p_2 < \varepsilon$ for player 2's best reply to p^k and p_2^k cannot be below $1/3$ by more than ε. This shows that p^* cannot be the limit equilibrium point of a test sequence.

10. A DECENTRALIZATION PROPERTY OF PERFECT EQUILIBRIUM POINTS

In this section it will be shown that the question whether a given behavior strategy combination is a perfect equilibrium point or not, can be decided locally at the information sets of the game. The concept of a local equilibrium point will be introduced which is defined by conditions on the local strategies. As we shall see, in perturbed games these local conditions are equivalent to the usual global equilibrium conditions. On the basis of this result a decentralized description of a perfect equilibrium point will be developed.

Notational Convention

Let Γ be an extensive game and let b_i be a behavior strategy of a personal player i in Γ. Let b'_{iu} be a local strategy at an information set u of player i. The notation b_i/b'_{iu} is used for that behavior strategy which results from b_i if the local strategy assigned by b_i to u is changed to b'_{iu} whereas the local strategies assigned by b_i to other information sets remain unchanged. Let $b = (b_1, \ldots, b_n)$ be a behavior strategy combination. The notation b/b'_{iu} is used for the behavior strategy combination b/b'_i with $b'_i = b_i/b'_{iu}$. The set of all local strategies at u is denoted by B_{iu}.

Local Best Replies

Let $b = (b_1, \ldots, b_n)$ be a behavior strategy combination for an extensive game Γ and let $\tilde{b}_{iu}$ be a local strategy at an information set u of a personal player i. The local strategy $\tilde{b}_{iu}$ is called a *local best reply to b in* Γ if we have

$$H_i\left(b/\tilde{b}_{iu}\right) = \max_{b_{iu} \in B_{iu}} H_i(b/b'_{iu}). \tag{43}$$

Local best replies in a perturbed game $\hat{\Gamma} = (\Gamma, \eta)$ are defined analogously: $\tilde{b}_{iu}$ is a *local best reply to be in* $\hat{\Gamma}$ if we have

$$H_i\left(b/\tilde{b}_{iu}\right) = \max_{b'_{iu} \in \hat{B}_{iu}} H_i(b/b'_{iu}) \tag{44}$$

where $\hat{B}_{iu}$ is the set of all local strategies at u for $\hat{\Gamma}$.

Conditional Realization Probabilities

Let $\hat{\Gamma} = (\Gamma, \eta)$ be a perturbed game of an extensive game Γ with perfect recall. For every information set u of a personal player i and every behavior strategy combination $b = (b_1, \ldots, b_n)$ for $\hat{\Gamma}$ we define a *conditional realization probability* $\mu(x, b)$:

$$\mu(x, b) = \frac{\rho(x, b)}{\sum\limits_{y \in u} \rho(y, b)}. \tag{45}$$

Obviously $\rho(x, b)$ is the conditional probability that x is reached by the play if b is played and u is reached. Since $\rho(x, b)$ is positive for every vertex x, the conditional realization probability $\mu(x, b)$ is defined for every vertex x. Let x be a vertex and let z be an endpoint after x. We define a second type of *conditional realization probability* $\mu(x, z, b)$ which is the probability that z will be reached if b is played and x has been reached. Obviously we have

$$\mu(x, z, b) = \frac{\rho(z, b)}{\rho(x, b)}. \tag{46}$$

Conditional Expected Payoff

For every information set u of a personal player i in a perturbed game $\Gamma = (\hat{\Gamma}, \eta)$ of an extensive game Γ with perfect recall we define a *conditional expected payoff* function H_{iu} for player i at u:

$$H_{iu}(b) = \sum_{x \in u} \mu(x, b) \sum_{z \text{ after } x} \mu(x, z, b) h(z). \tag{47}$$

$H_{iu}(b)$ is the conditional expectation of player i's payoff under the condition that b is played and u is reached by the play.

LEMMA 5: *Let* $b = (b_1, \ldots, b_u)$ *be a behavior strategy combination for a perturbed game* $\hat{\Gamma} = (\Gamma, \eta)$ *of an extensive game* Γ *with perfect recall. The conditional realization probabilities* $\mu(x, b)$ *do not depend on* b_i.

PROOF: In a game with perfect recall the information sets u of a personal player i have the property that the same choices of player i are on every path to a vertex $x \in u$. Therefore $\mu(x, b)$ does not depend on b_i.

LEMMA 6: *Let* $b = (b_1, \ldots, b_n)$ *be a behavior strategy combination for a perturbed game* $\hat{\Gamma} = (\Gamma, \eta)$ *of an extensive game* Γ *with perfect recall and let* $\tilde{b}_{iu}$ *be a local strategy for* $\hat{\Gamma}$ *at an information set* u *of a personal player* i. *The local strategy* $\tilde{b}_{iu}$ *is a local best reply to* b *in* $\hat{\Gamma}$ *if and only if the following is true*:

$$H_{iu}(b/\tilde{b}_{iu}) = \max_{b'_{iu} \in \hat{B}_{iu}} H_{iu}(b/b'_{iu}). \tag{48}$$

PROOF: The assertion of the lemma follows from the fact that the local strategy at u does not influence the realization probabilities of endpoints which do not come after vertices of u.

LEMMA 7: *Let $b = (b_1, \ldots, b_n)$ be a behavior strategy combination for a perturbed game $\Gamma = (\Gamma, \eta)$ of an extensive game Γ with perfect recall and let $\tilde{b}_i$ be a behavior strategy for a personal player i in $\hat{\Gamma}$. The behavior strategy $\tilde{b}_i$ is a best reply to b in $\hat{\Gamma}$ if and only if for every local strategy $\tilde{b}_{iu}$ assigned to an information set $u \in U_i$ by $\tilde{b}_i$ the local strategy $\tilde{b}_{iu}$ is a local best reply to $b/\tilde{b}_i$ in $\hat{\Gamma}$.*

PROOF: Suppose that for some $u \in U_i$, the local strategy $\tilde{b}_{iu}$ is not a local best reply to $b/\tilde{b}_i$ in $\hat{\Gamma}$. Let b'_{iu} be a local best reply to $b/\tilde{b}_i$ at u in $\hat{\Gamma}$. According to the definition of a local best reply $b'_i = \tilde{b}_i/b'_{iu}$ yields a higher payoff for player i than $\tilde{b}$, if the other players use their strategies in b. Therefore $\tilde{b}_i$ cannot be a best reply to b in $\hat{\Gamma}$. It follows that $\tilde{b}_{iu}$ is a local best reply to $b/\tilde{b}_i$ in $\hat{\Gamma}$.

Assume that every $\tilde{b}_{iu}$ is a local best reply to $b/\tilde{b}_i$ in $\hat{\Gamma}$ and that $\tilde{b}_i$ is not a best reply to b in $\hat{\Gamma}$. The theorem is true if this assumption leads to a contradiction. Let b'_i be a best reply to b in $\hat{\Gamma}$ and let b'_{iu} be the local strategies assigned by b'_i to the information sets $u \in U_i$. Let V_i be the set of all information sets $u \in U_i$, where b_{iu} is different from b'_{iu}. Obviously V_i is not empty.

In a game with perfect recall an information set $u \in U_i$ either comes *after* another information set $v \in U_i$ in the sense that every vertex $x \in u$ comes after a vertex $y \in v$ or u contains no vertex x which comes after a vertex of v. Therefore V_i contains information sets v such that no information set $u \in V_i$ has vertices after vertices of v. Let v be an information set of this kind.

We can assume without loss of generality that $b''_i = b'_i/\tilde{b}_{iv}$ is not a best reply to b in $\hat{\Gamma}$. Should b''_i be a best reply to b in $\hat{\Gamma}$, then we can use b''_i instead of b'_i for the purpose of this proof. If the same problem arises again, we can repeat the procedure if necessary several times until finally we find a best reply of player i in $\hat{\Gamma}$ which suits our purpose. Now we assume that b''_i is not a best reply to b in $\hat{\Gamma}$. With the notation $b/b'_i/\tilde{b}_{iv}$ for b/b''_i we can write

$$H_i\left(b/b'_i/\tilde{b}_{iv}\right) < H_i(b/b'_i). \tag{49}$$

In the following we shall show that b_{iv} is a local best reply to b/b'_i in $\hat{\Gamma}$. This is a contradiction to (49).

It follows by Lemma 5 that we have

$$\mu(x, b/b'_i/b_{iv}) = \mu\left(x, b/\tilde{b}_i/b_{iv}\right) \tag{50}$$

for every $x \in v$ and every local strategy b_{iv} of player i at v. Moreover the information set v has been selected in such a way that b'_i and $\tilde{b}_i$ assign the same probabilities to choice at information sets u after v. Therefore we have

$$\mu(x, z, b/b'_i/b_{iv}) = \mu\left(x, z, b/\tilde{b}_i/b_{iv}\right) \tag{51}$$

for every local strategy b_{iv} at v and for every $x \in v$. (47) together with (50) and (51) yields

$$H_{iv}(b/b'_i/b_{iv}) = \left(H_{iv}(b/\tilde{b}_i/b_{iv})\right). \tag{52}$$

Since $\tilde{b}_{iv}$ is a local best reply to $b/\tilde{b}_i$ it is a consequence of Lemma 6 and equation (52) that $\tilde{b}_{iv}$ is a local best reply to b/b'_i. This contradiction to (49) completes the proof of Lemma 7.

Local Equilibrium Points

A behavior strategy combination $b^* = (b^*_1, \ldots, b^*_n)$ for an extensive game Γ is called a *local equilibrium point* for Γ or for a perturbed game $\hat{\Gamma}$ of Γ if every local strategy b^*_{iu} which is assigned to an information set u by one of the b^*_i is a local best reply to b in Γ or $\hat{\Gamma}$, resp.

LEMMA 8: *A behavior strategy combination* $b^* = (b^*_1, \ldots, b^*_n)$ *for a perturbed game* $\hat{\Gamma} = (\Gamma, \eta)$ *of an extensive game* Γ *with perfect recall is an equilibrium point for* $\hat{\Gamma}$, *if and only if* b^* *is a local equilibrium point for* $\hat{\Gamma}$.

PROOF: The lemma is an immediate consequence of Lemma 7.

Local Limit Equilibrium Points

Let $\hat{\Gamma}^1, \hat{\Gamma}^2, \ldots$ be a test sequence for an extensive game Γ with perfect recall. A behavior strategy combination $b^* = (b^*_1, \ldots, b^*_n)$ for Γ is called a *local limit equilibrium point* of the test sequence $\hat{\Gamma}^1, \hat{\Gamma}^2, \ldots$ if every $\hat{\Gamma}^k$ has a local equilibrium point b^k such that for $k \to \infty$ the sequence of the b^k converges to b^*.

THEOREM 3: *A behavior strategy combination* $b^* = (b^*_1, \ldots, b^*_n)$ *for an extensive game* Γ *with perfect recall is a perfect equilibrium point of* Γ, *if and only if for at least one test sequence* $\hat{\Gamma}^1, \hat{\Gamma}^2, \ldots$ *for* Γ *the behavior strategy combination* b^* *is a local limit equilibrium point of the test sequence* $\hat{\Gamma}^1, \hat{\Gamma}^2, \ldots$.

PROOF: The theorem is an immediate consequence of Lemma 8 and the definition of a perfect equilibrium point.

11. THE AGENT NORMAL FORM AND THE EXISTENCE OF A PERFECT EQUILIBRIUM POINT

In this section the concept of an agent normal form will be introduced. The players of the agent normal form are the agents of the information sets described by H. W. Kuhn [1953] in his interpretation of the extensive form. An agent receives the expected payoff of the player to whom he belongs. The agent normal form contains all the information which is needed in order to compute the perfect equilibrium points of the extensive game. With the help of the agent normal form one can prove the existence of perfect equilibrium points for extensive games with perfect recall.

The Agent Normal Form

Let Γ be an extensive game and let $u_1, \ldots, u_N$ be the information sets of the personal players in Γ. For $i = 1, \ldots, N$ let ϕ_i be the set C_{ui} of all choices at u_i. In the following we shall define a normal form $G = (\phi_1, \ldots, \phi_N, E)$ where the players $1, \ldots, N$ are thought of as agents associated with the information sets $u_1, \ldots, u_N$. This normal form is called the *agent normal form of* Γ.

Let ϕ be the set of all pure strategy combinations $\varphi = (\varphi_1, \ldots, \varphi_n)$ for G. For every $\varphi \in \phi$ the expected payoff vector $E(\varphi) = (E_1(\varphi), \ldots, E_n(\varphi))$ is defined as follows: Let $\pi = (\pi_1, \ldots, \pi_n)$ be the pure strategy combination for Γ whose components assign the choice $\varphi_j \in \phi_j$ to every information set u_j. For this π we have

$$E_i(\varphi) = H_j(\pi) \quad \text{for } u_i \in U_j. \tag{53}$$

The expected payoff function E is extended to the mixed strategy combinations $q = (q_1, \ldots, q_N)$ of G in the usual way.

Induced Strategy Combinations

Let $b = (b_1, \ldots, b_n)$ be a behavior strategy combination for Γ and let $q = (q_1, \ldots, q_N)$ be a mixed strategy combination for the agent normal form G of Γ. We say that q is *induced* by b on G and that b is *induced* on Γ by q if for $i, \ldots, N$ the mixed strategy q_i is the same probability distribution over R_i as the local strategy assigned to u_i by the relevant component of b. Obviously this use of the word "induced" defines a one-to-one mapping between the behavior strategy combination b of Γ and the mixed strategy combinations q of G.

Perturbed Agent Normal Forms

Let G be a normal form $G = (\phi_1, \ldots, \phi_N, E)$ and let η be a function which assigns positive minimum probabilities η_c to every $c \in \phi_i$ with $i = 1, \ldots, N$, subject to the restriction

$$\sum_{c \in \phi_i} \eta_c < 1. \tag{54}$$

The pair $\hat{G} = (G, \eta)$ is called a *perturbed normal form* of G. A mixed strategy q_i for G is a *mixed strategy for* $G = (G, \eta)$ if q_i satisfies the following condition:

$$q_i(c) \geq \eta_c \quad \text{for every } c \in \phi_i. \tag{55}$$

A mixed strategy combination $q = (q_1, \ldots, q_N)$ is called a *mixed strategy combination for* $\hat{G} = (G, \eta)$ if for $i = 1, \ldots, N$ the mixed strategy q_i is a mixed strategy for $\hat{G}$. The set of all mixed strategies q_i of player i in $\hat{G}$ is denoted by $\hat{Q}_i$.

Let Γ be an extensive game and let G be the agent normal form of Γ. Obviously a behavior strategy combination for the perturbed game $\hat{\Gamma} = (\Gamma, \eta)$ is induced on Γ by every mixed strategy combination for the perturbed normal form $\hat{G} = (G, \eta)$ and vice versa. We call $\hat{G}$ the *perturbed agent normal form* of $\hat{\Gamma}$.

Equilibrium Points

A mixed strategy $\tilde{q}_i$ of a player i in a perturbed normal form $\hat{G} = (G, \eta)$ is called a *best reply to* $\hat{G}$ to the mixed strategy combination $q = (q_1, \ldots, q_N)$ for $\hat{G}$ if we have

$$E_i(q/\tilde{q}_i) = \max_{q_i' \in \hat{Q}_i} E_i(q/q_i'). \tag{56}$$

A mixed strategy combination $\tilde{q} = (\tilde{q}_1, \ldots, \tilde{q}_N)$ is called a *best reply to* q *in* $\hat{G}$, if every $\tilde{q}_i$ in $\tilde{q}$ is a best reply to q in $\hat{G}$. A mixed strategy combination q^* for $\hat{G}$ is called an *equilibrium point of* $\hat{G}$, if q^* is a best reply to itself in $\hat{G}$.

LEMMA 9: *Let $\hat{G} = (G, \eta)$ be the perturbed agent normal form of the perturbed game $\hat{\Gamma}(\Gamma, \eta)$ of an extensive game Γ with perfect recall. An equilibrium point of $\hat{\Gamma}$ is induced on Γ by every equilibrium point of $\hat{G}$ and an equilibrium point of $\hat{G}$ is induced on G by every equilibrium point of $\hat{\Gamma}$.*

PROOF: It is clear that a local best reply in $\hat{\Gamma}$ corresponds to a best reply in $\hat{G}$. Therefore the assertion follows by Lemma 8.

Perfect Equilibrium Points

A *test sequence* $\hat{G}^1, \hat{G}^2, \ldots$ for a normal form $G = (\phi_1, \ldots, \phi_N, E)$ is a sequence of perturbed normal forms $\hat{G}^k = (G, \eta^k)$ of G such that for $k \to \infty$ the sequence of the η_c^k converges to 0 for every c in the sets R_i. A *limit equilibrium point* q^* of a test sequence $\hat{G}^1, \hat{G}^2, \ldots$ is a mixed strategy combination for G, such that there is at least one sequence $q^1, q^2, \ldots$ of equilibrium points q^k for $\hat{G}^k$ which for $k \to \infty$ converges to q^*. A *perfect equilibrium point* of G is a mixed strategy combination q^* for G which is a limit equilibrium point of at least one test sequence $\hat{G}^1, \hat{G}^2, \ldots$ for G.

LEMMA 10: *A limit equilibrium point q^* of a test sequence $\hat{G}^1, \hat{G}^2, \ldots$ for a normal form G is an equilibrium point of G.*

PROOF: The proof is omitted here since it is completely analogous to the proof of Lemma 3.

THEOREM 4: *Let Γ be an extensive game with perfect recall and let G be the agent normal form of Γ. A perfect equilibrium point of Γ is induced on Γ by every perfect equilibrium point of G and perfect equilibrium point of G is induced on G by every perfect equilibrium point of Γ.*

PROOF: It follows by Lemma 9 that a one-to-one relationship between the test sequences for Γ and for G can be established where a perturbed game of Γ corresponds to its perturbed agent normal form. Therefore a limit equilibrium point of one of both sequences induces a limit equilibrium point of the other one.

Existence of Perfect Equilibrium Points

In the following it will be shown that every extensive game Γ with perfect recall has at least one perfect equilibrium point. In order to prove this, we make use of Theorem 4.

THEOREM 5: *Every normal form G has at least one perfect equilibrium point.*

PROOF: A perturbed normal for $\hat{G} = (G, \eta)$ satisfies well-known sufficient conditions for the existence of an equilibrium point in mixed strategies [see, e.g., Burger, 1958, p. 35, Satz 2]. Therefore every perturbed normal form $\hat{G}^k$ in a test sequence $\hat{G}^1, \hat{G}^2, \ldots$ for G has an equilibrium point q^k. Since the set of all mixed strategy combinations is a closed and bounded subset of a euclidean space, the sequence $q^1, q^2, \ldots$ has an accumulation point q^*. The sequence $q^1, q^2, \ldots$ has a subsequence which converges to q^*. The corresponding subsequence of the test sequence $\hat{G}^1, \hat{G}^2, \ldots$ is a test sequence with the limit equilibrium point q^*. Therefore q^* is a perfect equilibrium point of G.

THEOREM 6: *Every extensive game Γ with perfect recall has at least one perfect equilibrium point.*

PROOF: In view of Theorem 5 the agent normal form of Γ has a perfect equilibrium point. It follows by Theorem 4 that Γ has a perfect equilibrium point.

12. CHARACTERIZATION OF PERFECT EQUILIBRIUM POINTS AS BEST REPLIES TO SUBSTITUTE SEQUENCES

In this section it will be shown that the definition of a perfect equilibrium point as a limit equilibrium point of a test sequence is equivalent to another definition which is more advantageous from the point of view of mathematical simplicity. In view of Theorem 4 we can restrict our attention to perfect equilibrium points for normal forms. It is sufficient to analyze the agent normal form if one want to find the perfect equilibrium points of an extensive game with perfect recall. It is important to point out that it is not sufficient to analyze the ordinary normal form. This will be shown in Section 13 with the help of a counterexample.

Substitute Sequences

Let $G = (\Pi_1, \ldots, \Pi_n; H)$ be a game in normal form. A mixed strategy q_i of player i is called *completely mixed* if for every $\pi_i \in \Pi_i$ the probability $q_i(\pi_i)$ assigned to π_i by q_i is positive. A mixed strategy combination $q = (q_1, \ldots, q_n)$ is called completely mixed if q_i is completely mixed for $i = 1, \ldots, n$. Let $\bar{q} = (\bar{q}_1, \ldots, \bar{q}_n)$ be a mixed strategy combination for G. An infinite sequence of mixed strategy combinations $q^1, q^2, \ldots$ is called a *substitute sequence* for $\bar{q}$ if q^k converges to $\bar{q}$ for $k \to \infty$ and every q^k is completely mixed. A strategy q_i or a strategy combination q is called a *best reply* to the substitute sequence $q^1, q^2, \ldots$ if q_i or q, resp. is a best reply to every q^k in the sequence.

Substitute Perfect Equilibrium Points

A mixed strategy combination $q^* = (q_1^*, \ldots, q_n^*)$ for a normal form G is called a *substitute perfect equilibrium point* of G if q^* is a best reply to at least one substitute sequence for q^*.

LEMMA 11: *A substitute perfect equilibrium point of a normal form G is an equilibrium point of G.*

PROOF: Let q^* be a best reply to the substitute sequence $q^1, q^2, \ldots$ for q^*. For $k = 1, 2, \ldots$ and for $i = 1, \ldots, n$ we have

$$H_i(q^k/q_i') = \max_{q_i \in Q_i} H_i(q^k/q_i). \tag{57}$$

In view of the continuity of H_i and the continuity properties of the maximum operator it is clear that (57) remains valid if on both sides we take limits for $k \to \infty$. This shows that q^* is an equilibrium point.

Associated Perturbed Normal Forms

Let $G = (\Pi_1, \ldots, \Pi_n; H)$ be a normal form, let $q = (q_1, \ldots, q_n)$ be a completely mixed strategy combination for G, and let ε be a positive number such that for $i = 1, \ldots, n$ we have $q_i(\pi_i) > \varepsilon$ for every $\pi_i \in \Pi_i$. For every triple (G, q, ε) of this kind we define an associated perturbed normal form $\hat{G} = (G, \eta)$, where the minimum probabilities of the pure strategies for G are as follows:

$$\eta_{\pi_i} = \begin{cases} q_i(\pi_i) & \text{if } \pi_i \text{ is not a best reply to } q \text{ in } G \\ \varepsilon & \text{if } \pi_i \text{ is a best erply to } q \text{ in } G \end{cases} \tag{58}$$

for $i = 1, \ldots, n$ and for every $\pi_i \in \Pi_i$. Obviously η satisfies the condition that the minimum probabilities for all pure strategies of a player sum up to less than 1.

LEMMA 12: *Let $\hat{G} = (G, \eta)$ be the associated perturbed normal form for the triple (G, q, ε). The strategy combination q is an equilibrium point of $\hat{G}$.*

PROOF: A mixed strategy is a best reply to q in $\hat{G}$ if the pure strategies which are not best replies to q in G are used with their minimum probabilities. In view of (58) this is the case for every component q_i of q.

LEMMA 13: *A substitute perfect equilibrium point of a normal form G is a perfect equilibrium point of G.*

PROOF: Let $q^* = (q_1^*, \dots, q_n^*)$ be a substitute perfect equilibrium point for G and let $q^1, q^2, \dots$ be a substitute sequence for q^* such that q^* is a best reply to $q^1, q^2, \dots$. Let $\varepsilon_1, \varepsilon_2, \dots$ be a sequence of positive numbers with $\varepsilon_k \to 0$ for $k \to \infty$, such that for $k = 1, 2, \dots$ and for $i = 1, \dots, n$ we always have $q_i^k(\pi_i) > \varepsilon_k$ for every $\pi_i \in \Pi_i$. Since every q^k is completely mixed we can find a sequence $\varepsilon_1, \varepsilon_2, \dots$ of this kind. Let $\hat{G}^k = (G, \eta^k)$ be the perturbed normal form associated with the triple (G, q^k, ε_k).

In the following it will be shown that $\hat{G}^1, \hat{G}^2, \dots$ is a test sequence for G. Obviously for $k \to \infty$ those minimum probabilities which are equal to ε_k converge to 0. Consider a pure strategy $\pi_i \in \Pi_i$ which is not a best reply to q^*. For this pure strategy we must have $q_i^*(\pi_i) = 0$. Therefore for $k \to \infty$ the minimum probabilities of pure strategies which are not best replies to q^* converge to 0, too. Consequently, $\hat{G}^1, \hat{G}^2, \dots$ is a test sequence of G.

The sequence $q^1, q^2, \dots$ is a sequence of equilibrium points q^k for the perturbed game $\hat{G}^k$ in a test sequence $\hat{G}^1, \hat{G}^2, \dots$ for G. This follows by Lemma 12. Moreover the sequence $q^1, q^2, \dots$ converges to q^*. Therefore q^k is a limit equilibrium point of the test sequence $\hat{G}^1, \hat{G}^2, \dots$. Consequently q^k is a perfect equilibrium point of G.

THEOREM 7: *A mixed strategy combination $q^* = (q_1^*, \dots, q_n^*)$ is a perfect equilibrium point of G, if and only if q^* is a substitute perfect equilibrium point of G.*

PROOF: In view of Lemma 13 it remains to be shown that a perfect equilibrium point q^* of G is substitute perfect. Let $\hat{G}^1, \hat{G}^2, \dots$ be a test sequence for G, such that q^* is a limit equilibrium point of $\hat{G}^1, \hat{G}^2, \dots$. Let $q^1, q^2, \dots$ be a sequence of equilibrium points q^k for $\hat{G}^k$ which converges to q^*. The definition of a perfect equilibrium point requires that such sequences $\hat{G}^1, \hat{G}^2, \dots$ and $q^1, q^2, \dots$ exist.

Let T_i^k be the set of all those pure strategies of player i which appear with more than minimum probability in q^k, i.e., π_i is in T_i^k, if and only if we have $q_i^k(\pi_i) > \eta_{\pi_i}^k$ for player i's component q_i^k in q^k. Obviously a pure strategy $\pi_i \in T_i^k$ is a best reply q^k in G but T_i^k may not contain every pure best reply to q^k in G. Since the q^k converge to q^* and the $\eta_{\pi_i}^k$ converge to 0, there must be a number m, such that for $k > m$ every pure strategy π_i with $q_i^*(\pi_i) > 0$ is in T_i^k for $i = 1, \ldots, n$. Without loss of generality we can assume $m = 0$ since otherwise we can use subsequences of the original sequences $\hat{G}^1, \hat{G}^2, \ldots$ and $q^1, q^2, \ldots$ for the purpose of this proof.

Since every π_i with $q_i^*(\pi_i) > 0$ is in T_i^k and every $\pi_i \in T_i^k$ is a best reply to q^k in G, the mixed strategy q_i^* is a best reply to g^k for $k = 1, 2, \ldots$. The q^k are completely mixed and $q^1, q^2, \ldots$ converges to q^*. The sequence $q^1, q^2, \ldots$ is a substitute sequence for q^* and q^* is a best reply to this sequence. q^* is a substitute perfect equilibrium point.

13. TWO COUNTEREXAMPLES

One might be tempted to think that a perfect equilibrium point of the normal form G of an extensive game Γ with perfect recall always corresponds to a perfect equilibrium point of Γ. If this were the case one would not need the agent normal form. In the following we shall present two counterexamples. The first one is quite simple but less satisfactory than the second one.

The First Counterexample

The extreme game of Figure 2 has exactly one perfect equilibrium point, namely the pure strategy combination (Rr, L). Here Rr refers to that pure strategy of player 1 where he chooses R at the origin and r at his other information set. The fact that this is the only perfect equilibrium point follows immediately by the subgame perfectness of perfect equilibrium points. (See the corollary of Theorem 2 in Section 8).

In the normal form (Rr, L) is a perfect equilibrium point, too but not the only one. Since the strategies Rl and Rr are equivalent (Rl, L) is just as perfect in the normal form as (Rr, L). In a perturbed game of the extensive form the strategies Rl and Rr are not equivalent but this information is lost in the normal form and cannot be regained by the construction of perturbed normal forms.

The first counterexample is not quite satisfactory since one may be content with the fact that among the two equivalent perfect equilibrium points of the normal form, there is one which is perfect in the extensive

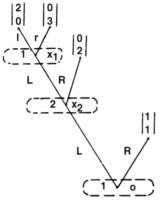

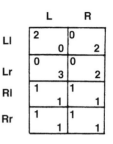

FIG. 2. Extensive form and normal form for the first counterexample. The conventions of the graphical representation of the extensive form are explained at figure 1.

form. One may take the point of view that it is not important to distinguish between these two equilibrium points.

The Second Counterexample

Consider the equilibrium points (Rl, L_2, R_3) and (Rr, L_2, R_3) of the game of Figure 3. As we shall see both of these equilibrium points are perfect in the normal form but they fail to be perfect in the extensive form.

Perfectness in the Normal Form

It is sufficient to show that (Rl, L_2, R_3) is a perfect equilibrium point of the normal form; if this is the case the same must be true for (Rr, L_2, R_3) since in the normal form Rr is a duplicate of Rl.

In order to show the perfectness of (Rl, L_2, R_3) we construct the following substitute sequence $q^1, q^2, \dots$: In q^k every pure strategy which does not occur in (Rl, L_2, R_3) is used with a small probability ε_k. The pure strategies Rl, L_2, and R_3 are used with probabilities $1 - 3\varepsilon_k$, $1 - \varepsilon_k$, and $1 - \varepsilon_k$ resp. The ε_k are selected in such a way that $\varepsilon_1, \varepsilon_2, \dots$ is a decreasing sequence which converges to 0. Of course ε_1 must be selected sufficiently small, say $\varepsilon_1 < 1/100$.

It can be verified easily that (Rl, L_2, R_3) is a best reply to this substitute sequence. We omit the computational details. It follows by

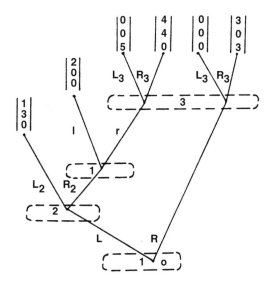

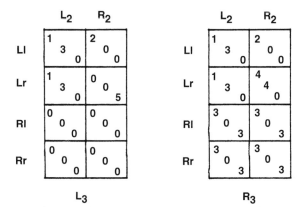

FIG. 3. Extensive form and normal form for the second counterexample. The normal form is described by two trimatrices, one for player 3's choice L_3 and one for his choice R_3.

Theorem 7 that (Rl, L_2, R_3) and (Rr, L_2, R_3) are perfect equilibrium points for the normal form of the game of Figure 3.

Imperfectness of (Rl, L_2, R_3) in the Extensive Form

Let $b^1, b^2, \ldots$ be a sequence of behavior strategy combinations with completely mixed local strategies for the game in Figure 3 such that b^k

converges to (Rl, L_2, R_3) for $k \to \infty$. We may call $b^1, b^2, \ldots$ a *substitute sequence* for (Rl, L_2, R_3). According to Theorems 4 and 7 the equilibrium point (Rl, L_2, R_3) cannot be perfect unless there is at least one such sequence with the property that the choices in (Rl, L_2, R_3) are local best replies to every b^k. In order to see this one just has to translate the terminology of the agent normal form into that of the extensive form.

For every ε there is a number $m(\varepsilon)$ such that for $k > m(\varepsilon)$ the probabilities prescribed by b^k to the choices selected by (Rl, L_2, R_3) are greater than $1 - \varepsilon$. It can be seen immediately that for sufficiently small ε the local best reply for player 1 is r, not l. This shows that the sequence $b^1, b^2, \ldots$ cannot be such that (Rl, L_2, R_3) is a best reply to every b^k. Consequently (Rl, L_2, R_3) fails to be a perfect equilibrium point of the extensive game of Figure 3.

Imperfectness of (Rr, L_2, R_3) in the Extensive Form

In the same way as before let $b^1, b^2, \ldots$ be a sequence of behavior strategy combinations with completely mixed local strategies which converges to (Rr, L_2, R_3). Here, too, for sufficiently big k the choices in (Rr, L_2, R_3) have probabilities greater than $1 - \varepsilon$ in b^k. It can be seen immediately that for sufficiently small ε player 2's best reply to b^k is R_2. Therefore the sequence $b^1, b^2, \ldots$ cannot be such that (Rr, L_2, R_3) is a best reply to every b^k. Consequently (Rr, L_2, R_3) fails to be a perfect equilibrium point of the game of Figure 3.

Interpretation

In the following we shall try to give an intuitive explanation for the phenomenon that an equilibrium point which is perfect in the normal form may not be perfect in the extensive form.

In order to compare the normal form definition with the extensive form definition, we shall look at a perturbed game $\hat{\Gamma}$ of an extensive game Γ with perfect recall and at a perturbed normal form $\hat{G}$ of the normal form G of Γ. Let the behavior strategy combination $b^k = (b_1^k, \ldots, b_n^k)$ be an equilibrium point for $\hat{\Gamma}$ and let the mixed strategy combination $q^* = (q_1^*, \ldots, q_n^*)$ be an equilibrium point for $\hat{G}$.

A choice c in Γ is called *essential* for b^* if the relevant local strategy selects c with more than the minimum probability for c required by $\hat{\Gamma}$. A choice which is essential for b^k must be a local best reply to b^k in Γ.

A pure strategy π_i is called *essential* for q^k if $q_i^*(\pi_i)$ is greater than the minimum probability for π_i required by $\hat{G}$. A pure strategy which is essential for q^* must be a best reply to q^* in Γ.

Both b^* and q^* reach all parts of the extensive form in the sense that the realization probabilities of all vertices are positive. Nevertheless there is a crucial difference between b^* and q^*. This difference concerns the conditional choice probabilities $\mu_i(c, u, q_i^*)$ which have been defined with the help of Lemma 1 and Lemma 2 in the proof of Kuhn's theorem. In the case of q^* these conditional choice probabilities are defined for every personal information set.

It may happen that player i's pure strategies which are essential for q^* are such that a given information set u is not reached by q^*/π_i for every one of these essential strategies π_i; the realization probabilities $\rho(x, q^*/\pi_i)$ are 0 for every $x \in u$. An information set u of this kind will be called inessentially reached by q^*.

If an information set u of player i is inessentially reached by q^*, then the conditional choice probabilities $\mu(c, u, q^*)$ will be exclusively determined by those pure strategies of player i which are inessential for q^*. Therefore the $\mu_i(c, u, q^*)$ may be very unreasonable as a local strategy at u.

The crucial difference between b^* and q^* is as follows: Whereas every local strategy in b^* is reasonable in the sense that the essential choices are local best replies, q^k may lead to unreasonable conditional conditional choice probabilities at those information sets which are inessentially reached by q^*.

As an example let Γ be the game in Figure 3 and let q^* be such that only the pure strategies in the equilibrium point (Rr, L_2, R_3) are essential for q^*. The information set of player 1, where he chooses between l and r is inessentially reached. Therefore the conditional choice probabilities for l and r are not determined by Rr but exclusively by the minimum probabilities for Ll and Lr which may be such that l is selected with a high conditional choice probability.

In an extensive game, where every player has at most one information set, it cannot happen that the information set of a player i is not reached by q^*/π_i for one of his pure strategies π_i. His strategy does not influence the realization probabilities of the vertices in his information set. The agent normal form corresponds to an extensive form where every player has at most one information set. Therefore no difficulties arise in the agent normal form.

REFERENCES

Burger, E.: Einführung in die Theorie der Spiele, Berlin 1958.

Kuhn, H. W.: Extensive Games and the Problem of Information, in: H. W. Kuhn and A. W. Tucker (eds.): Contribution to the Theory of Games, Vol. II, Annals of Mathematics Studies, 28, pp. 193–216, Princeton 1953.

Nash, J. F.: Non-cooperative Games, Annals of Mathematics **54**, 155–162, 1951.

Neumann, J. v., and O. Morgenstern: Theory of Games and Economics Behavior, Princeton 1944.

Selten, R.: Spieltheoretische Behandlung eines Oligopolmodells mit Nachfrageträgheit, Zeitschrift für die gesamte Staatswissenschaft **121**, 301–324 and 667–689, 1965.

——: A Simple Model of Imperfect Competition, where 4 are Few and 6 are Many, International Journal of Game Theory **2**, 141–201, 1973.

Received September, 1974

LIST OF CONTRIBUTORS

ROBERT J. AUMANN
  Institute of Mathematics and Computer Science
  The Hebrew University
  Jerusalem, 91904 Israel

DAVID BLACKWELL
  Department of Statistics
  University of California at Berkeley
  Berkeley, Calif. 94720 USA

GERARD DEBREU
  Department of Economics
  University of California at Berkeley
  Berkeley, Calif. 94720 USA

H. EVERETT (deceased)

THOMAS S. FERGUSON
  Department of Mathematics
  University of California at Los Angeles
  Los Angeles, Calif. 90095 USA

JOHN C. HARSANYI
  The Walter A. Haas School of Business
  University of California at Berkeley
  Berkeley, Calif. 94720 USA

HAROLD W. KUHN
  Department of Mathematics
  Princeton University
  Princeton, NJ 08544 USA

MICHAEL MASCHLER
Department of Mathematics
The Hebrew University
Jerusalem, 91904 Israel

JOHN F. NASH, JR.
Department of Mathematics
Princeton University
Princeton, NJ 08544 USA

BEZALEL PELEG
Institute of Mathematics and Computer Science
The Hebrew University
Jerusalem, 91904 Israel

JULIA ROBINSON (deceased)

HERBERT E. SCARF
Department of Economics
Yale University
New Haven, Conn. 06520 USA

REINHARD SELTEN
Department of Economics
University of Bonn
D-53113 Bonn, Germany

LLOYD S. SHAPLEY
Departments of Mathematics and Economics
University of California at Los Angeles
Los Angeles, Calif. 90095 USA

MARTIN SHUBIK
Department of Economics
Yale University
New Haven, Conn. 06520 USA

F. B. THOMPSON
Department of Computer Science
California Institute of Technology
Pasadena, Calif. 91105 USA